Tribology in Environmental Design 2000

Conference Chairman

Sir Gordon Higginson
Pro-Chancellor, Bournemouth University, UK

Honorary Committee

Professor Stathis Ioannides
SKF Group Technical Development,
The Netherlands

Professor Gillian Slater
Vice-Chancellor, Bournemouth University, UK

Gordon Waddington
Research and Technology, Rolls Royce plc, UK

Professor Peter Hogarth
Bournemouth University, UK

Professor Esteban Fernández
Universidad De Oviedo, Spain

Professor H Peter Jost
President, International Tribology Council, UK

Professor Peter Hills
Institution of Engineering Designers, UK

International Scientific Advisory Committee

Dr Rehan Ahmed
Heriot-Watt University, UK

Professor Leo Alting
Technical University of Denmark, Denmark

Dr Ing Jonathan Borg
University of Malta, Malta

Dr Robin Cundill
SKF, The Netherlands

Dr Mark Hadfield
Bournemouth University, UK

Dr Kenkichi Sato
Chiba University, Japan

Dr Peter Smith
Castrol International, UK

Professor Shogo Tobe
Ashikaga Institute of Technology, Japan

Professor John Tripp
SKF, The Netherlands

Dr Ricardo Vijande
Universidad De Oviedo, Spain

Dr John Williams
Cambridge University, UK

Conference Co-ordinator

Ing Christopher Ciantar
Bournemouth University, UK

Conference Secretariat

Mrs Christine Thwaites
Bournemouth University, UK

Keynote Speakers

Dr Ricardo Vijande
Dr Ing Jonathan Borg
Dr Rehan Ahmed
Dr John Williams
Professor Esteban Fernández
Dr Kenkichi Sato

1st International Conference on

Tribology in Environmental Design 2000

The Characteristics of Interacting Surfaces – a Key Factor in Sustainable and Economic Products

Held on the 4–6 September 2000
Bournemouth University, UK

Edited by

M Hadfield and **C Ciantar**
School of Design, Engineering, and Computing
Bournemouth University, UK

Organized by

Sustainable Product Engineering University Research Centre
Bournemouth University, UK

Professional Engineering Publishing

Professional Engineering Publishing Limited
Bury St Edmunds and London, UK

First Published 2000

ISBN 1 86058 266 4

A CIP catalogue record for this book is available from the British Library.

Printed by The Cromwell Press, Trowbridge, Wiltshire, UK

The Publishers are not responsible for any statement made in this publication. Data, discussion, and conclusions developed by authors are for information only and are not intended for use without independent substantiating investigation on the part of potential users. Opinions expressed are those of the Authors and are not necessarily those of the Institution of Mechanical Engineers or its Publishers.

Related Titles of Interest

Title	Editor/Author	ISBN
Guide to Wear Testing for Industry	M Neale	1 86058 287 7
IMechE Engineers' Data Book – Second Edition	C Matthews	1 86058 248 6
History of Tribology	D Dowson	1 86058 070 X
New Directions in Tribology	I M Hutchings	1 86058 099 8
Lubrication and Lubricant Selection	A R Lansdown	1 86058 029 7
Tribology Casebook	J D Summers-Smith	1 86058 041 6
Introductory Guide to Industrial Tribology	J D Summers-Smith	0 85298 896 6
Analysis of Rolling Element Bearings	W Changsen	0 85298 745 5
Gearing, Transmission, and Mechanical Systems	Edited by D Su	1 86058 260 5
Gear Technology	IMechE Seminar	1 86058 280 X
Fluid Film Bearings	IMechE Seminar	1 86058 278 8

For the full range of titles published by Professional Engineering Publishing contact:

Sales Department
Professional Engineering Publishing Limited
Northgate Avenue
Bury St Edmunds
Suffolk
IP32 6BW
UK

Tel: +44 (0)1284 724384
Fax: +44 (0)1284 718692

Contents

Surface Engineering

Lubricants

Test Methods

Preface

Tribology in Environmental Design 2000

In welcoming you all to this major event in the year 2000 at Bournemouth University, I note first with great pleasure that the authors of the papers alone come from 15 countries. This is a truly international event and I know that we shall learn a great deal from each other, not only in the formal sessions but also in the small clusters that will gather in spare moments throughout the three days we are together.

The first major event in this country in the field of tribology was held in 1937 by the Institute of Mechanical Engineers, a four-day conference entitled 'General discussion on lubrication and lubricants'. There are over 100 papers from far fewer countries than we have, and numerically they were predominantly from the UK. Of course, in those days, the title tribology was not used; it first appeared in 1966 as one of many outcomes of the Jost Report, the landmark publication of the work of a national committee chaired by H Peter Jost, whom we shall meet during the conference.

Our conference considers aspects of tribology almost unheard of, and certainly not researched, in those earlier days. The emphasis on environmental effects is especially novel. I hope you will find benefit and enjoyment during your few days in Bournemouth.

Sir Gordon R Higginson
Bournemouth
June 2000

Life-oriented Products

Influence of tribology on machine design methodology

R VIJANDE, E FERNÁNDEZ, M CADENAS, and **J M SIERRA**
Department of Construction and Engineering of Production, University of Oviedo, Gijón, Spain

ABSTRACT

This article is about the design of a tribometer that has been developed since the announcement of TED 2000. The different steps of interest are described, from the design to the calculation, manufacturing and application of the tribometer, making special reference to the implications that tribology has in each of these stages. It was kept in mind that in the last few years, new technologies, like plasma spraying and laser, have gradually been incorporated into the field of production and have now become of common place. Mechanical engineers today need to consider tribological technology and, of course, the fundamental principals of tribology, from the very first stage of their projects.

1 INTRODUCTION

The company TN CORP., on November 2nd 1999, communicated to the group of Design and Tribology of the Area of Mechanical Engineering of the E.T.S.I.I. and I.I. of Gijón, the interest they had in the study of tribological-mechanical behaviour of some model wheels, figure 1. They wanted to compare these wheels with the ones that they were already using with a view to substituting them.

The interest resides in the possible substitution of the traditional wheels for the new ones, which could lead to a significant saving, both economically and in the energy required (less friction), without apparent damage to other benefits.

The tests will reproduce, as far is possible, the real operating conditions of the rollers, comparing the response of these new wheels, manufactured with an innovative plastic material that has little contrasted tribological characteristics, with that of those used up to now.

In order to carry out these tests, it was agreed to design, calculate and build a tribometer able to carry out wheel-rail tests of the rollers. The surface of the rollers would be made from a plastic material that would allow them to rotate without slipping under certain load conditions and a determined speed.

The objective is to obtain some results that allow one to determine if it is convenient to substitute the old rollers for the new ones in the future.

2 PHASES OF DESIGN

2.1. Specifications

To begin to carry out the design, some basic specifications were set. So that the tests simulated, as far as possible, the operation of the rollers currently in use. These initial specifications are:

- Geometric dimensions of the rollers: external diameter, ∅= 75mm and wide = 24mm
- Rollers of a polymer material
- Normal loads applied on each roller: 1850 N
- Angular speed of the roller: 425rpm
- Roller on plane rolling (plane rail of steel with some specific superficial characteristics of roughness.)

2.2. Specification analysis

First of all the rolling action was studied. The type of rolling that it was required to study was that of a cylinder on a plane.

An important function for the design of the tribometer was to be able to carry out the tests by means of two cylinders of parallel axes; one would be the wheel or roller to be tested and the other to substitute the rail. We understand that it is this way, due to the uniformity of the rotational movement (constant speed) between cylinders, and the non-uniformity one (big peaks of acceleration) in that of rotation-translation between the cylinder and the rail as our experience has shown us in other mechanisms.

2.2.1 First tribological problem

Here, the first important question of the design arises: is it possible to simulate the behaviour of a cylinder-plane with another cylinder-cylinder?

2.2.2 First tribological solution

Following Hertz theory [1], and as Halling points out [2], when two cylindrical bodies of radius R_1 and R_2 are pressed to each other with their axes parallel, exercising a force P for unit of longitude, the general problem becomes two-dimensional, figure 2. The contact occurs in a fringe along the length of the cylinders and with a thickness of *2a*, the value of "*a*" being:

$$a = \sqrt{\frac{4PR}{\pi E^*}} \quad \{1\},$$

$$\frac{1}{E^*} = \frac{1-\nu_1}{E_1} + \frac{1-\nu_2}{E_2}$$

E Young's module, ν Poisson module and $1/R=1/R_1+1/R_2$ represent the relative curvature between both cylinders and R is the equivalent radius of curvature in the main plane I (the plane that is perpendicular to the axes of the cylinders).

For a given material, {1} can be expressed as:

$a \propto \sqrt{R}$, {2}.

Moreover, in the case of the contact of a cylinder on a plane (plane rolling), the radius of the plane is infinite, that is to say, $R=R_1$, then:

$a \propto \sqrt{R_1}$ {3}.

On the other hand, $x = R_1/R_2$ can be a good measure to check the correspondence of the case of rolling between cylinders to that of cylinder-plane rolling (x=0). Substituting $R_2 = R_1/x$ into the expression $1/R=1/R_1+1/R_2$ you arrive at the expression: $R = R_1/(1+x)$; a function that, if it is evaluated by points (this can be seen in Table 1), gives the deviation between R_1 and R as x varies.

In this table R1 has been given a value of 100 units in order to directly obtain the percentage of variation of R.

As can be appreciated, for x values very close to zero (0,01), the agreement is very high (99%) between the contact of cylinder-cylinder and cylinder-plane, although it demands a drum 100 times bigger than the roller. It is necessary to come to a compromise, sacrificing some accuracy to facilitate the construction and design. Taking x=0,15, as Niemman recommends [3] where possible, results in R=0,87R1.

For these conditions, the size of the contact area *a* is given by {2} and would take a value a':

$a' \propto \sqrt{0{,}87R_1} \Rightarrow a' \propto 0{,}933\sqrt{R_1}$.

Therefore, the difference between one contact area and another is less than 7% which is an acceptable margin of error for this kind of simulation.

As a consequence, the testing machine can simulate with good precision, a cylinder-plane or cylinder-rail rolling contact if the relationship between the radii of the roller to test and the drum that faces is smaller or close to 0.15

2.3. Pre-design

Next, we began to sketch a prototype by means of successive outlines, to outline and improve the design until arriving at the definitive one, shown in Figure 3. Among others, the following considerations were kept in mind:

The chosen system to transmit the load should be simple, cheap, comfortable and effective, which greatly reduces the number of possibilities, discarding pneumatic systems, hydraulic systems, electronic systems, etc. The solution considered is that of gravitational masses, since it is relatively simple, applies a constant pressure and the mechanism is very cheap and functional. It was also interesting that it multiplies the weight of the masses with the advantage that the user does not have to handle them. (With 200 N of weight, a force of 2000 N is obtained in the rollers).

To apply the normal load to the rollers, a mechanism was investigated that, also distributes the forces symmetrically to achieve a good stability of the machine. Therefore, a mechanism that presses the rollers on the drum in diametrically opposed positions was chosen.

The time factor was considered very important, since these tests are 1600 hours long, which means a lot of time to obtain results of a significant sample. To mitigate this circumstance, the machine tests three couples (6 rollers) at the same time.

2.4. Final design

Once the kinematic, dynamic and necessary resistance analyses had been carried out to decide the geometric forms and the dimensions of the elements that would compose the machine, we arrived at the definitive design. For it, we used computer simulation programs of mechanisms and structural calculations.

The machine finally consists of a revolving drum (1) of steel upon which rotate (driven by friction) six rollers (2) (material to test). The position of the rollers has been made in two diametrically opposed rows, one on each side of the drum and with three rollers in each one (Figure 3 and 6).

The rollers are mounted in the centre of the vertical independent arms (3) that are articulated at their lowest end. At the top end of the arm, the load of the weights is applied (5) supported by the lever plate (4). Each arm is joined with the one that occupies a diametrically opposed position, regarding the drum, by means of a union rod (6). All these elements compose the mechanism of load transmission.

To transmit the power to the drum, a geared motor with frequency converter was prepared (8) joined to the drum by means of the flexible coupling (9).

The set-up was completed with the control system (here it is not described) for such functions as: cycle counting, rpm, temperature, automatic stop at end of test, etc.

To overcome the friction couple of the rollers on the drum, an electric motor of alternating current with frequency converter was chosen, that permits modification of the test speed and to make the machine more versatile.

2.4.1 Second tribological problem

Here a second important tribological query arises: how to achieve or to simulate the real friction? That is to say, how to give the appropriate superficial characteristics to the cylinder figure 3 (1) that simulates to the rolling rail?

2.4.2 Second tribological solution

Initially it was observed that there were three important aspects to consider:

- The average roughness Ra, must remain constant throughout 5.5 microns;
- The hardness, must remain constant throughout 60 Rc;

- The duration in time, without loosing the early characteristics, so that repeated tests can be carried out without the necessity for maintenance on the surface of the drum fig. 3 (1);

Thermal spraying is one of the most common methods to improve the surface characteristics of materials. It consists in spraying the substrate with molten substances at high speed. As a result, a coating is formed whose thickness may vary from a few hundredths to several millimetres according to the characteristics of the substance, the spraying technique used and the results required. Plasma spraying and HVOF are clear examples. The second gets more adherence, a similar roughness and a little less porosity. Also, a layer of projected plasma, later treated with laser, achieves very high benefits [4, 5, and 6], Figure 4 and 5.

It was decided to split the drum in to three fringes, each by pair of rollers and covered them by means of plasma spraying. Tungsten carbide powder was used (CW-Co12%) for its high resistance to wear, its good adherence and a final roughness (selecting the appropriate half size of grain) valid for the test, without later necessity of any machining process. This way the high level of hardness required is obtained ($HV_{0.3} \approx 1100$) and, especially, the long duration of these characteristics.

It is worth highlighting that a later treatment with laser on the coatings of plasma was considered necessary in order to achieve excellent protection of the steel drum against corrosion [6], since the wheels may be tested in some corrosive atmospheric surroundings.

2.5 Problems in the production

2.5.1 Third tribological problem

Also, the manufacturing process is seen to be affected by these techniques, which, at the same time can be used with other specific tribological objectives, such as those noted above.

When manufacturing part of the components, different tools may be employed: lathe, milling machines, water cutting, oxicut, laser cutting, etc.

It becomes frequently difficult to decide which to use due to a lack of normalised information with comparative data for the varying techniques.

The components 3 (Arms), 4 (Lever Plate) and 7 (Bench) were cut by means of laser. Here the Third Tribological Problem was presented and we pick it up in this article: faulty final finish of the holes carried out in 3, 4 and 7, especially noticeable in those that are used as kinematic couples between bars.

The phenomenon is known and several authors had explained it, for example George Chryssolouris [7] whose words will be quoted:

"During laser machining, high power densities are introduced on the surface of the workpiece in order to raise the temperature of the volume to be removed to the melting or vaporisation point".

In the laser cutting process, a kerf is created through relative motion between the laser beam and the workpiece surface.

When material is removed through melting, a molten layer forms at the erosion front. When laser cutting metals, several factors can affect cutting surface quality. The relationship between kerf quality and cutting parameters was investigated for laser-gas-cutting of stainless steels.

Laser cutting has several advantages:

- For most industrial materials with workpiece thickness up to 10 mm, laser cutting produces a significantly higher material removal rate than mechanical cutting or shearing.
- Laser cutting produces kerf widths, which are narrower than those achievable with mechanical cutting.

The drawbacks of laser cutting in comparison to conventional methods are:

- Laser cutting effectiveness reduces as the workpiece thickness increases. Workplaces greater than 15mm in thickness generally cannot be cut effectively by modern industrial lasers.
- Laser cutting produces a tapered kerf shape, compared to the straight vertical kerf walls achievable by conventional methods. The kerf taper is a result of the divergence of the laser beam and becomes more pronounced as the workpiece thickness increases. The kerf taper can be reduced by adjusting the focal point of the laser beam to the interior of the workpiece instead of on the workpiece surface.

2.5.2 Third (desired) tribological solution

To our understanding, at this time important publications of studies over concrete aspects related with the laser technology exist [8, 9, 10, 11, and 12]. It would be opportune to use a standardisation of operative procedures that includes aspects together, such as: equipment power, laser type, appropriate parameters related to material, feasible tribological operations, feasible automated operations, degree or level of the operation (roughness, hardness, adjustment, tolerances, etc)

3. CONCLUSIONS

From these reflections, it can be concluded:

1. The importance that tribology has, to give solutions and to pose queries in connection with the mechanical design.

2. The importance that tribology has to give solutions and to outline queries in connection with the production processes.

3. The importance that tribology has to give solutions to lengthen the life-cycle of the machines, as derived from the above (longest duration of the surface characteristics found by the thermal projection, p.e.)

4. It is necessary to popularise these technologies (plasma, laser, etc.), especially for the surface characteristics that are obtained, to eliminate barriers and to facilitate their employment for the designer.

5. What about the environment that TED 2000 invites us to consider?
Allow us to finish with a rhetorical question; does the contribution of Tribology not lead us to the optimum solution of a problem, before it is born?

REFERENCES:

[1] H. Hertz Z. reine angew. Math. 92, (1881), 155 or see s.P. Timoshenko and J. N. Goodier. Theory of elasticity. McGraw-Hill, N. Y., (1970), p 409.
[2] J. Halling and al., Principes of tribology, edited by J. Halling, H. K. 1983
[3] G. Niemann, Elementos de máquinas, Labor,V 1, Barcelona,1987
[4] Vijande R.; Belzunce, J.; Fernández, J.E. "Wear and microstructure in fine ceramic coatings" Wear, 148 (1991) 221-233
[5] Cuetos, J.M.; Fernández, J.E.; Vijande, R. "Plasma-sprayed coatings treated with lasers: tribological behaviour of Cr_2O_3" Wear, 169 (1993) 173-179
[6] Fernández, J.E.; Rodriguez, R.; Ynglong Wang; Vijande, R.; Rincón, A. "Sliding wear of plasma-sprayed Al_2O_3 coating" Wear 181-183 (1995) 417-425
[7] G. Chryssolouris, Laser Machining. Theory and Practice. Ed. Spriger-Verlag. 1991.
[8] Hatefeuille L., Beylat L., Carron D. and Puig T. "Amelioration des propiétés de surface par refusion laser". Proc. Conference Laser de puissance et traitements des matériaux. Ecole de Printemps. France. 1991. pp. 27-31.
[9] Wang K., Lin J. and Han D. "Microstructure and properties of laser remelted cobalt based and nickel based plasma srayed coatings". Surface Engineering. 1996. 12(3). pp. 235-238.
[10] Mateos J., Cuetos J.M., Cadenas M. and Montes H. "Tratamiento con laser de CO_2 de recubrimientos de carburo de tungsteno proyectados por plasma". Anales de Ingeniería Mecánica. 1997. 4. pp. 113-119.
[11] Fernández E., Cuetos J.M., Vijande R. and Rincón A. "Comparison of wear behaviour of plasma sprayed Al_2O_3 coatings with and without laser treatment". Tribology International. 1996. 29 (6). pp. 477-485.
[12] Mateos J., Cuetos J. M., Fernández E. and Cadenas M. "Plasma sprayed tungsten carbide coatings after laser surface treatment". Proc. 1998 Powder Metallurgy World Congress & Exhibition. 1998. Spain. 1. pp. 531-537.

ACKNOWLEDGMENTS

Antonio Calleja Palacio by his collaboration
T.N. SA and FICYT by their support

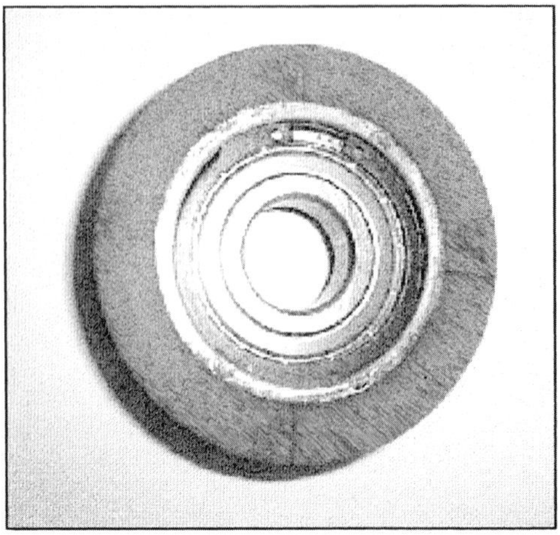

FIGURE 1. Roller

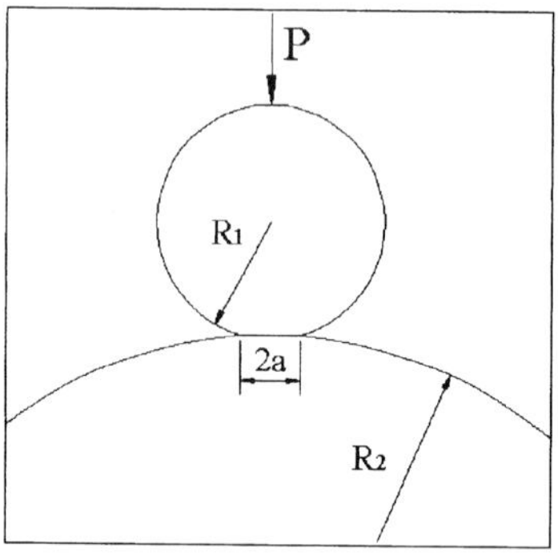

FIGURE 2. Contact of two cylinders

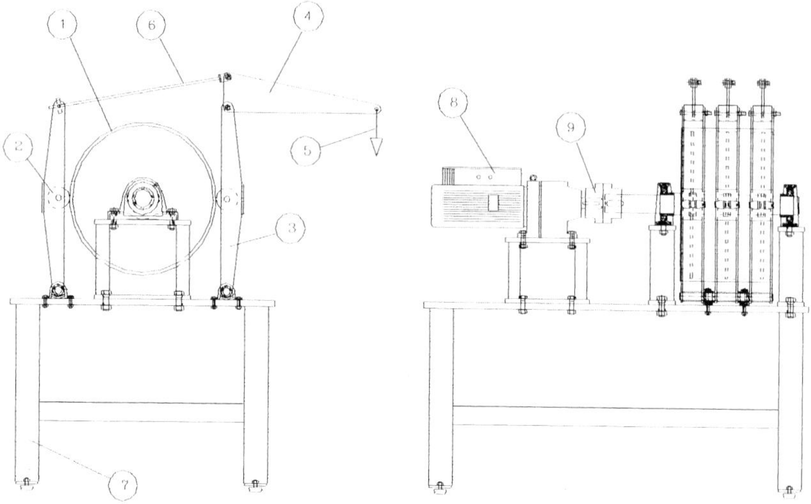

FIGURE 3. Tribometer outline

(1) Cylinder or drum, (2) Rollers (wheels), (3) Arms, (4) Lever plate, (5) Weight, (6) Arms union rod, (7) Bench, (8) Geared motor with speed control module, (9) Flexible coupling.

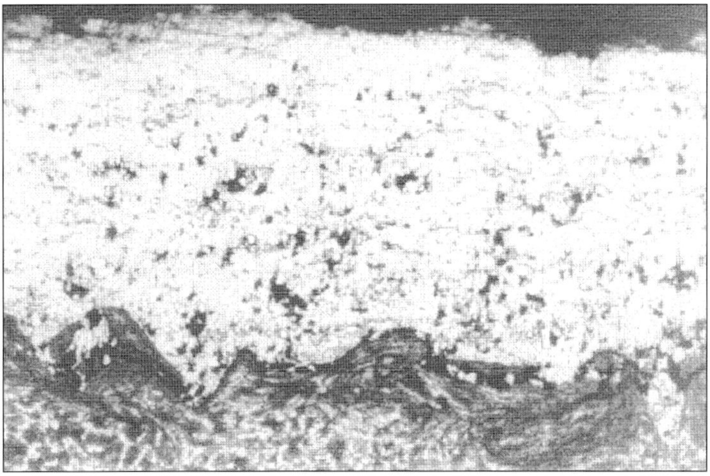

FIGURE 4. Plasma sprayed coating

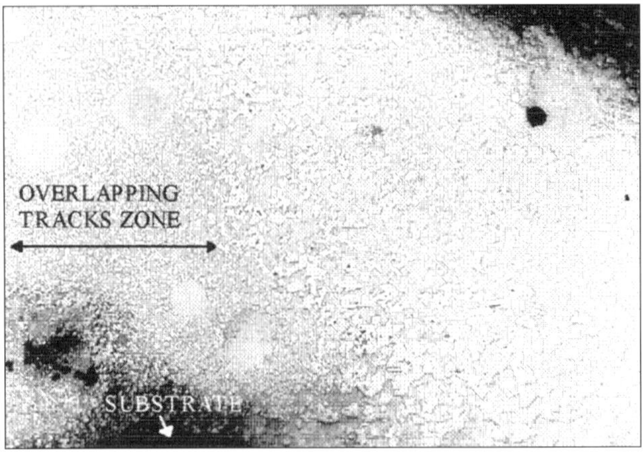

FIGURE 5. Laser remelted coating

FIGURE 6. Tribometer

TABLE 1. Values of R in % in function of x

R1	x=R1/R2	R=R1/(1+x)	R1	x=R1/R2	R=R1/(1+x)
100	0,01	99,01	100	0,13	88,50
100	0,02	98,04	100	0,14	87,72
100	0,03	97,09	100	0,15	86,96
100	0,04	96,15	100	0,16	86,21
100	0,05	95,24	100	0,17	85,47
100	0,06	94,34	100	0,18	84,75
100	0,07	93,46	100	0,19	84,03
100	0,08	92,59	100	0,2	83,33
100	0,09	91,74	100	0,21	82,64
100	0,1	90,91	100	0,22	81,97
100	0,11	90,09	100	0,23	81,30
100	0,12	89,29	100	0,24	80,65

Search and evaluation heuristic for ecological optimized products

R ZÜST
ZHW, Winterthur, Switzerland

ABSTRACT

This contribution discusses and evaluates case examples of industrial and commercial products from the area machine construction and Electro technique. The aim is to support the analysis and search process in the early phase of product development through a search heuristic.
The presented search heuristic primarily builds on the perception that there are both active and passive as well as locally moved and immobile objects. From this perception various work hypotheses can be derived - both with regard to the situation analysis and to the solution search. The heuristic can be refined and be made more concrete through further parameters.

1 INTRODUCTION

1.1 Context

An organization has activity scopes with regard to designing its activities, products and services which exceed the legal frameworks, such as laws, ordinances and standards. This entrepreneurial scope should be regarded in the following.
In the early planning phase important decisions are made with regard to the quality of the solution. Therefore, methods and procedures supporting an actual 'business innovation' in the early planning phase are vital. Obviously, the ecological improvement potential increases with the expansion of innovation perspectives beyond mere process and product innovations (see e. g. [1]and [2]).
While process innovations can still content themselves with technical solutions (possibly even in the sense of 'end-of-pipe-solutions'), the ecological design scope already considerably increases with the product innovations. The product life-phases are optimized in the meaning of a 'life-cycle engineering'.

Additional ecological improvement potentials arise with the transit to function and need innovations. Here, there is a chance to satisfy given needs through totally new products ideas, e. g. "benefit instead of selling goods".
With need-oriented innovations the given needs are also questioned themselves. This also shows that the solutions increasingly go beyond company limits by approaching the ecologically really interesting innovation perspectives. Finally they even represent society projects (see [2]).

This contribution aims at the processing of methodical bases which support the product, function and need innovation in the early product-development phase.

1.2 Objects

In the early development phase the use of checklists and guidelines is not or only partially possible ([3]). This is especially due to incomplete and/or inexact existing information. Further, calculations as e. g. within life-cycle-assessments or a simulation are only possible, if there are many assumptions made. These assumptions however, lead again to great uncertainties with regard to the evaluation and interpretation of the results. Therefore the question arises to which extent these solution approaches make sense for the present problem task.
Therefore this contribution discusses case examples of industrial and commercial products of the areas machine construction and Electro technique and evaluates them with regard to a corresponding heuristic. The aim is to develop a search heuristic supporting the analysis and search process within an innovation plan in the early product development phase.

1.3 Aspects

Within the heuristic, primarily ecology-oriented aspects are regarded. Therefore, the energy and material flows caused by company-own products and services are supplemented by a corresponding environmental relation (see [4]). The input stream is subdivided into materials, energy and resources flows. While materials and forms of energy go through preceded functions, resources are drawn directly from the Eco-sphere. Output, in addition to the product itself, involves undesirable flows which may be categorized in terms of by-products and emissions. The term "emissions" signifies several output flows delivered directly into the Eco-sphere. Products and by-products undergo existing functions (figure 1).

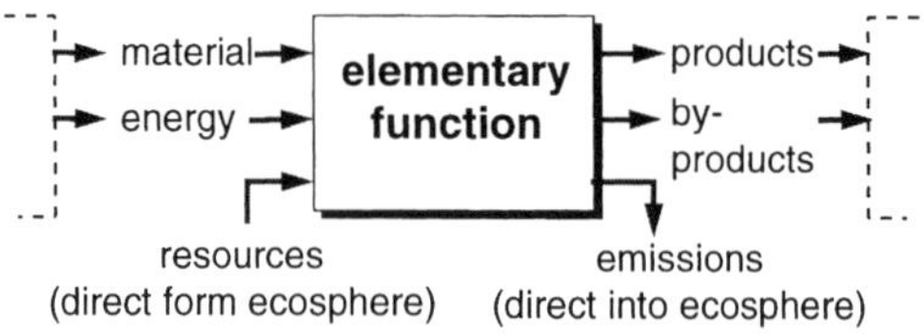

Fig. 1 Input and output flows ([4]

Many input and output streams may be specified in an in-depth manner. In this way products, by-products and emissions re-emerge as materials, energy or resources. By joining the corresponding elementary functions one after the other, a total life-cycle can be depicted (see [4]and [5]).

Here it is important that the direct environmental relation is visible in form of environmental impacts. This corresponds to the direct resource extraction and to releasing emissions. Registration and depicting the environmental impacts forms the basis for a systematic environmental evaluation.

2 EXAMPLES OF ENVIRONMENTAL PRODUCT EVALUATIONS

Within a search project about "environmental information-systems" the so-called "important environmental aspects" according to ISO 14001 (see [6]) had to be determined with representative product groups. As the industrial partners mostly determine the functions as well as the individual characterizations of the products, the whole life-cycle from material and energy supply until disposal was regarded. The product registration and evaluation was performed analogous to the previous chapter. For the quantitative evaluation both existing Eco-inventories (e. g. [7]) and evaluation methods (e. g. [8]and [9]) were used. The intermediary results as well as the individual evaluation steps were continuously interpreted and evaluated - as described in ISO 14040 [10].

2.1 Example 1: Products of the Sanitary Field

In this example products from the sanitary field were regarded. On the one hand, these were products in the field of supply and disposal and active products such as flushing systems or the urinal electronic to steer the flushing. (For the distinction between "active" and "passive" products see [11]). On the other hand, installation systems (= Vorwand-Installations-Systeme) were analyzed (see figure 2).

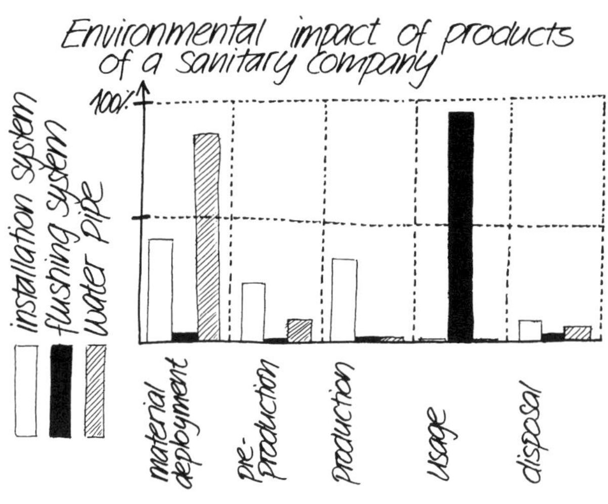

Fig. 2 Relative environmental impacts of various products of the sanitary field (data basis [12], published in [13])

The results show clear differences. So, it can be seen that the total local environmental impacts related to the whole production life-cycle is small to meaningless. There are enormous differences, however, in the usage phase. Passive products cause no or only irrelevant environmental impacts during their usage. Active products however, such as flushing systems, cause 95% and more of the entire environmental impacts during the usage phase. The production and disposal phases are irrelevant here.

The presented results must be applied in an effective environmental management. For product-developers work becomes easier with such information. So, with the flushing systems the flushing function must primarily be improved. Here, the designer does not need environmental knowledge in its narrow sense. He can focus on the flushing function. Further, the customer benefit is easy to communicate. In the future, he can reduce the use of water and thus also his costs.

2.2 Example 2: Various Cable Types

The next example describes the relative environmental impacts of various cable types with various ways of usage. On the one hand, these are stationary performance and signal cables. On the other hand, cables that are moved in the usage phase, as in cars, railway wagons or planes, are examined. The three examples are shown in figure 3 in an overview.

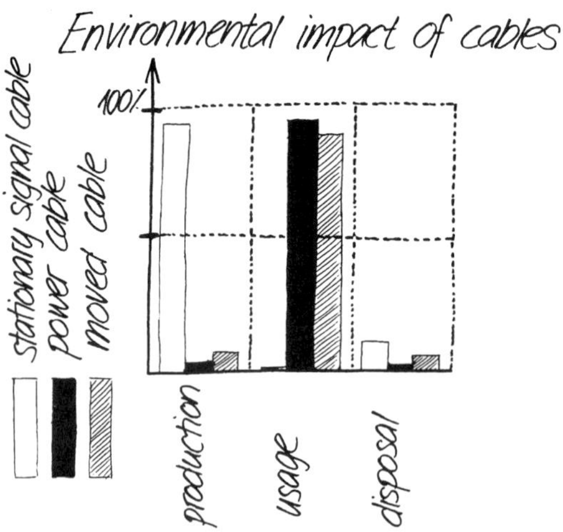

Fig. 3 Relative environmental impacts of various cable types (Data basis [14], published in [15])

Here, we have a similar picture as before. Passive products, such as signal cables, cause important environmental impacts by their material intensity (choice of material and processing process). The loss performance caused by the Ohm resistance leads to low ecological impacts during usage. With energy cables this is just the opposite. The loss performance during usage is rather high. The only possibility to reduce the transfer losses is an increase of the cross-section and thus an increase in material intensity. Further, if the cables are moved, the moved mass plays an important role. In this case the material intensity must be reduced despite simultaneous increase of the loss performance.

Both examples show different situations for similar products or for products of the same branch. The solution strategies must be adapted correspondingly. The example of the stationary energy cables also shows that a material-intensity reduction for cables, as demanded in many publications, is counter-productive. This would even worsen the environmental performance of the energy transfer. Here it should be considered to which extent the cable cross-section could be increased.

2.3 Conclusions

To conclude, six statements are made which can be deduced from the formerly presented basics and examples. The following hints should help to recognize environmental weaknesses more effectively in practical application (see also [16]):

1. The planning processes have a high environment relevance. There, the later in various places occurring environmental impacts are largely determined. It is irrelevant whether physical products or services are designed. With both, the planning-decision consequences on the later life phases must be regarded.

2. Analyses, considering the production site only, do not make sense normally (also see the example in chap. 4) because decisive problem and solution fields remain unrecognized.

3. Environment-critical materials can cause significant environmental impacts already in small amounts. Input-output analyses regarding only the large material flows are not suitable in this case.

4. With so-called passive products, i.e. products causing no further environmental impact during use, the material intensity often represents an important environmental aspect (for distinction between "passive" and "active" products see [11]). If the examined company has a high production depth, its production processes are also relevant. Here, environmental improvements can be achieved e.g. by a modified product design (choice of material, good use of material, design ...) or by other production sequences and processes (less rejects and less waste products caused by processes) (also see [17], [5]).

5. With active products, i.e. products causing significant environmental impacts by use, the way of function often plays an important role. Cars, electronic devices but also cables (energy transfer) and flushing systems, as shown before, represent active products.

6. With mobile objects, e.g. use of railway wagons, light ways of construction also should be regarded. The way and manner of material deployment and waste disposal is often subordinated with regard to a holistic improvement of environmental performance. (also see example before).

3 SEARCH HEURISTIC FOR ECOLOGICAL IMPROVEMENT POTENTIALS

Within ecology-oriented product design there is a repeated demand for a long life duration. Depending on the way and usage of the product, this characteristic can even be counter-productive to the utmost. Therefore, in the following the question is examined how differentiated demands can be raised with regard to 'function', 'light construction' and 'recycling ability'.

From the environmental perspective two features are central, based on the previous considerations. These are:

- Way of usage: the already regarded objects can have an 'active' or 'passive' behavior.
- Place of usage: during usage the objects remain at the same place, i. e. 'stationary' or they are regularly moved and are thus 'moved objects'.

3.1 Active Product, locally immobile

Active products are characterized by causing further environmental impacts during the usage phase through the usage of the intended functions and services. Here, the desired function can only be reached if additional processes can be introduced. Examples for active and immobile products are e. g. current meters in a building, TV sets in a room, refrigerators in a kitchen combination, flushing system in a toilet and stationary energy cables (see e. g. in chapter 2).

For active and locally immobile objects the following solution strategies emerge:

- improved function and lower losses: this demand primarily concerns the product functionality.
- strategies for material-re-integration: immobile objects can have a heavier construction. With a material-intensive construction the material choice must be made with regard to the material provision, processing and assembly / disassembly / material recycling.
- life duration: if no important product function-related technical innovations are to be expected a long life duration should possibly be achieved. If important technical changes are to be expected, the demand for a long life duration must be reconsidered.

3.2 Active Product, regularly moved

Active objects can also be regularly moved. Examples for active and moved objects are e. g. cooling aggregations in a vehicle, railway (transportation of people and goods) or TV sets in a plane.

For active and locally moved objects the following solution strategies are developed, contrary to immobile / active objects:

- improved function and lower losses: this demand primarily concerns the product functionality.
- light construction / miniaturization: reduced mass enables easier acceleration and slowing down.
- life duration: demand for long life duration is rather secondary. Here, a (too) long life duration can prevent the introduction of new product technologies as well as the demand for reduced mass.
- Strategies for material-re-integration: demand for recycling ability rather secondary because thus an improved function principle and a light construction is prevented.

3.3 Passive Product, locally immobile

In the usage phase, passive products cause no further environmental impacts by using the provided functions and services. Examples for passive and immobile products are e. g. furniture and facilities in a building.

Home worker and household devices can be close to passive products because of their rare usage (partly below 2 hours), although they do cause further environmental impacts during usage.
For passive and close to passive and locally immobile objects the following solution strategies result:

- long life duration: As no further environmental impacts result through usage and as the object is not regularly moved, the life duration must clearly be optimized by reconsidering production and disposal.
- strategies for material-re-integration: immobile objects can have a heavier construction. For a material-intensive construction the material choice must be made with regard to processing and assembly / disassembly / material recycling.
- Light construction and miniaturization are rather meaningless as they shorten the life duration and reduce the environmental performance.

3.4 Passive Product, regularly moved

Passive objects can also permanently be moved. Examples for passive and simultaneously moved objects are seats and desks in a vehicle (plane, car, railway) and re-usable package and transportation materials.

For passive and locally moved objects there are the following solution strategies:

- light construction / miniaturization: reduced mass enables easier acceleration and slowing down.
- life duration: demand for long life duration is rather secondary. Here, a (too) long life duration can prevent the demand for reduced mass.
- Strategies for material-re-integration: demand for recycling ability rather secondary because thus a light construction is prevented.

4 DISCUSSION

The examples in chapter 2 have clearly shown the great differences within the same trade and within the same products with regard to the whole life-cycle. At the beginning of the planning effective strategies are required both in the analysis phase and within a first solution search for Eco-effective product development. One possibility consists in the application of the previously described search heuristic. First structure and estimations can be made at an early planning stage. Thus, sensible work hypotheses can be derived and the development process can be arranged into sensible steps.

The search heuristic can be supplemented by further dimensions (also see [18]and [15]). These are e. g.:

- kind of materials or substances: materials have an influence on the environmental performance of an object based on the providing, processing and recycling processes.
- mixture grade: usually, materials are available in a 'mixed' form. Therefore, it must be examined how the various materials are mixed or alloyed with each other.

- geographical distribution: the individual objects can have a local, regional or global distribution. Thus various distribution service and / or recycling concepts emerge. This again influences the environmental performance or services of a product.
- ownership: Products often change the owner after the production phase. Therefore manipulation and changes made by the new owner are possible. Thus, taking-back concepts or 'up-grading' possibilities are only limited.
- At present, an attempt is made to apply the previously presented search heuristic in development projects. The search heuristic should further be specifically supplemented and be evaluated with regard to the effectiveness.

5 LITERATURE

[1] Schneidewind, U., "Mit COSY – Company oriented Sustainability – Unternehmen zur Nachhaltigkeit führen", Diskussionspapier Nr. 15, IWÖ-HSG, St. Gallen, 1994.

[2] Minsch, J. et. al., "Mut zum ökologischen Umbau", Springer Verlag Basel, Boston, Berlin, 1996.

[3] Luttropp, C., Züst, R., "Eco-effectiv products: Holistic view point" (Keynote-Paper), in 5th International Seminar on Life Cycle Engineering, Proceedings Life Cycle Design'98, KTH Stockholm, 1998, pp. 45-55.

[4] Caduff, G., Züst, R., "Increasing Environmental Performance via Integrated Enterprise Modelling", in ECO-Performance - 3rd International Seminar on Life Cycle Engineering CIRP, edited by Züst, R. et al., IO Verlag Zürich, 1996, pp 39-46.

[5] Caduff, G., "Umweltoreintierte Leistungsbeurteilung", Gabler Verlag Wiesbaden, 1998.

[6] n.n., "EN ISO 14001 Umweltmanagementsysteme - Spezifikation mit Anleitung zur Anwendung, CEN-Zentralsekretariat (Publisher) Brüssel, 1996.

[7] n.n., "Ökoinventare für Verpackungen", Schriftenreihe Umwelt Nr. 250 (Band I und II), Bundesamt für Umwelt, Wald und Landschaft (BUWAL) Bern, 1996.

[8] Ahbe, S., Braunschweig, A., Müller-Wenk, R., "Methodik für Ökobilanzen auf Basis der ökologischer Optimierung", Schriftenreihe Umwelt Nr. 133, Bundesamt für Umwelt, Wald und Landschaft (BUWAL) Bern, 1990.

[9] Goedkoop, M., "The Eco-Indicator 95 - Weighting Method for environmental effects that damage ecosystems or human health on a European scale", Amersfoort (NL), Pre consultants, 1995.

[10] n.n., "EN ISO 14040 Environmental Management Systems - Principles and framework", CEN-Zentralsekretariat (Publisher) Brüssel, 1998.

[11] De Winter, A.D., Kals, J.A.G., "A methodic approach to the environmental effects of manufacturing", in CIRP 2nd international seminar on Life Cycle Engineering, Feldmann, K. (Publisher),: Meisenbach Verlag Bamberg, 1994, pp.287-301.

[12] Gerber, R., "Bestimmen der bedeutenden Umweltaspekte der Produkte der Firma Geberit", Unveröffentlichte Diplomarbeit, Betriebswissenschaftliches Institut (BWI) der ETH Zürich, 1997.

[13] Frei, M., Züst, R., "Die öko-effektive Produktentwicklung - Die Schnittstelle zwischen Umweltmanagement und Design", in Markt- und Kostenvorteile durch Entwicklung umweltverträglicher Produkte (VDI-Berichte, Nr. 1400), VDI Verlag Düsseldorf, 1998, pp. 51-71.

[14] Seipelt, D., "Bestimmen der bedeutenden Umweltaspekte von Energie- und Signalkabel der Firma Huber+Suhner", Unveröffentlichte Diplomarbeit, Betriebswissenschaftliches Institut (BWI) der ETH Zürich, Zürich, 1998.
[15] Frei, M., "Öko-effektive Produktentwicklung", Gabler Verlag Wiesbaden, 1999.
[16] Züst, R., "Principles of Environmental Assessment in the Context of Design Process" (Keynote-Paper), in 5th International Design Conference - Design 98, edited by D. Marjanovic, WDK Zürich, 1998, pp. 53-63.
[17] Mannhart, M., "Umweltmanagement für KMU", Unveröffentlichte Projektarbeit, Betriebswissenschaftlichen Institut der ETH Zürich, Zürich, 1996.
[18] Frei, M., Caduff, G., Züst, R. "Eco Effectiveness - Systematic Inclusion of Ecological Aspects in Product Development", in NordDesign '96, Helsinki, pp. 130-140.

Design for tribology and design for environment – two goals, one way

B ROSEMANN, C SCHAUFLER, and **H MEERKAMM**
Lehrstuhl für Konstruktionstechnik, Friedrich-Alexander-Universität, Erlangen-Nürnberg, Germany

SYNOPSIS

Solving tribological problems, traditionally, means the usage of lubricants e.g. grease or oil, which normally are harmful for the environment. Thin film coatings with PVD solid lubricant films can help to solve the tribological problem on the one hand and enable us to 'Design for the Environment', on the other. Some characteristics of machine elements, which have interacting surfaces standing in relative motion, have impacts both on the environment and on the life cycle of the product. A competent knowledge of the effects of wear and friction between these interacting surfaces as well as knowledge concerning the effects of lubrication are prerequisites when determining function, quality and cost of many products in our life. Such knowledge is the pre-condition for the realisation of products with tribological elements that exhibit excellent properties in both fields. For a holistic optimisation it is very important to emphasis fundamental considerations in the early stages of a methodical product design process and to manufacture appropriate tribological systems by electing to use environmentally friendly technologies. Products, which use tribological construction elements, have different environmental consequences. For example, lubrication has to be used to guarantee functionality, but leaking lubrication into the environment could cause the pollution of ground water. Recycling of such parts calls for the used lubrication materials to be removed in expandable and partly difficult processes. An appropriate technology, which could help to avoid or reduce these environmental problems in the whole life-cycle, is the thin-film-technology, more specifically the PVD-process (Physical-Vapour-Deposition). This is a non-polluting method for producing thin films (100nm up to 5 μm) on high precision mechanical parts with rolling/sliding contact, used to retain reliability. Different coating systems e.g. low-friction for wear protection or soft solid lubricant films can be created. Thin-film-technology avoids using toxic materials e.g. by application of carbon or molybdenum disulphide as lubricants. In recycling, no additional steps or special processes are necessary. Examples for solid lubricated mechanical parts are deep groove ball bearings for x-ray tubes, ball joints, uniball bearings and linear guidance systems that are briefly presented.

1 PREFACE

Today, in Technical Systems, which at least partly consist of mechanical functions, there are various elements that have interacting surfaces in relative motion. In the phase of conceptual and embodiment design, it is the task of the product designer to select and dimension appropriate mechanical components in such a way, that an optimal function is achieved for the required life span.

In the course of the product life, an increasing wear of the contacting surfaces occurs, as a consequence of use, until the product cannot fulfil the required function any longer. In the Federal Republic of Germany approx. 5% of the gross national product is lost due to the mechanisms of wear. This corresponds to a value of approximately 170 milliard DM (approx. 87 milliard €) [17].

Usually friction is directly connected with wear and is respectively, one of the main causes of wear. Increased friction accelerates the wear and leads to an increase of the waste heat and, with it, to thermal load of the parts as well.

In order to achieve a sufficient solid surface separation and subsequently a minimisation of the wear, lubricants such as oil and grease are used. To guarantee their function as well as the request of required life span, these lubricants are supplemented with additives. Fluid oil based lubricants and especially additives are probably the most ecologically questionable part of the production. This is as true for their use as well as in their disposal and they cause high costs.

Resulting from this, there is the desire to reduce the application of such additives as well as the prescribed amount of lubricant, which is used. However, through this reduction, the danger of system failure, caused by insufficient of lubrication, increases.

2 DESIGN FOR THE ENVIRONMENT

Design for the Environment is a subject in the field of Design for X (DFX). The necessity for more intense consideration of this aspect as one part of the Design for X has its roots not only in actual problems of waste management and environmental pollution, but also in the shortages of natural resources.

If we consider the whole life-cycle of products, whether PC's, household appliances, ball bearings, linear guideways etc., we determine, that these products produce a variety of different environmental impacts arising from processes during the production and the usage and finally during recycling and disposal. These environmental burdens are mainly defined during the product development process. Therefore it is the task of the designer to consider all phases of the life-cycle in order to design an environmentally friendly product instead of relying on so called end-of-pipe technologies, which filters residues or renders it harmless.

The design methodology offers the designer a problem-oriented procedure to reach new and innovative solutions with less iterative steps. This methodology also offers the framework of the Design for the Environment (DFE). After the definition of requirements, the designer starts to describe relevant functional requirements to his product and groups them in a function structure. The general requirement to ecological products is the holistic minimisation of environmental burdens like production of emissions, waste heat, noise waste materials or

products, etc., whereas holistic means, with regard to the hole life-cycle. After searching for principal physical effects, evaluation and selecting principle combinations and the first topological relations are generated by building up working faces and a working structure. For these solutions, machine elements are selected and dimensioned in the next step. Missing or incomplete areas of the embodiment design are detailed.
In addition to the conventional way, the designer needs a lot of information about the impacts of his design drafts and associated processes to the environment on the one side, and a assistance to evaluate the consequences of different design decisions on the other hand (see Fig. 1). Different tools to support this processes in DFE are presented in more detail in [16],[22].

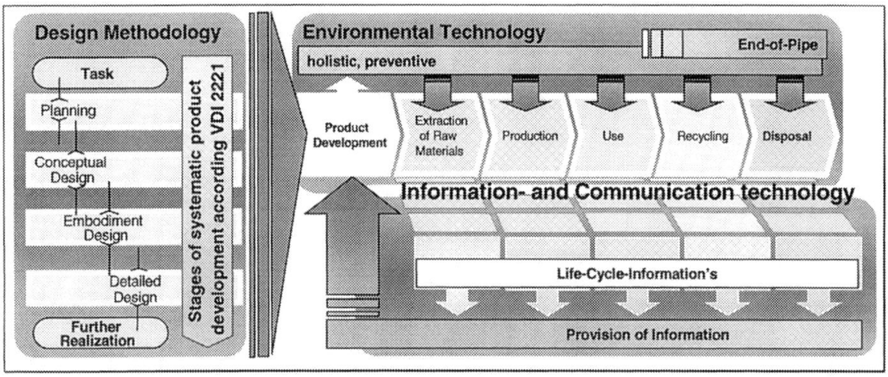

Fig. 1 Life-Cycle-Information's needed for the development of environmentally sound products in the framework for systematic product design according to VDI 2221 [25]

Considering the life-cycle of state of the art solutions in tribology to identify environmental burdens in this field, we get to the result that one important influence is the use of fluid lubricants in most cases of bearing points. The problems arising with conventional lubrication oriented on the three phases manufacturing, use and recycling/disposal are discussed in detail in Section 3. In Section 4, we present different approaches to meet tribological requirements and posses decreasing environmental burdens.

3 MAJOR PROBLEMS

Tribology is the science and technology of interacting surfaces, which are in relative motion, as well as related topics and methods (Jost, 1966). Friction and wear, as well as lubrication also belong to the thematic complex of tribology. With respect to a tribo-technological system one differentiates between the sizes of the stress collective and the tribological system structure as well as between the friction and wear characteristics [5].

3.1 Today's state of the art

If two items are in contact, depending on the state of the contacting surface, we have to differentiate between *dry friction*, *mixed friction*, *liquid* and *gas friction* [4]. The elementary mechanisms of friction, which are adhesion, plastic deformation, crenation, flexible hysterises and absorption are usually overlaying in real systems (see Fig.2).

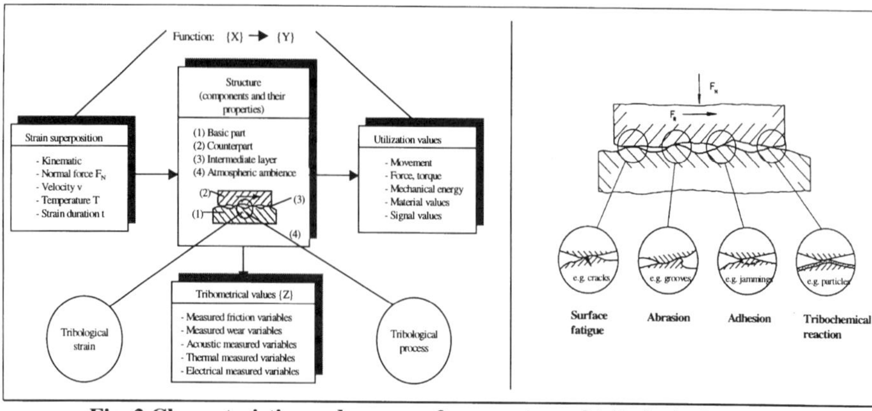

Fig. 2 Characteristics and groups of parameters of tribological systems and mechanisms of wear [1].

By the relative motion of the bodies there will be a progressive waste of material from the surface of one or both friction partners. It is the function of the lubricant to separate these contacting surfaces and thereby to minimise wear and ensure the function of the assembly. The lubricant can be present in liquid (oils, grease) and in fixed form (solid lubricants such as C, MoS_2).

State of the art, today, is the application of liquid lubricants with the associated pro and cons. During hydrodynamic lubrication, components have to be sealed according to a high constructional expenditure, to prevent a leaking of the lubricants into the environment. Furthermore, complex lubrication systems are necessary. However, the use of oils dramatically decreases wear and frictional heat, which develops as result of friction contact, will be lead away.

3.1.1 Liquid Lubricants (Oils and Grease)

Lubricants are made from basic oils, thickener and additives. In the evaluation of their hazard potential, these lubricants are analysed in "new" condition with no additives [6]. Additives are normally those substances in lubricants, which are toxic and environmentally hazardous.

Due to the thermal and mechanical load of the lubricants, in operation, a modification of the material could occur through chemical and physical processes. Apart from a negative effect on the function of the component, sometimes this leads to chemical accruement of toxic reaction products. Today, however only small knowledge exists about the status of used lubricants, since toxicological analyses are expensive and time-consuming and are thus only rarely performed. Therefore, the application of biologically degradable lubricants without additives in quantities as small as possible is to be aimed.

3.1.2 Solid Lubricants

Fixed lubricants possess the same functions as liquid lubricants. The following table shows an overview of substances used as solid lubricants.

Table 1 Overview of solid lubricant substances

lubricant types	lubricant	application
joinings with layer grid structure	graphite, MoS_2, WS_2	roller bearing, vacuum technology (not graphite)
metalls	gold, silver, lead	elektric,thermal conductance, x-ray tube bearings
polymers	PTFE, Polyamid, Polyimid	sliding contact bearings, cages
comnpouds	PTFE + glas fiber	roller bearing cages

If we compare fixed lubricants and liquid lubricants in regards to the stability against external influences and in regards to loads and costs (Table 2), we can conclude that fixed lubricants, are substantially more resistant. Presently, the part life is still low, when pure dry lubrication is used and pure dry lubrication has comparatively high coating costs. This is due to equipment technology still being expensive.

Table 2 Comparison of the properties of liquid and solid lubricants [23]

requirements/ properties	consis-tance	vacuum	low/high temp.	flamm-able	radi-ation	high velocity	life	cost
solid lubricants	+	++	++	++	+	-	-	-
fluid lubricants	-	--	--	-	--	+	+	+

3.2 Life-cycle oriented problems in the use of liquid lubricants

The use of liquid lubricants compared with fixed lubricants, however, is accompanied by disadvantages related to the product life cycle.

3.2.1 Manufacturing

The raffination process within the production of lubricants is very resource intensive, but we will not describe this process in more detail. For a more technical description, we refer to appropriate literature [1],[2].

The plasma supported thin film technologies for the production of wear-protection layers, such as PVD technique, are generally known as very environmental friendly processes. Contrary to the galvanic techniques (electrochemical coating), no poisonous substances result in the coating and purification processes [18].

3.2.2 Use

- Depending upon the field of application, in relation to the ageing of assigned fluid lubricates it is necessary to change the lubricant in maintenance intervals in order to ensure a safe operation of functional bearing points.
- A further cause for necessary oil changes is the need to remove rubbed off particles, which will be collected in the lubricant with increasing working time. This particles promotes additional wear.
- From lubrication leakage, large quantities of lubricating oil are introduced into the environment. Estimated values from the year 1987 assume that, annually in the FRG approx. 30,000 to 35,000 tons of oil will contaminated the soil and air. Due to the increase in use of goods in the last decades, we have to assume that a rise in lubrication oil leakage will occur [17]. Examples for leakage lubrication are:
  - o two-stroke engines (lawnmower, chainsaw, outboard motors, etc.)
  - o sawing chain lubrication

- o lubrication of switch points
- o applying of (mould) release agent (conversion technology)
- o wire lubrication
- o lubrication of unhoused gears
- o etc.

3.2.3 Disposal

- The Federal Government of Germany expects an increase in waste oil of approximately 332,000 tons per year until the year 2005. Industrial lubricants produce approx. 40% of this waste. Since the predominant part of these materials is not biologically degradable, they require special disposal. The subsequent burning process is connected with high emissions [24].
- Recycling processing procedures for waste oil cannot be implemented without pre-treatment. The results of these processes are mineral oil emulsions, which are likewise burned. Thereafter, the residual substances are deposited [17].
- Before a material utilisation of used functional elements, a purification process must be implemented. These processes often operate with cleaning agents, which is again environmentally problematic.

4 APPROACHES FOR IMPROVEMENT

An ideal solution for avoiding environmental impacts would be the complete denouncement of fluid lubricants. Since this is largely not possible today in the fulfilment of the function of moved components, we should aspire to use the smallest quantity of lubricant as possible and preferable in solid form.

4.1 Approaches in R&D

4.1.1 Minimal Quantity Lubrication

A basic approach is the lubrication with an just sufficient amount of oil. If you look at the friction torque and the temperature of cylindrical roller bearings versus the used amount of oil, it can be determined, that the friction torque, the temperature and also the wear is minimised by a reduced amount of oil. But if the oil falls below a critical quantity, the roughness peaks of the material couple make contact (solid friction) that results in an increase of the wear. Lubrication with an optimal amount of oil is called Minimal Quantity Lubrication (MQL). With a minimum volume of lubrication and also lubrication costs a good tribological behaviour will be reached, and environmental protection is guaranteed.
Machine elements with MQL normally show a lower damping and a higher acoustic emission. If cylindrical roller bearings operate in the area of MQL temperature, friction and wear are obviously reduced.

4.1.2 Function Relocation

Another approach in avoiding environmental harmful lubrications is to remove the toxic additives out of the oil or grease and put their function (corrosion inhibitor, oxidation inhibitor, wear resistant, friction reducer, adhesive strength conditioner etc.) into a thin film which is coated on the surface. In the special research project 442 "ecological tribological systems" the goal is, on the one hand the development of economically beneficial cooling lubrication for cutting tools and, on the other hand the development of rolling bearings, linear guidance systems and warm gear spindles with lubrication optimised operation. Both goals should be achieved by using of a combination of special additive-free fluids and PVD-coatings. The function of the additives is substituted by new PVD-coatings. For further information refer to [21].

4.1.3 Dry Lubrication

The use of thin layer coatings, e.g. PVD coatings, such as wear-protection layers on tools of the conversion technique as well as the cutting manufacturing is already state of the art. They are also used increasingly as fixed lubricant layers on rolling/sliding elements. For example: Coatings on engine parts (plunger) or on fuel injection system parts in automobiles. Such thin films can effectively protect precision and highly loaded components from wear, due to their high hardness, good sliding qualities and dimensional accuracy.

In pollution-sensitive areas, like food or textile industry, unlubricated operation with solid lubricant layers represents an ideal solution. Other areas such as air and space technology or medical technology do not get by without solid lubrication, since the extreme conditions (vacuum, extreme temperatures) do not permit the use of grease or oil in this application (degassing, decomposition). Layer systems, which are able to fulfil the fore mentioned specified requirements, are films based on carbon or MoS_2. In the following section, such layer systems, which were developed at our institute, will be represented.

4.2 Process Technique

The PVD process used at the institute belongs to the group of the plasma-supported thin film procedures. It operates according to the principle of the d.c.-cathodic sputtering. The layer material will be transferred from the fixed phase into the gaseous phase, without the necessity of a minimum temperature. The material vapour condensed onto the surface of the components and forms a thin layer. Based on the low process temperatures, it is possible to coat temperature-sensitive materials such as roller and ball bearing steels, synthetic material or plastics.

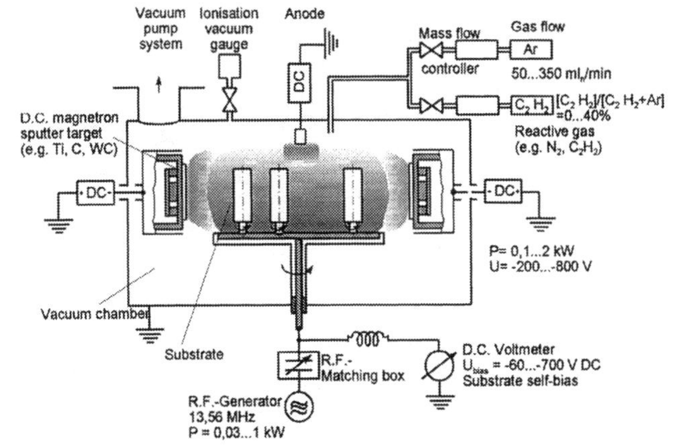

Fig. 3 Schematic diagram of a CC800 coating system for the deposition of low-friction and wear protecting coatings.

Such films were prepared at the Chair of Engineering Design (KTmfk) using a CC800 PVD coating system (CemeCon GmbH, Germany). This Coating system is illustrated in Fig. 3. The coating machine was equipped with two water-cooled unbalanced dc magnetron cathodes and a capacitively coupled r.f. discharge (13.56 MHz) system, which is connected to the rotating table for the purpose of argon plasma etching of the substrates. Before the sputter coating starts the integrated etching unit cleans the substrates and adhesive strength is thereby improved. Because of the fourfold substrate rotation, even mechanical parts of complicated shape can be coated fairly uniformly.

4.3 Technical implementation

In the following sections we will discuss two coatings which were developed at our institute.

4.3.1 Diamond like carbon coatings

The application of solid lubricant coatings is useful when reliability and life cycle of mechanical parts, working under special operating conditions (e.g. extreme environmental conditions), is important. PVD coatings increase the lifetime of machine parts under critical friction conditions (mixed lubrication and solid-body friction at poor lubrication and under dry running conditions). A decisive criterion for the functionality and lifetime of precision parts are often their frictional properties. The deposition of a thin, low-friction PVD coating gives the treated parts the necessary tribological lubrication properties. The low dry friction coefficient $\mu \leq 0.25$, when sliding against steel, provides a protection even for the uncoated sliding partner against wear.

Carbon coatings, produced in the PVD machine CC800, exhibit, in spite of their high hardness and wear resistance, low-friction coefficients. Another advantage is the low deposition temperature of about 100 to 150 °C.

The amorphous, hydrogen free carbon coating (a-C) is produced by magnetron sputtering from graphite targets and shows low intrinsic stresses. Its hardness is about 1,400 HV [8].

The metal containing amorphous carbon coatings (Me-C:H) is a further development of the very hard diamond like carbon coatings (DLC, hardness 2,500 - 3,000 HV). They are deposited by a combination of magnetron sputtering of metals (e.g. Ti, Cr or WC) and plasma decomposition of carbon hydrogen gas (e.g. methane, acetylene). This recently developed coating exhibits improved friction and wear properties. Because of its lower hardness (1,500 - 2,000 HV), its intrinsic stresses are lower and, therefore, the adhesion on the substrate is improved. In contrast to the DLC coating, the tribological properties of Me-C:H show a greater independence on the environmental humidity.

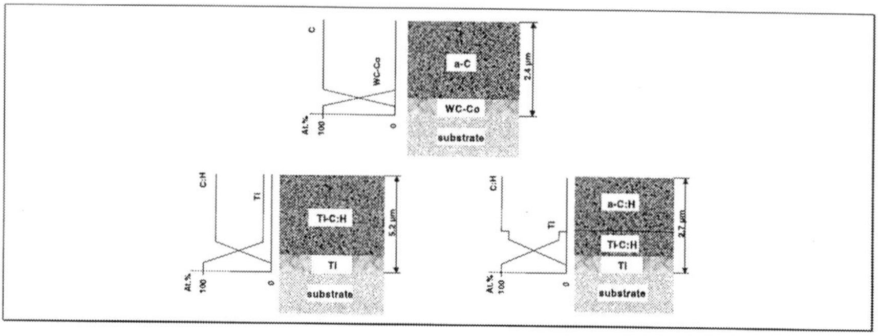

Fig. 4 Structure of different carbon layers

4.3.2 Metal containing molybdenum sulphide coatings

When surfaces of mechanical parts running under extreme application conditions (aerospace, nuclear and vacuum technology) have to be deposited with a low-friction wear resistant coating, MoS_2 containing PVD sputter coatings are appropriate. Because of the sensitivity of MoS_2 to humidity, the application possibilities of this soft lubricant under certain environmental conditions are limited. In vacuum and dry air, however, the friction coefficient against steel can fall to values of about 0,01. With specially developed PVD process, the tribological properties of the deposited coatings could be optimised. Co-sputtering of metals or cemented carbides (WC-Co) give the soft MoS_2 an improved tribological and wear resistant property, in particular, at ambient conditions.

The metal doped MoS_2:Me films are deposited with a coating thickness of 0.5 - 1.5 μm and exhibit a good uniformity and adhesive strength, even on parts of complex geometry. Because of the low deposition temperature (≤ 120 °C), the process doesn't effect a distorting of the parts and, moreover, no loss of hardness of the bearing steel takes place. In contrast to bonded MoS_2 films (or similar solid lubricant films), PVD sputtered MoS_2:Me contains no organic binders that are sensitive to high energetic radiation (e.g. x-rays) and shows a negligible degassing in vacuum.

4.4 Application Areas

As a result of the combination of wear-resisting quality and good friction characteristics, application fields for such layers on tribological stressed components arise. Therefore, for typical parts of application could be find in the field of mechanical engineering.

For example, application fields and typical parts (machine elements) in product design are:

- bearing points
    - sliding bearings,
    - ball bearings
    - linear bearings
- drive elements
    - gear wheels,
    - shafts or axes,
    - chains,
- pairings of pistons and cylinders,
    - motor pistons
    - anti-shock pads
    - valves
- etc.

In many cases, a clear reduction of the assigned amounts of solvent lubricants is possible. In some cases a lubricant-free operation can be achieved.

In the field of production engineering, primary, we distinguish two application areas, in which, similar to application fields of vapour deposition we are able reduce environmental burdens of solvent lubricants:

- conversion technologies (non cutting) and
- cutting technologies.

5 EXEMPLARY RESULTS OF RESEARCH

In the following sections of the paper, we will present examples of our research.

5.1 Ball Joint

One of the fields of research at our Institute is the improvement of the tribological behaviour of ball joints (Fig. 5). This field is generally marked by breakaway and lunging problems. The causes for these effects are often the used grease that gives rise to adhesion and sticks during the motion.

In contrast to the ball joint greased steel/plastic couple, which exhibits enormous noticeable break-away at starting motion the coatings on the plastic bowl achieve a clear reducing of these unwished stictions at starting [9], [14].

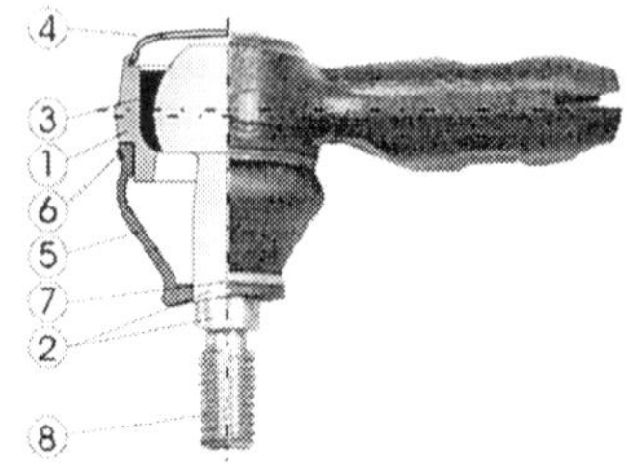

1 joint case
2 ball-ended spindle, steel
3 bowl, plastic
4 steel plate cover
5 sealing
6 stretching ring
7 stretching ring
8 drilled for cotter pin

Fig. 5: Assembly of a Ball Joint

5.2 Uniball bearing

Another machine part, which was examined, is the uniball bearing (Fig. 6). The aims hereby were to reduce the grease consumption, the bearing clearance and to improve the bearing endurance by using a low friction wear resistant PVD-coated inner ring.

A variety of different coatings were tested (hard nitrides, MoS_2 and hard carbons) in special uniball bearing test rigs. Good results were obtained with a Cr-C:H coating (2000 HV, thickness 10 – 15 μm) to extend the re-lubrication intervals. Fig. 6 shows the number of stress cycles until an increase of a bearing clearance of 60 μm as well as the grease consumption for a PVD-coated bearing and a standard bearing. The PVD-coated uniball bearing was re-lubricated with grease after each 200 000 stress cycles, whereas the standard bearing after each 125 000 stress cycles [13],[15].

The coated bearing had an endurance of 500 000 stress cycles and 9 grams of grease were consumed, the standard bearing had an endurance of 750 000 stress cycles and the grease consumption was 12 gram per 500 000 stress cycles (25 % of grease consumption).

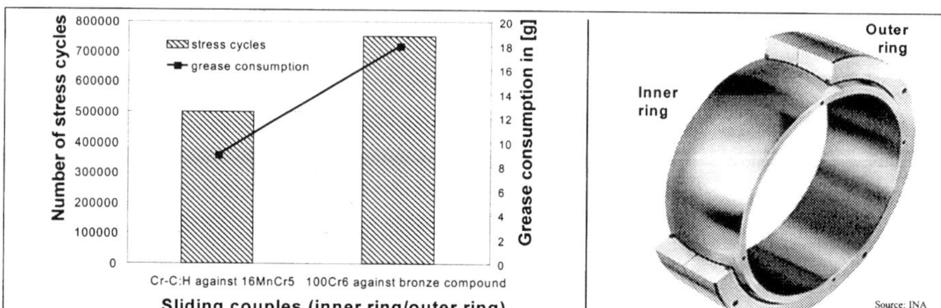

Fig. 6 Number of stress cycles up to an increase of a bearing clearance of 60 μm of uniball bearings

5.3 Linear Roller Bearings and Guideway Assemblies

Certain branches of industry require linear guidance with ball rotating (see Fig. 7) units, which have to be used in unlubricated operation (machine tool manufacture, paper -, textile -, and food industry). In oil-lubricated linear bearing units, an increased wear danger exists due to snubbing and slowing down at the oscillating operation and, therefore, exhibits no building up of a hydrodynamic lubrication. By lubrication intervals a contamination risk exists for the environment.

At the institute - in the research program FOROB II - drying lubricant films, which are based on carbon and molybdenum disulfide are developed for ball rotating units, which meet the described criteria. The film systems are to be applied on the balls, since these are in contact with all components of the tribological system. The fixed lubricants are to be distributed by transfer lubrication with assistance of the balls on cars and rail. Ball coating process economic advantages, since these can be coated as bulk material [19].

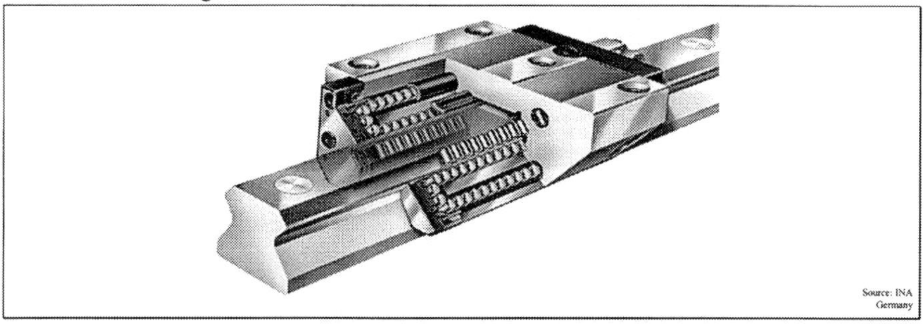

Fig. 7 Typical linear guideway system

5.3 Bearing of X-Ray Tubes

An application in which a solid lubrication is stringently required is deep groove ball bearings applied to rotating-anode X-ray tubes (see Fig. 8), which are used in a typical 10^{-4} Pa vacuum, at a rotating speed of about 9,000 rev min^{-1} and a maximum temperature of 450 °C (see Fig. 8).

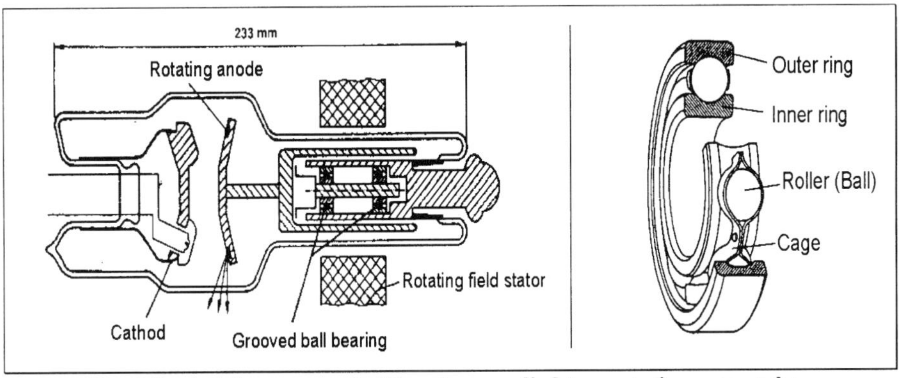

Fig. 8 Coated deep groove ball bearings applied to a rotating x-ray tube

Here soft metal coatings like silver or sulphides like MoS_2 behave fairly well as solid lubricants in bearings running under these severe environmental conditions. In a project of our institute special multi-layer coatings were developed which outranged the state of the art coatings in a pin-on-disk model test rig. The coefficient of friction after 10 km sliding distance could be lowered about 15- 35%, by a reduction of 90% of the ball wear rate and 20-50% of the coating and substrate wear rate. Field tests will follow [12].

6 SUMMARY

Liquid lubricants today often smear tribological systems. In many cases, these liquid lubricants are the source of much environmental problems, which could be reduced by use of coating technologies.

A possible solution is represented by the application of modern thin-film-technologies. With these processes, it is possible to coat thin, low friction as well as wear resistant solid lubrication film onto surfaces. Such layers are appropriate due to their small thickness - within the range of 100nm to 10 µm - even on highly precise components. With high adhesive strength then applied, these layers together assume the function of separating rubbing surfaces. The advised unlubricated operation, in conventional terms, holds a high reduction potential of costs, e.g. by a shortened need of maintenance and restoration (through life span lubrication), by simplifications in the design of bearing points (through omission of lubrication mechanisms, seals, etc.), and by reduction of energy use and material use (through production and application of the lubricants). Typical solid lubricants, such as carbon in different modifications (amorphous hydrogen-free carbon a-C, metalliferous hydrocarbon layer Me-C:H as well as diamond-similar carbon layer DLC) and molybdenum sulphide layer systems (pure or metal-doped) are completely ecologically harmless. When such layer systems are applied with a pollution free coating process, e.g. by means of plasma-supported PVD procedures (Physical Vapour Deposition), this combination then represents a distinguished possibility of ensuring the function of components in a very pollution free manner.

7 REFERENCES

[1] Bartz. W. J., „Tribologie und Schmierungstechnik", Kompendium Teil I, January 1997.
[2] Bartz. W. J., „Tribologie und Schmierungstechnik", Kompendium Teil II, January 1997.
[3] Czichos, H., Habig, K.-H., „Tribologie Handbuch", Vieweg-Verlag, Berlin, 1992.
[4] DIN 50281, „DIN 50281: Reibung in Lagerungen", Beuth Verlag, Berlin, Oktober 1977.
[5] DIN 50320, „DIN 50320: Verschleiß", Beuth Verlag, Berlin, Dezember 1979.
[6] Dornhöfer, G., „Lebensdauerschmierung von Kfz-Aggregaten unter Berücksichtigung der Umweltschonung", Tribologie und Schmierungstechnik, 41. Jahrgang, 2/1994
[7] Hambrecht, R.; Meerkamm, H., „Smearing in Grease Lubricated Roller Bearings", Tribology 2000 - Plus, 12th International Colloquium Tribology, Stuttgart/Ostfildern, January, 2000.
[8] Krumpiegl, T., Meerkamm, H., Fruth, W., Schaufler, C., Erkens, G., Böhner, H., „Amorphous carbon coatings and their tribological behaviour at high temperatures and in high vacuum", Surface and Coatings Technologies 120-121, 1999, P. 555-560.
[9] Meerkamm, H.; Fruth, W., „Reibungs- und Verschleißminderung in Kugelgelenken durch PVD-/PCVD-Beschichtungen für die Kfz-Industrie", FOROB, 2. Statusseminar, Erlangen, December 1996.
[10] Meerkamm, H.; Fruth, W.; Krumpiegl, T., "Reibungsarme Verschleißschutzschichten für ungeschmierte Lagerkomponeneten", VDI-Werkstofftag '97, Neu-Ulm, June 1997.

[11] Meerkamm, H., Fruth, W, Krumpiegl, T., Schaufler, C., „Mechanical and tribological properties of PVD and PACVD wear resistant coatings", International Journal of Refractory Metals & Hard Materials 17, 1999, 201-208.
[12] Meerkamm, H.; Fruth, W.; Schaufler, C,. „Verbessertes Lagerlaufverhalten in medizinischen Röntgenröhren", Beitrag für den Uni-Kurier, Sonderheft Medizin und Technik, Erlangen, July 1997.
[13] H. Meerkamm, T. Krumpiegl, „Vermeidung von Schmierstoffen in Gelenklagern durch PVD-Beschichtungen", BMBF-Abschlussbericht, Teilvorhaben: Untersuchungen zum tribologischen Verhalten beschichteter Gelenklager, FKZ 13N6220, Erlangen, 1997.
[14] Meerkamm, H.; Krumpiegl, T.; Fruth, W.; Schaufler, C., „Einsatz von PVD-Verschleißschutzschichten bei Stahl / Stahl- und Stahl / Kunststoff-Gleitpaarungen als Schmierstoffersatz sowie zur Reduzierung von Stick-Slip-Effekten", in VDI-Berichte 1380 Gleit- und Wälzlagerungen - Gestaltung, Berechnung und Einsatz, Veitshöchheim, March 1998.
[15] Meerkamm, H.; Krumpiegl, T.; Fruth, W., „Tribologische Optimierung von ungeschmierten Lagerkomponenten durch reibungsarme PVD-Verschleißschutzschichten", Tribologische Fachtagung 1997 - Reibung, Schmierung und Verschleiß, Göttingen, 30 September/October 1997.
[16] Meerkamm, H., Schweiger, W., Rosemann, B., „ Sustainable Development as a Basis for the Reduction of Waste", Proceeding of 2nd international BayFORREST Conference on Development and Application of Waste Technology, Munich, July 1-3,1998.
[17] Möller, U. J., „Umweltaspekte beim Einsatz von Schmierstoffen", in VDI Berichte 680, Das Öl als Konstruktionselement, VDI, Cologne, 1988, Page 321-336.
[18] Simon, H.; Thoma, Marin, "Angewandte Oberflächentechnik für metallische Werkstoffe: Eignung, Verfahren, Prüfung, Hanser-Verlag, Wien, 1985.
[19] N.N., „Trockenlaufende und verschleissreduzierte Kugelumlaufeinheiten in Linearführungen durch Entwicklung und Einsatz keramischer Werkstoffe und PVD-Festschmierstoff-Multilayer-Schichten", in 1. Ergebnisbericht FOROB II, Bayerischer Forschungsverbund für Oberflächentechnik, Erlangen 1999, Page 13-32.
[20] N.N., „GfT Gesellschaft für Tribologie e. V. – Ziele, Aktivitäten", Moers, 1998.
[21] N.N., "SFB 442 – Umweltverträgliche Tribosysteme", http://www.rwth-aachen.de/ifas/Ww/sfb442/sfb_home.html, RWTH-Aachen, 2000.
[22] B. Rosemann, H. Meerkamm, St. Trautner, K. Feldmann, „Design for Recycling, Recycling Data Management and optimal End-of-Life Planning based on Recycling-Graphs", Proceeding of ICED 99, Munich, August, 1999.
[23] Schröer, A.; Savan, A; Pflüger, E., „Reibungs- und Verschleissminderung durch Festschmierstoffe", Tribologie und Schmierungstechnik, 46. Jahrgang, 1/1999.
[24] N.N., „Umwelt: Rückgang von Altöl für 2005 erwartet", Woche im Bundestag, Heft 16, 07.10.97.
[25] N.N., „VDI-Guideline 2221 - Systematic approach to the development and design of technical systems and products", Berlin, 1993.

The development of surface topography during wear of matt finish femoral stems

L T BROWN, L BLUNT, and **J HOWELL**
Centre of Precision Technologies, University of Huddersfield, UK

ABSTRACT

In the UK around 50,000 Total Hip Replacements are performed annually, of these up to 10% are revision operations [8]. Revision operations are required when a primary prosthesis fails. The premature failure of prostheses has been attributed to the production of debris from the wear of the femoral stem and head. There have been many studies to qualitatively characterise the wear of femoral stems, however few quantitative studies have been performed. In this semi-quantitative study 25 explanted stems were examined and the wear characterised in relation to the extent of damage on the matt surface.

INTRODUCTION

In the UK around 50,000 Total Hip Replacements are performed annually [8]. Total Hip replacements are performed to give better quality of life to patients with various conditions including osteoarthritis; chronic rheumatoid arthritis; and fractures of the hip.
During total hip replacement the head of the femur is removed and replaced with a prosthetic head, normally the head articulates in an ultra high molecular weight polyethylene (UHMPE) acetabular cup. This head is positioned on a femoral stem, which is inserted into the medullry canal of the femur. There are two main types of prostheses, cemented and uncemented, uncemented are made from porous materials which allow for boney ingrowth to occur within the mudullary canal. The cemented prosthesis is fixated into the femur using Acrylic bone cement. This study concentrates on the wear of cemented total hip replacement. The expected life of a successful total hip replacement is 10+ years, however of the operations that are performed, up to 10 % can be revision operations to replace prostheses that have failed prematurely. Revision operations are costly and have a lower chance of success compared with Primary Total Hip Replacement. There has been an emphasis in research on determining the

reason for premature failure of the implants. A great deal of revisions of femoral components are due to aseptic loosening of the stem. Premature failure of the femoral stem is often attributed to the production of metallic debris, which damages body tissue causing bone resorption and loosening of the component within the cement mantle.

Initially all cemented femoral stems had a highly polished surface finish, subsequently there was a move towards the use of matt finish femoral stems with a bead blasted surface, which were used in the period 1976 – 1982. The rationale behind matt surfaces was greater fixation of the stem in the cement, however the result was a greater incidence of aseptic loosening in matt finished stems.
In a study by Howie et al (1998), 40 Total Hip replacements were performed, 20 polished and 20 matt finished stems. Apart from the surface finish the stems were identical Exeter type prostheses (Stryker Howmedica). At a nine-year follow up, four of the matt finished stems had been revised for loosening whereas none of the polished stems had shown signs of loosening. The study concluded that: "The matt surface appears to be responsible for increased loosening in the double-tapered stem design."[6]
Tentative suggestions indicate that the wear of femoral stems in vivo may be due to fretting wear. "Most of these repairs are performed on limbs and these can undergo a large number of stress reversals in one day; the conditions are therefore those in which fretting corrosion can be expected."[13]. Fretting wear is the result of small oscillatory micromovements between two contacting surfaces, the characteristic signs of fretting wear are the production of metal oxide debris and the formation of surface wear scars. Studies have shown these characteristics to be present on explanted stems, both matt and polished and of different materials [4,10,12]. There does appear to be an additional wear mechanism involved in the wear of matt stems. Before fretting takes place it is suggested that there is polishing of the matt surface where the rough asperities are removed [2].

Although there have been many studies, in vivo and in vitro, to qualitatively characterise the wear on femoral stems, there has been little work done on quantitative characterisation. The aim of the present study is to examine the development of the surface wear of femoral stems in the hope that understanding the development of the surface topography will lead to better knowledge of the wear mechanisms involved.

THE STUDY

25 Explanted Matt finish femoral stems were examined, The degree of wear varied on each one however the position of wear on explanted femoral stems does appear to follow a pattern due to the loading regime in vivo [2,4].
The stems were examined and a visual grading system was used to categorise the wear in different areas of the stem [5]. This grading system contains 5 categories, pol 0 denotes no wear, with an increase in wear denoted by pol 1 through to pol 4, pol 4 being an indicator of high wear. (i.e. a high degree of polishing)
Of the 25 explanted stems examined there were 25 stems showing areas of no wear (Pol0), this surface measurement was taken from an area which lies outside of the cement mantle, resulting in no contact hence no possibility of wear. 18 of the stems showed areas graded as pol 1; 18 of the stems had areas graded as pol 2; 9 of the stems had areas graded as pol 3 and 5 of the stems showed areas of high wear and were graded as pol 4. *

The surface topography of the visually graded areas was measured using 3D surface measurement.

The specimen surfaces were then measured using a Wyko NT 2000 optical interferometer operating in the vertical scanning mode. Various measurement areas were examined with a vertical magnification of between 25x and 100x. Vertical resolution of this system in the VSI mode is 1nm.

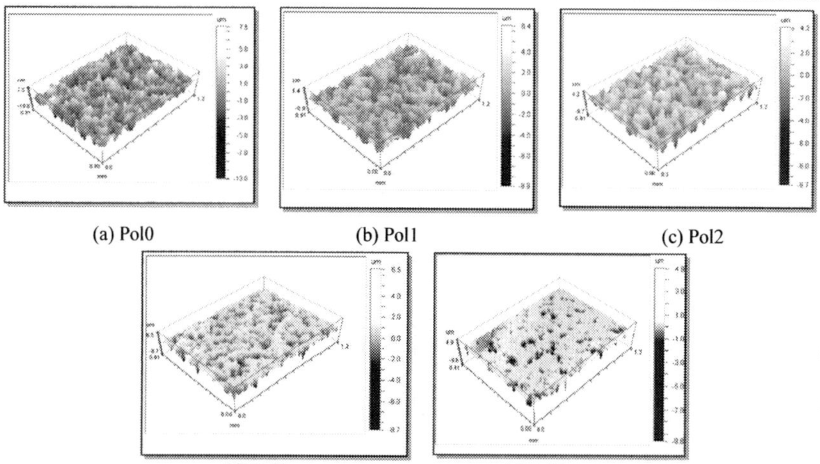

(a) Pol0 (b) Pol1 (c) Pol2

(d) Pol3 (e) Pol4

Fig.1- Axonometric projections showing the change in surface texture with increase in wear.

From the axonometric plots (fig.1), it can be seen that the rough asperities caused by the bead blasting process are progressively worn away, and the surface becomes smooth. Debris is generated, through a form of contact wear between the stem and the bone cement, and is a large factor contributing to aseptic loosening of the stem resulting in failure of the prostheses.

Traditionally quantitative surface measurement was carried out using 2D surface parameters. Unfortunately as many as 105 different surface parameters were developed in an attempt at improving functional surface description, the large number of 2D surface parameters has been referred to as a "Parameter rash"[13]. Most of these parameters were of limited use in describing surface interaction. With developments in 3D surface measurement technology, a need was recognised for the introduction of 3D surface parameters to characterise surface topography more accurately.

In recent years researchers have developed a primary set of 17 surface parameters which seem to describe the 3D surface topography more comprehensively [11]. The 17 parameters can be found summarised in table 1.

Table.1 3D Parameters

Family	Nomenclature	Parameter
Amplitude	Sq	Root-mean-square deviation of the surface
	Sz	Ten point height of the surface
	Ssk	Skewness of topography height distribution
	Sku	Kurtosis of topography height distribution
Spatial	Sds	Density of the summits of the surface
	Str	Texture aspect ratio of the surface
	Sal	The fastest decay autocorrelation length
	Std	Texture direction of the surface
Hybrid	SΔq	Root-mean-square slope of the surface
	Ssc	Arithmetic mean summit curvature of the surface
	Sdr	Developed interfacial area ratio
Functional Index	Sbi	Surface bearing Index
	Sci	Core fluid retention
	Svi	Valley fluid retention
Functional Volume	Sm	Material volume of the surface
	Sc	Core void volume of the surface
	Sv	Valley void volume of the surface

Many authors have expressed the importance of these 3D parameters in characterising surfaces and manufacturing processes[3]. With the complicated nature of surface topography a one to one numerical representation is impossible [11]. Therefore it is necessary to represent surfaces numerically by parameters.

In the present study, each parameter was calculated for each data set taken from the worn stems. To gain an overall picture of the suitability of the parameters for characterisation of the development of surface topography during the wear of femoral stems an average (see previous section *) was taken for each parameter for each grade of wear. The Average parameter values were plotted against visual grade of wear. From these results it was possible to determine which parameters were suitable for characterisation of the development of topography during wear of matt finish femoral stems.

RESULTS

From the results, the most useful parameters for showing fretting wear of explanted replacement hip femoral stems were determined, the strong relationship between increase in visual grade of wear and parameter can bee seen in fig.2 (a-c)

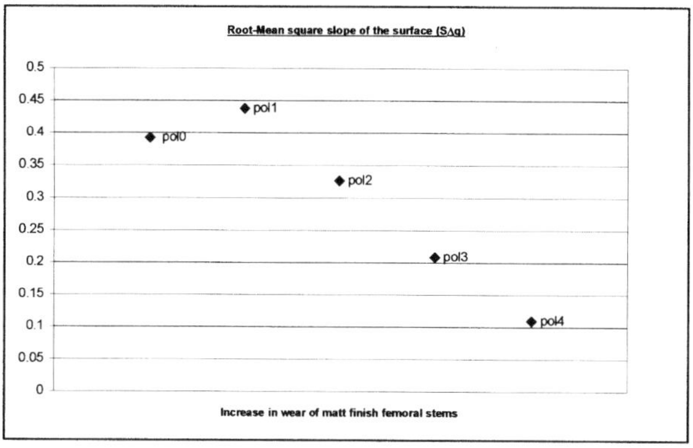

Fig 2. (a) Average hybrid parameter, showing
SΔq – Root mean square slope of the surface

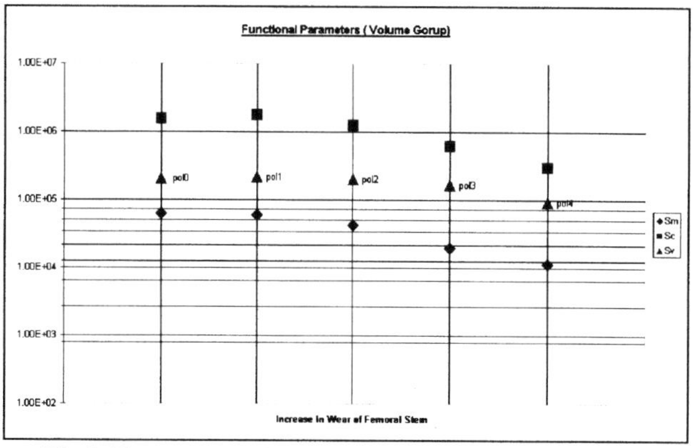

Fig.2 (b) – Average functional volume group, showing:
(I) Sm Material volume of the surface
(II) Sc Core void volume of the surface
(III) Sv Valley void volume of the surface

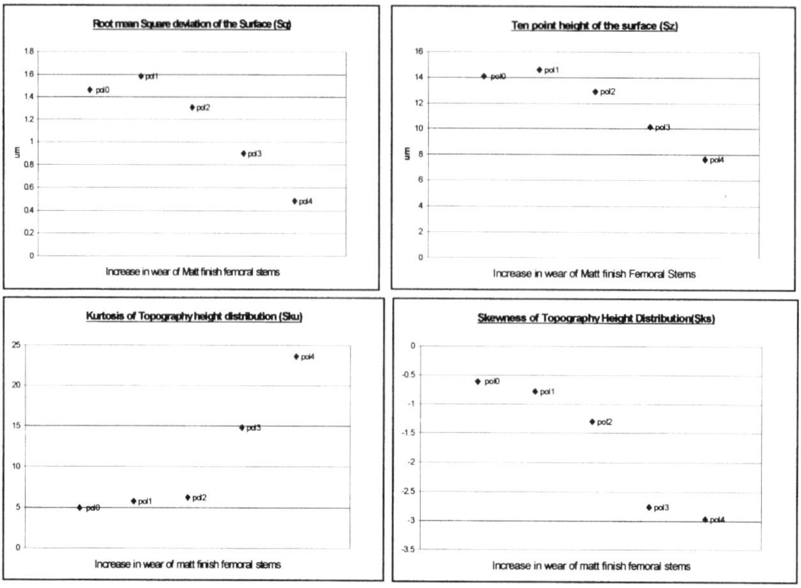

<u>Fig 2(c) -Average Amplitude parameters, showing:</u>

(I) Sq – Root mean Square deviation of the surface
(II) Sz – Ten point height of the surface
(III) Sku – Kurtosis of topography height distribution
(IV) Ssk – Skewness of topography height distribution

From these graphs (Fig.2 a-c) the relationship between increase in wear can be clearly seen. The parameters that are believed to be the most suitable for this application are those that show the change in surface amplitude and the change in surface volume with increase in wear:

Fig2a shows the data for the hybrid parameter rms slope of the surface $S\Delta q$, although the first data point Pol0 does not follow the pattern, the other data points poll – pol4, decrease almost linearly.

$S\Delta q$ represents the root mean square slope of the surface, the decrease in $S\Delta q$ shows the steepness of the slopes of the asperities decreases with wear. This is to be expected as the asperities are worn down and the surface becomes smoother.

Fig.2b shows the functional – volume family. Again the data points for pol0 show discontinuity with the other results. However the data shows Sm, Material volume of the surface to decrease, this is to be expected as it shows the volume of material to be decreasing as the asperities of the matt surface are worn away with increase in wear grade.

Sc, Core void volume of the surface also decreases with increase in wear grade, though to a lesser degree than the Material volume of the surface.

Sv, Valley void volume of the surface indicates the change in the valleys with increase in wear, from the data shown in fig 2b it can be seen that Sv undergoes only slight change with increase in wear. This shows, with the previous volume family parameters that although the surface is changing with increase in wear the valleys remain relatively unchanged. The pattern would indicate that the rough asperities of the bead blasted surface are worn away leaving the valleys intact.

The other set of parameters found to be useful are shown in fig.2c, these are the amplitude family of parameters. Sq, is the root mean square deviation of the surface, it is a widely used parameter, in statistical terms it is the sample standard deviation of the surface. Again ignoring the data for pol0 there is linear decrease in Sq with increase in wear, showing the overall surface roughness to be decreasing with increase in wear.

Sz, The ten point height of the gives the average value of the heights of the 5 highest peaks and the depths of the 5 deepest valleys. Sz shows the same relationship with increase in wear, as did the data for Sq. This shows the distance between the highest peak and the deepest valley's is decreasing, this also shows that the asperities are being worn away.

Sku, Kurtosis of topography height distribution is a measure of the peakedness of the surface height distribution, and characterises the spread of the height distribution. The results show that the spread of the height distribution becomes smaller as the value for Sku increases with increase in wear. "A centrally distributed surface has a kurtosis value larger than 3 whereas the kurtosis of a well spread distribution is smaller than 3"[11].

Ssk, Skewness of topography height distribution can effectively describe the shape of the topography height distribution. This parameter gives an indication of the structure of a surface. The decrease in Ssk with increase in wear shows the peaks on the surface being removed by the wear process. This correlates with the visual inspection of the axonometric projections where it is apparent that the peaks created by the bead blasting process are worn away to produce a more polished surface.

As the mechanism behind the wear of Matt finish femoral stems is not fully understood, a move towards quantifying the amount of wear of stems in vivo is crucial to determining the wear mechanism. An attempt has been made at a semi quantitative analysis of the wear in addition to the use of 3D parameters to characterise the surface.

From the 25 stems, 4 were chosen which show the full range of visually graded wear, bearing area curves were produced for the most worn plateaus for each stem, the plateau on the bearing area curve gives the average height of the worn areas of the surface. The height of this worn plane was used as a baseline to determine how much material has worn away in terms of volume per mm^2. By plotting the volume of lost material from pol0 to pol4 a representation of the material loss with increase in wear can be determined.

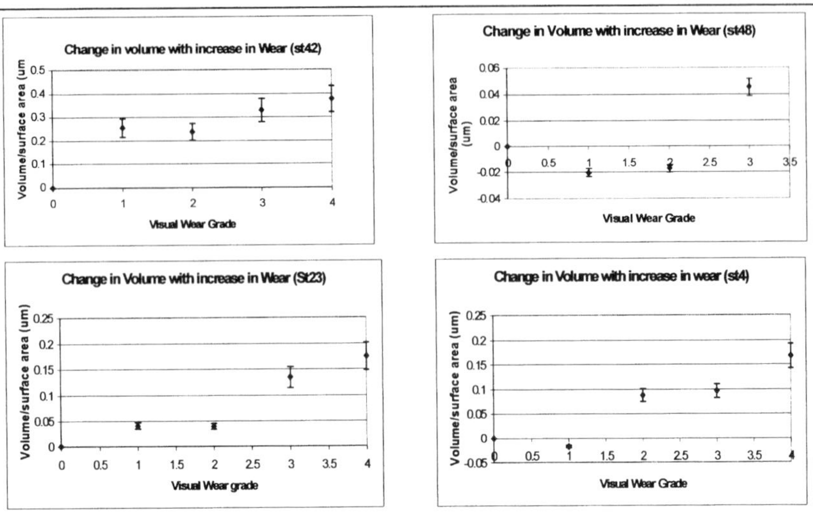

Stem Details: st42, Charnley Cobra 316L matt
St48, Charnley Roundback 316L matt
St23, Exeter 316L matt
St 4, Exeter 316L matt

Fig.3 Normalised volume loss with increase in visual wear grade for 4 matt finish femoral stems

There seems to be some confusion between Pol1 and Pol2 of the visual grading, as in all 4 cases there is a discontinuity with the results in this area. There does however appear to be a strong pattern suggesting an initial rapid wear rate between pol0 and pol2 becoming a more steady increase in volume lost with pol3 and pol4. This does correlate with the pattern of wear associated with fretting wear however it also follows the regime of most forms of sliding wear. "It must always be borne in mind that the boundary between different types of wear is not always a rigid one." [7].

The plots for st48 and st4 show negative values for the volume of material lost, showing there to be some discontinuity with results. This is attributed to the measurement of the values for pol0 being taken under the collar of the femoral head where no wear due to absence of contact has been incurred. It is believed that the initial surface roughness due to bead blasting may differ under the collar, to that of the rest of the stem due to the complex nature of the geometry of the stem resulting in an inconsistent roughness after blasting. This theory is also shown in the work on use of 3D parameters to characterise the surface, where an average of the 25 stems was taken.

DISCUSSION

Fretting wear is the mechanism presently attributed to the wear of femoral stems in vivo. The term Fretting denotes a small oscillatory movement between two solid contacting surfaces [7]. The factors that differentiate fretting from other forms of sliding wear are the magnitude of movement and the nature of the damage caused.

Fretting is concerned with amplitudes of less than 25um and certainly not greater than 130um[13]. From this study of matt finish femoral stems it is apparent that although the wear could be classified as fretting in terms of the magnitude of micromovements between the stem and the bone cement, due to the stress reversals of the limbs. The damage observed on the stem does not exactly mimic that which would be expected of fretting. According to Waterhouse R B (1972) "It is a general observation that the higher the degree of surface finish the more serious is the fretting damage"[13]. When looking at the semi-quantitative results obtained in this study it is apparent that the Matt stems undergo considerable material loss although the surface finish is of a rough nature." There is possibly another form of wear involved, which initially polishes the surface of the matt stems before classic fretting takes place. The semi-quantitative results indicate that this could be due to a number of types of wear as the data shows a better correlation with ideal abrasive and sliding mechanisms, Referring to ideal abrasive wear Rabinowicz (1993) illustrates, "Blunting of the abrasive surface after wear"[9].

The wear occurs between the PMMA bone cement and the stainless steel matt finish stem. Third body wear of metallic oxide particles could also play a factor as metallic oxide debris has been found present at revision operations, not only with stainless steel implants but also Titanium and Cobalt Chrome [1,4,10].

It is considered that matt finish stems have a higher failure rate than that of polished stems from considerable clinical evidence. From the study of the matt finish femoral stems and from evidence to show the wear of polished stems to be due to fretting, it is apparent that the wear of matt and polished stems is due to two different wear mechanisms. The authors consider that the initial wear of matt stems is of an abrasive form, which, when the matt stems become more polished then changes to a more classical fretting nature.

Further work is needed taking in to account the anomalies with the data for pol0 and using alternative data as a measure of unworn surface. A method of relocation markings on the surface of a stem which undergoes simulation to determine the material loss in relation to number of cycles may be a better way of determining the mechanism behind the wear of both matt and polished femoral stems. The visual grading system also needs to be clarified to establish definite differences between the grades of wear. There appears to be some overlap between the grades of pol1 and pol2, it may be more beneficial to cut the number of grades to 4 showing no wear slight wear heavy wear and severe wear.

There is also scope for performing simulated pin on disk tests, to examine the way PMMA bone cement and stainless steel perform under oscillation.

An additional area of study would be to model the initial wear using a simple truncation wear technique. This would involve digitally wearing a virgin surface using a truncation model. Essentially the surface is sliced away at given depths and at each depth the parameters can be calculated to establish an ideal modelled surface with which the experimental data can be compared.

CONCLUSIONS

- There are 3D parameters, which are useful for characterising the wear of explanted femoral stems.
- The use of semi-quantitative models may go some way to determining the wear mechanism behind the failure of femoral components during total hip replacement.
- The study has provided a good basis for continuing work on quantification, and has provided some insight into the possible wear mechanisms involved.
- The wear mechanisms of matt stems appear to be initially different to that of the more successful polished stems.

REFERENCES

[1] Bischoff UW, Freeman MAR, Smith D, Tuke MA, Gregson PJ, "Wear Induced by Motion Between Bone and Titanium or Cobalt-Chrome Alloys.",
The Journal of bone and Joint Surgery 76-B No 5, 1994, pp713-715

[2] Bridson D, "Fretting Tests on Deliberately Losened Exeter Hip Replacement Stems"
Unpublished data, 1999

[3] Butler D L, "The Topographic Characterisation of Cylinder Liner wear",
Birmingham University, PhD Thesis, 1999

[4] Cook J E, "Fretting Wear of Total Hip Replacement Femoral Stems",
Exeter University, PhD Thesis, 1998

[5] Howell J, Blunt L, Gie G, Timperly A, Ling RMS, "An Investigation of the Fretting Wear Seen on Explanted Hip Replacement Femoral Stems."
The Journal of Bone and Joint Surgery, 81-B Suplement III, 1999

[6] Howie DW, Middleton RG, Costi K, "Loosening of matt and polished cemented femoral stems", The Journal of Bone and Joint Surgery, 80-B No 4, 1998, pp573-576

[7] Hutchings I M, "Tribology: Friction and Wear of Engineering Materials",
Edward Arnold Publishers UK, 1992

[8] Murray DW, Carr AJ, Bulstrode CJ, "Which Primary Total Hip Replacement"
The Journal of Bone and Joint Surgery, 77-B No 4, 1995, pp520-526

[9] Rabinowicz E, "Friction and Wear of Materials", John Wiley and sons, USA, 1965

[10] Shahgaldi BF, Heatley FW, Dewar A, Corrin B, "In Vivo Corrosion of Cobalt-Chromium and Titanium Wear Particles",
The Journal of Bone and Joint Surgery 77-B No 6, 1995, pp962-966

[11] Stout K J, Sullivan P J, Dong W P, Mainsah E, Luo N, Mathia T, Zahouani H.
"The Development of Methods for the Characterisation of Roughness in Three Dimensions",
Commission of the European Communities Report,
University of Birmingham Press, UK, 1993

[12] Walczac J, Shahgaldi F, Heatley F, "In Vivo corrosion of 316L stainless-steel hip implants: morphology and elemental compositions of corrosion products"
Biomaterials, 19, 1998, pp229-237

[13] Waterhouse R B, "Fretting corrosion", Pergamon Press, Hungary, 1972

[14] Whitehouse D J, "The Parameter Rash is there a Cure", Wear 83, 1982, pp75-7

Increase of life cycle due to application of surface coating methods

S MITROVIC and **Z ZIVIC**
Mechanical Engineering Faculty, University of Kragujevac, Yugoslavia

SYNOPSIS

It is natural that elements of a tribo-mechanical system (TMS) are exposed to various types of wearing processes during the exploitation period. Worn machine parts, in most cases, are considered to be a disposable material when replaced with new ones. The present need for savings in every aspect of the economy and the environment forces many producers to start thinking about regeneration of worn and damaged machine elements, especially with older high-priced machines. In these cases an appropriate coating method can provide desired surface properties and bring the machine back to work avoiding expensive new parts. The welding reparational method of profile wire production rolls, which represents the parts of rolling stand, is presented here as an example of the effective application of coating methods in this area. The choice of the corresponding method of welding and the regulation of the complete reparational technology are also considered. Results have shown that the increase of life cycle is significant. Behavior of regenerated rolls showed all exploitation characteristics that had original ones.

KEYWORDS: reparation, welding

1. INTRODUCTION

Modern trends of world economy, in general, assume savings. Savings should be done in every single domain in order to reach this global aim. Therefore, in every engineering discipline exists the imperative of lowering the costs of production process. From this aspect, prolonging the life cycle of a machine element is very important and essentially reflects this requirement. If we consider the tribo-mechanical system and its elements, there are many solutions to this problem from the standpoint of tribology as a science and technology of friction and wear. The method of regeneration is considered in this paper.

The wearing of tribo-mechanical system elements is unavoidable. In recent years, machine elements that had been worn or damaged otherwise during their work time, were mostly ordered from their manufacturer, without even considering the possibility of their reparation. Nowadays, many producers think about regeneration of these elements, especially with older high-priced machines. In these cases, appropriate coating methods can provide desired surface properties and bring machine back to work avoiding more expensive new parts.

Machine elements' reparation can overcome problems with supply of new parts such as: difficulties with product technology, high costs of manufacturing process, long term production, complicated administration if the parts are to be imported etc. [1]. The reparation process economics are determined depending on many factors some of which are: ratio of a new part – regenerated part price, comparison of time needed for a production of a new element to a time needed for an old one and also the life cycle of a new part compared to the repaired one. Every production plant should systematically investigate possibilities of worn and damaged element regeneration within its own maintenance services, if possible. Or to look for a specialized firm that deals with the regeneration processes of machine parts. As an example of the above, the welding reparational method of profile wire production rolls which represents the parts of rolling stand, is presented here. Results showed the effect of regenerative welding, as a surface coating method, since properties of welded parts fulfilled all exploitable demands. Therefore this method can be adopted in increasing the life cycle of a machine element.

2. REPARATIONAL WELDING

Surface coating methods represent the powerful tool in improving characteristics of various metal surfaces (hardness, toughness etc.) in area of thin surface layers where tribological processes are located. The cause of machine parts wearing out in a working regime can be of many different sources, such as: cavitation, corrosion, abrasion, erosion, increase of temperature level, dynamic loads etc. Methods of welding and metallisation are the usual regeneration methods applied in cases of worn machine elements. The reparational welding method is the process of applying metal layers onto the working surfaces of some parts, in order to increase their wear resistance, resistance to impact pressure and resistance to other forms of loads. It can be successfully used to fix parts broken during work regime and also ones made with defects. It is usually applied onto the specific surfaces of working parts, which are exposed, to the greatest wear level. It is possible, sometimes, to apply this method more than once before the old part must be exchanged for the new one.

The selection of the technology for reparational welding must include not only the weldability of the base material and additional material options, but also the economics of the selected technology. It should be noted that reparational welding could be done with almost every machine part and almost every material. There are some difficulties that appear with the gray iron which is unclean (residue of sand at casting, presence of inclusions) and with the gray iron which has its structure changed due to a long working regime with high temperature levels in organic oil environment, for example. Repaired machine parts in most cases regain their first function completely. Welding methods are powerful tools in obtaining the longer life cycle with various constructions. In general, regenerative welding is used for:

- Regeneration of machine parts and devices by applying metal onto surfaces damaged by friction, cavitation, corrosion or breakage,

- Gaining the requested physical and chemical properties for surfaces of specific elements or their segments. It can be hardness, abrasion resistance, corrosion resistance etc.

Work life for many machines and devices are limited due to a wearing of their specific parts only. Therefore, for machine elements, different materials have been chosen in order to obtain approximately the same life cycle for all of them. Until now, this was done by improvement of working surface quality, by thermal treatment (quenching, cementation, and nitriding) and lately by reparational welding which successfully replaces some of these processes. Advantage of welding is the fact that the metal used for welded layers can be of different chemical composition than the one for the base material. Therefore, desired properties of working surfaces can be obtained by depositing thin layers of expensive materials onto the less expensive base material thus saving good quality materials and lowering the final price of a machine part at the same time. The thickness of a welded zone is usually up to 3 mm. For bigger thickness required, it should be performed with several passes.

Experience in welding technique application encourages its use in various manufacturing applications. Nowadays, with the level of its development, it is more economic to organize special welder's workshops, while cutting down the production of new spare parts. Investigations of many welded parts showed that they had bigger wear resistance than new ones, and their price was lower at the same time.

3. PROFILE WIRE PRODUCTION ROLLS - DESCRIPTION, WORKING CONDITIONS, CONSTRUCTION

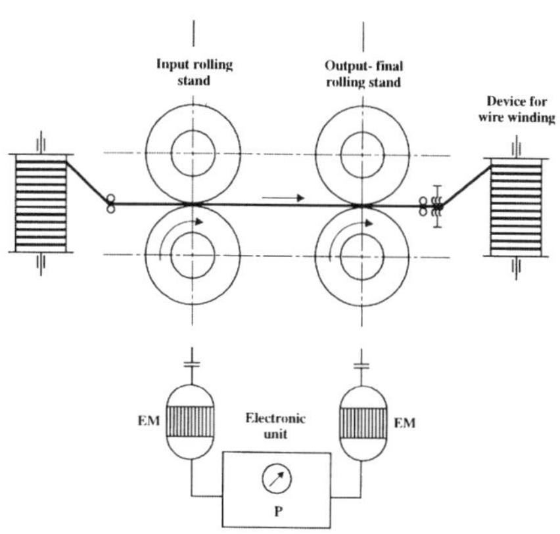

Fig. 1 Schematic representation of rolling of profile wire

Profile wires production rolls, which represents the parts of rolling stand [2], is presented in Figure 1. Manufacturer of these parts is "Marshal Roberts" – England, and it is installed in Factory "Kablovi", Jagodina, SR Yugoslavia. The machine has been primarily used for copper profile wire production but it can also be used for production of steel ones. The motion of the rolling stands is due to a power supply from electromotors of 15 kW power.

Synchronization, that is, regulation of rolled wire, is done by changing the revolution number of rolls through the potentiometer P and electronic unit that acts onto the direct current motor drives. Stretching of rolled wire must not be larger than 30 %, in this operation. Revolution number of rolls is from 0 – 225 °/min. Working temperature is below 100°C, for rolls. During its lifetime, in order to minimize friction, rolls are lubricated by an industrial grease. Rolls must be of the same diameter (D1 = D2), in order to achieve the good quality of produced wires. Surfaces for rolls must be very fine, that is, highly polished surfaces and of a hardness of 60 HRc. During rolling, the wire has to be without internal defects, porosity and macro and micro-cracks. Maximum width of the rolled profile is equal to the rolls' width, but the limiting factor is the power of drive motors, Profiles are mostly of the types shown in Figure 2 with characteristics given in Table 1.

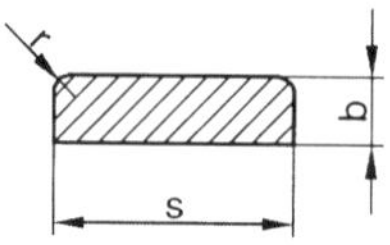

Fig. 2 Profile type

Table 1 Profile characteristics

b (mm)	s (mm)	r (mm)
5.6 ± 0.02	2.12 ± 0.015	0.80 ± 0.06
6.7 ± 0.035	2.5 ± 0.015	
7.1 ± 0.035	2.5 ± 0.015	
8.0 ± 0.035	2.5 ± 0.015	

Rolls on the rolling stand must fulfill some important characteristics in order to achieve good wire quality. Obtained wire surface must be [1]:

- smooth and clean,
- without dirt or residue of pull-out material,
- without surface defects as cracks, scaling, inclusions of metal or non-metal origin
- without any other surface defects that could affect the process of lacquering and therefore the quality of the produced wire.

Any of these defects on the rolled wire can be the first sign of rolls wearing out. In Figure 3, dimensions of one roll are presented.

4. SELECTION OF PROCEDURES AND TECHNOLOGY FOR ROLLS REPARATION

4.1. Material of rolls and its weldability

According to JUS standard, material for rolls is C7421 steel (DIN standard 25 MoCr 4). Chemical composition (in %) is given in Table 2. Tensile strength of this steel is from 400 to 490 MPa, and hardness is 217 HB. It is Mo-Cr cemented steel with higher level of carbon.

Table 2 Chemical composition for rolls' material (%)

C	Si	Mn	P_{max}	S_{max}	Cr	Mo
0.23-0.29	0.15-0.40	0.60-0.90	0.035	0.035	0.4-0.6	0.4-0.5

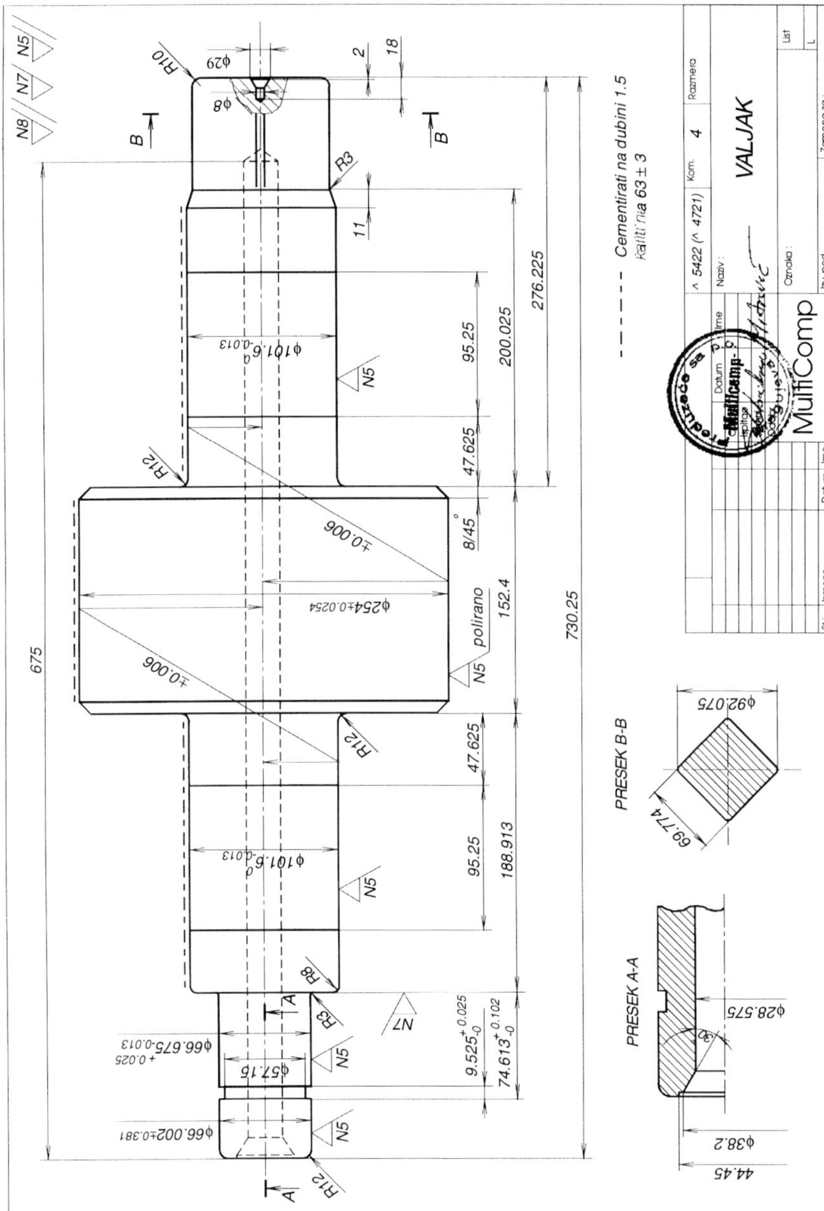

Fig. 3 Drawing of a roll

This steel is in the group of poor weldable steels. It is thus necessary to do preheating and then apply a more complex welding technology than that needed for good weldable types.

4.1.1. Weldability evaluation

As a base criteria for evaluation of welding for this steel the so called chemical equivalent carbon CE is used and calculated by:

$$CE = C + \frac{Mn}{6} + \frac{Cr + Mo + V}{5} + \frac{Ni + Cu}{15} \%$$

$$CE = 0.29 + \frac{0.90}{6} + \frac{0.60 + 0.50}{5} = 0.66$$

It can be seen, that for the steel of such a chemical composition, preheating has to be done.

4.1.2. Calculation of preheating temperature

Seferijan method of calculating the preheating temperature, has been applied:

$$T = 350\sqrt{|C| - 0.25} \ °C$$

$$|C| = |C|_H + |C|_D \ (\%)$$

where:

$|C|$ - total equivalent carbon

$|C|_H$ - chemical equivalent carbon

$|C|_D$ - equivalent carbon of thickness D [mm]

Calculation of equivalent carbon, chemical equivalent carbon and equivalent carbon of thickness D is done by formulas:

$$360\ |C|_H = 360C + 40\ (Mn+Cr) + 20Ni + 28Mo$$

$$360\ |C|_H = 360 \cdot 0.29 + 40\ (0.90+0.60) + 28 \cdot 0.50$$

$$360\ |C|_H = 0.496$$

$$360\ |C|_D = 0.005 \cdot s\ |C|_H$$

$$360\ |C|_D = 0.005 \cdot 250 \cdot 0.496 = 0.620$$

$$360\ |C| = 0.496 + 0.620 = 1.116\ \%$$

$$T = 350\sqrt{1.116 - 0.25} \ °C$$

$$T = 325 \ °C$$

In this way, the calculated preheating temperature considers the worst possible case, that is when the construction is big and material preheating is done only in specific area of welding. However, in this case roll is heated through the whole cross section and the criteria for temperature calculation can be somehow changed. Including this criteria, preheating temperature is T = 250°C. This results in large amounts of saved energy compared to calculated level of T = 325°C.

4.2. Selection of welding method and regime

As the welding method of reparation of rolls, the electric arc welding under powder has been chosen (EPP). Welding under powder has often been used in practice because of its high productivity and good quality of welded surfaces. Only massive parts in horizontal position is welded under powder and additional material can be in a form of a wire (one or two) or a ribbon.

The above stated form of welding is mostly applied in case of cylindrical surface regeneration, like steel rolls of a big diameter. Welding can be done along a generating line of roll or circles, and the process itself can be automated with rotary motion of roll and transitive motion of welding head along the axis of the roll. Welded layers in most cases have bigger hardness and wear resistance than the base material. This welding regime is adopted taking into account two opposite requirements:

- for larger productivity the current intensity should be as big as possible;
- for mixing rate, smaller current intensity is required.

Depending on the practice case, the optimal solution is selected. Welding velocity is from 10 to 30 kg/h, and the welding thickness is 4 mm.

4.3. Welding parameters

Welding cylinder reparation has been done with following parameters:

I_z = 680 – 700 A - welding current

U_z = 32 – 34 V - stress

V_z = 40 cm/min - welding velocity

d_e= 5 mm - electrode diameter

4.4. Additional material

Additional material has been chosen in combination with covering powder. Considering the fact that there is a small offer for domestic additional material, welding is done with electrode wire EPP Cr 6 and covering powder OP 100, which are products of steel factory "Jasenica" from Slovenia. Chemical composition of additional material is given in Table 3:

Table 3 Chemical composition of additional material

C	Si	Mn	Cr	Ni	Mo
0.1 %	0.25 %	0.7 %	7.0 %	0.35 %	0.55 %

However, direct application of this material onto the base material (C 7421 steel) of the roll showed some negative consequences like crack appearance. Therefore we decided the first layer is to be of an additional material EPP 2 in combination with EP 45 powder, and other layers with EPP Cr 6 wire. This combination:

- base material C 7421 steel;
- layer EPP 2 with powder EP 45;
- layers EPP Cr 6 with powder OP 100

showed the best possible working conditions and so it was adopted for the reparation process.

4.5. Welding technology

The reparation process for rolls consisted of five phases:

1. Surface preparation;

2. Preheating;
3. Welding;
4. Thermal treatment (annealing)
5. Final machine treatment (scraping, grinding and polishing)

The first reparation phase is the determination of the degree of damage of the roll and then preparation of the surfaces that are to be welded. The number of passing that should be applied onto the base material depends on a depth of these damaged layers. Surface preparation consisted of two steps:

- Smoothening of rolls, that is, removal of base material layers by appropriate machine treatment until all the cracks vanish, and
- Surface cleaning from corrosion.

EPP welding is done with starter plates as a method of prevention of end cracks. With rolls, starter plates are substituted by pipes that are welded at the ends of the roll and taken off during the final machine treatment. The second phase is the preheating of rolls. This process is carried out in a burning kiln made by fire resistant bricks, right next to the welding equipment (Figure 4). The heating is done with propane – butane burners. The preheating temperature should be kept at the constant level during the whole process of welding.

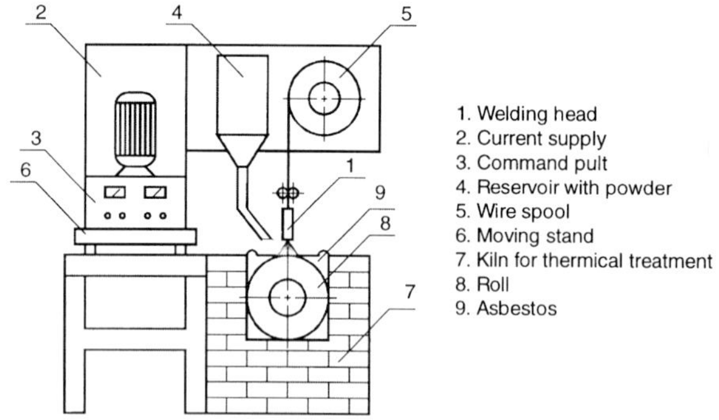

Fig. 4 Preheating of rolls

As soon as the preheating temperature is reached the welding is performed. In this phase the potential danger are deformations that could appear on the roll. Therefore, it should be noted that the order of welded layers is important. This layer should be placed diametrically opposite than the preceding one as shown in a Figure 5. This phase continues until the required dimension of the roll is reached.

The next phase is the annealing which is carried out at the temperature of 650 °C in order to reduce internal stresses resulting from the welding process in the previous phase. The heating is done in the same preheating kiln with a propane – butane flame. This process is schematically presented in Figure 6.

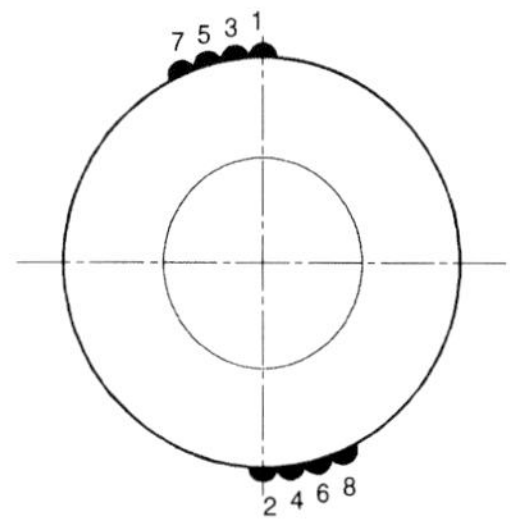

Fig. 5 Order of welded layers

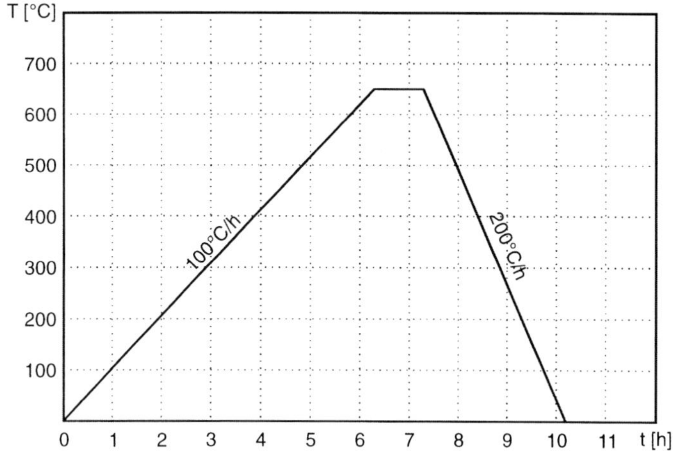

Fig. 6 Flowchart for thermal treatment - annealing

After this, rolls are subjected to the machine treatment (scraping and polishing) for the required dimension (∅254) and the quality of machined surface (polished surface of hardness 60 HRc) to be achieved.

5. CONCLUSIONS

Established results of the reparational method of rolls have proved that this method is very effective and rational in increasing the working life of these elements. Characteristics of reparated rolls satisfy all the exploitation demands same as the original ones did. The production cost for a new roll is 1800 DM, while the reparation cost per roll is 600 DM. This fact itself is enough to justify the application of the reparation process. The work conducted here represents a valuable experience in light of saving energy and money while obtaining the good results in the production of rolled wire.

REFERENCES

[1] Smith, D., "Skills and Technology of Welding", New York, 1995.
[2] Jovanovic, M., Adamovic, D., Lazic, V., "Tehnologija zavarivanja", Kragujevac, 1996.

Product Life Design Tools

Towards 'CAD for wear control'

J C BORG and **G C KINSELLA**
Department of Manufacturing, University of Malta, Malta

ABSTRACT

Besides being a phenomena that negatively influences an industry's production facilities, *wear* has also an affect on the 'use' performance of the artefacts developed. In order for firms to deliver competitive artefacts, improving the *cost* and *time* performance measures of their product development process is not enough. Manufacturing firms need to also deliver artefacts of high *quality* – one way is generate artefacts that are better in terms of their 'wear characteristics'. This paper, concerning on-going research dealing with developing *knowledge intensive CAD* tools supporting *'life-oriented' design*, focuses on the notion of how *wear issues* can be modelled to proactively guide designers in controlling component wear *during* the *synthesis* of conceptual mechanical design solutions.

1. INTRODUCTION

Design decisions can result in *unintended consequences* that have a *propagation effect* [3] across *multiple* life-phases such as manufacturing, use and disposal. Thus focusing on 'function' *during* synthesis leads to artefacts that are "not being designed for..." [6] a number of 'X-abilities' including *wear*. According to some estimates [2] [12], losses resulting from a lack of tribological 'know-how' are exceptionally high. Controlling wear *during* design can thus reduce operating and maintenance costs, increase utilization time and reduce disposal problems, the latter having a direct influence on environmental issues. Such an artefact 'life-oriented design' (LOD) approach is now necessary in an environmentally conscious society. However, only a few tools support 'Design for Wear Life' (DFWL).

These include the very simple notions such as, *use hard materials, maintain low contact pressure, use a lubricant* and *prevent high temperatures*. A promising CAD approach is that reported in [8] which focuses on providing designers with wear issues related to the disassembly of a product. A method widely employed to cater for wear during product development is that of *testing and simulation* [10]. A major limitation of this *costly* and *time-consuming* approach is that it used *after* a candidate solution has been physically realized, meaning that designers need to resort to re-design. Further, tests focus mainly with the *material selection* of the components, ignoring other designer definable characteristics [13] such as *structure, form* and *surface finish.* Thus, a lack of relevant tools to aid designers in *wear control* amplifies the problem of catering for wear issues *during* design.

By acquiring practical tribological expertise and exploiting the acquired domain knowledge *during* design to engage in *provident thinking* [1], designers can develop artefacts of a better usage reliability. However, due to its complexity [10], little of this knowledge is explicitly made available to designers during synthesis. At the same time, it is known that a lack of appropriate knowledge during early design decision making is known to result in undesired life-cycle consequences (LCCs) [3]. Given therefore the need of capturing wear phenomena expertise for utilization *during* conceptual solution synthesis, a feasible approach towards wear control is to embed relevant and properly structured wear domain knowledge into a *Knowledge Intensive CAD* tool [4]. KICAD provides a means of how captured *wear LCC* knowledge can be reused at fast processing speeds during new design situations. The development of the KICAD tool is based on the 'design tool' development framework [5] that states that computational tools should be based on *information models* derived from models of the *reality*, termed *phenomena models*. The specific phenomena concerned with in this paper is that design decisions give rise to life-cycle consequences including those of *wear*. The *knowledge model* is concerned with *what* elements should form part of the knowledge structure and with *how* these elements should be related to each other and organized, in order to result in codified knowledge. The *computer model* encompasses existing or new techniques to realize and manipulate the knowledge model for supporting the design reality. To develop tools allowing designers to foresee co-evolving *wear LCCs*, it is therefore necessary to understand *how* wear LCCs are generated from decision commitments.

2. RELATING 'ARTEFACT DESIGN' AND 'WEAR PHENOMENA'

A mechanical artefact can be considered as a decomposable system [7] consisting of a number of elements (sub-assemblies, components, component elements), termed in this paper *product design elements (PDEs)*, related to each other with *'part of'* relationships. The life-phases, *design (D), realization(R), use(U)* and *disposal(Di)* forming a mechanical artefact's life, involve the *re-use* of technical systems (e.g. manufacturing or maintenance systems) that realize the relevant transformation effects of that phase. These systems (e.g. servicing system) can also be decomposed into sub-systems (e.g. a holding device) termed in this paper as *life-cycle phase elements (LCPE)*. Such a decomposition supports life-phase system compositional modelling, where *LCPEs* are related to each other with *'part of'* relationships. As argued in [3], the *interaction* of an artefact with either *natural* (e.g ocean) or *artificial* (e.g. an elctro-chemical machine) life-phase systems (Figure 1) can give rise to a number of unintended life-cycle consequences (LCCs). A LOD approach therefore requires that designers *foresee* what life-phase systems will be met during the life of an artefact and that they also foresee the consequences of such interactions during design. Thus, designers need to *possess* and *utilize* a wide breadth of knowledge related to different *life-cycle* phases, to enable them to reveal interactions and hence possible wear LCCs. Therefore a LOD approach requires a vast amount of distributed LCC knowledge to be acquired, readily available and easy to access in order for it to be explicitly utilized during synthesis. Therefore, understanding the phenomena of how wear LCCs are generated during solution synthesis is essential for developing a KICAD tool for wear control,

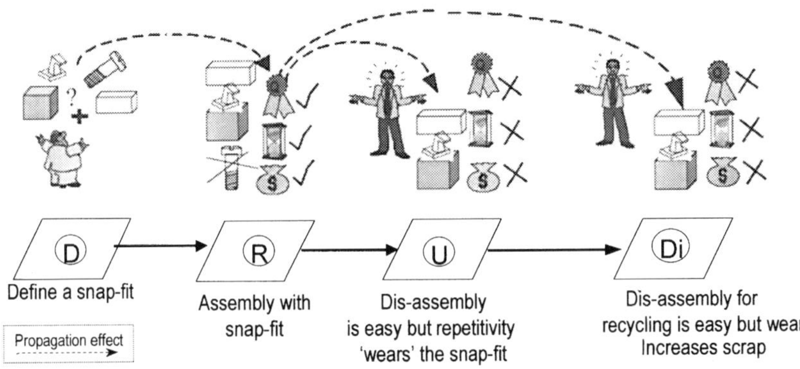

Figure 1 – Typical consequences with a snap-fit assembly feature

2.1 Wear Mechanisms

Wear concerns surfaces in relative motion. From a high level point of view, a study of wear mechanisms such as *adhesive, abrasive, fatigue erosion, corrosive* and *electrical-arc-induced wear* [2][11] [12] reflects that there are several factors which should therefore be considered during design, some being:

<u>Mating material pairs</u>

Hardness - in general increasing hardness decreases the wear of a material. Usually for adhesive wear, hardness above 700VPN tends to suppress severe wear and the system will operate in the mild wear regime. In abrasive wear [9], there is some evidence that the wear rate decreases with hardness. *Metallurgical Compatibility* - the tendency of the sliding metals to adhere to each other is indicated by their metallurgical compatibility [12], which is the degree of solubility when two metals are melted together. Using incompatible materials results in a lower wear rate. *Material Structure* - different material structures result in different wear properties e.g. hexagonal metals have a limited number of slip planes resulting in low wear and friction. *Material's Geometric Form* - the surface geometry may affect the wear rate. When mating a sharp surface against a flat surface (point-plane contact), the tendency is to have greater wear on the flat surface due to the high pressure at the point contact. Increasing the apparent contact area (Figure 2) results in a better load distribution, implying lower wear rates.

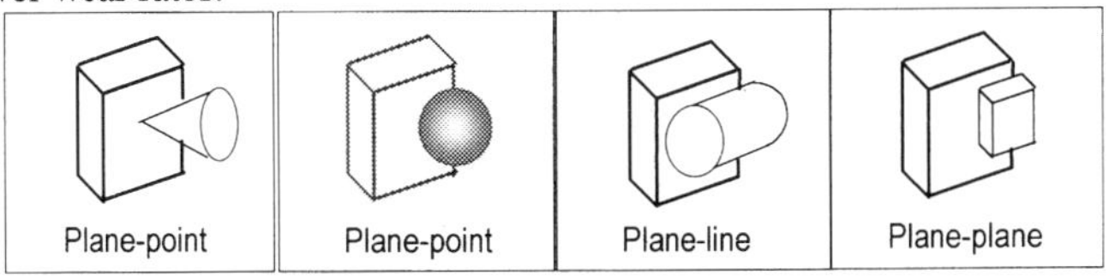

Figure 2 – Basic forms of contact area

Operating Conditions
Load - As the load increases, the wear rate also increases. There is a transition point at a certain load where the rate of wear increases faster with an increase in load. Below this particular load, the system is said to operate in the mild wear regime and above this point, the system is said to operate at the severe wear regime. Mild wear regime is usually a characteristic of oxide forming metals such as steel, nickel alloys and copper alloys. *Speed* - an increase in speed can lead to an increase or decrease in wear, depending on the effect of temperature on the surfaces. *Temperature* - an increase in operating temperatures tend to increase wear rate, because the materials involved become softer. However, an increase in temperature may sometimes cause a decrease in wear due to a higher oxidation rate, thus keeping the system in the mild wear regime.
Contact Surface Conditions
Surface Roughness – high surface roughness can cause abrasive wear especially with hard materials. Very smooth surfaces can also cause friction and hence wear. *Surface Preparation* - cleaning a surface or coating a surface can have different effects on wear. Clean surfaces exhibit high wear rates, due to higher adhesion. Hard undercoats such as nickel undercoats for use with high reliability electrical contacts, generally increases the wear resistance of the topmost coating or prime plate, since the composite hardness of the system has increased.

2.2 'Artefact Life' Design Decision-Making

This section presents a model of synthesis decision-making (Figure 4) upon which the *'KC' approach framework* [3] to LOD developed in this research is based. The space of alternatives available in a solution space, make the design process, decision intensive. In this paper, a decision is therefore assumed to exist due to a selection between a number of alternatives related to some domain. In the case of components, the designer encounters alternatives concerning the manipulable characteristics [13] that include reusable PDEs such as, form features, assembly features, material and surface textures (Figure 3). In addition, in artefacts that involve *relative motion* between different components, a *lubricant* is also considered as a PDE. Similarly, the synthesis of life-phase system models involves the reuse of LCPEs.

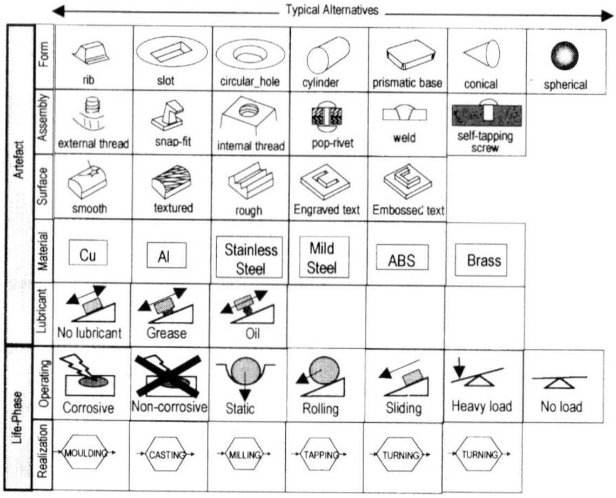

Figure 3 - Typical alternatives in mechanical artefact life design

In any case, the alternative selected, a result of the synthesis decision-making process, is termed here the *decision commitment.* A design decision is made to *intentionally* achieve a *desired* consequence. Thus, alternatives are interpreted by the decision-maker, in terms of their *expected* consequences. However, decision commitments also result in *unintended* consequences[11]. This, highlights why *explicitly* providing designers with knowledge of unintended LCCs resulting with their commitments, is beneficial to supporting them in selecting 'life-oriented alternatives'. These commitments, made during synthesis to an evolving solution, are termed *synthesis decision commitments* [3]. Basically, during synthesis, a designer *generates* a set {O} of possible solution options to the sub-problem being tackled. This set gives rise to a decision proposal D{O}. The set of options {O} can consist of reusable PDEs or LCPEs. As one of the *alternative* has to be, the designer engages in a decision-making process (DMP) to make a selection.

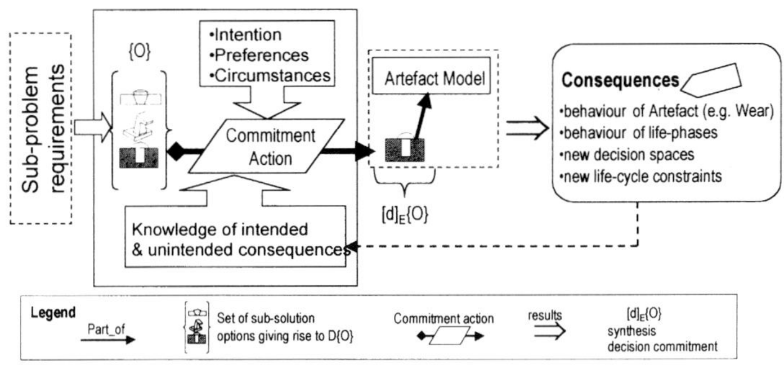

Figure 4 - Synthesis decision commitment model

During the DMP, the designer considers *intentions*, *preferences* and known *circumstances*. Following these considerations, the designer selects an alternative by making a synthesis decision commitment $[d]_E\{O\}$ to the evolving solution model (Figure 4). Due to consequences arising from the *interaction* of an artefact and life-phase systems, this paper argues that LOD requires that designers, concurrently generate and model the *artefact solution* and *life-phase system solutions*. These models collectively form an *artefact life model*. During such concurrent synthesis, decision commitments therefore concern both the *artefact model* and the different *life-phase models*. Life-phase synthesis decisions concern for instance the selection of technical processes (e.g. milling) and technical process parameters (e.g. feedrate) and operating conditions (e.g. loads, corrosive environments etc.).

2.3 Wear Life-Cycle Consequences Phenomena

A phenomena model [4] describing *how* LCCs are generated, has been developed, this summarized in Figure 5. Basically, this model highlights that LCCs are generated from two fundamentally different conditions:

i. individual, *non-interacting* synthesis decision commitments (Figure 5a) resulting in what are termed, non-interacting LCCs, i.e.$\mathbf{LCC_{ni}}$. For example, an LCC_{ni} resulting from the commitment of a rolling element bearing (O_1^1) to a mechanical structure is that 'rolling motion is possible in the use-phase'. A *wear* LCC_{ni} is that 'due to the repeated rolling motion during the use phase, the bearing will experience fatigue wear, therefore requiring replacement in the servicing phase;'

ii. multiple and *interacting* synthesis decision commitments (Figure 5b) resulting in what are termed, interacting LCCs i.e. **LCC_i**. For example, an LCC_i resulting from the commitment of a rolling element bearing (O_1^1) *and* the commitment of a corrosive operating environment (O_2^2) in the use phase, is that 'corrosive wear is likely to occur'.

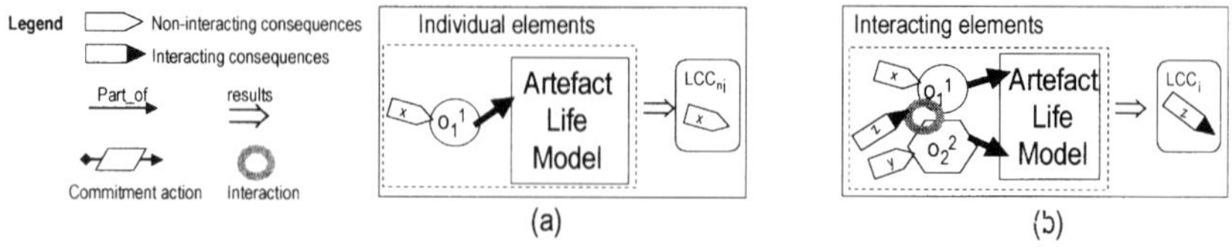

Figure 5 - LCC generation phenomena model

This LCC phenomena model discloses that knowledge about wear LCCs is *implicitly* co-evolving with synthesis decision commitments being made to the artefact life solution. It also highlights that concurrent synthesis is a necessity if *wear LCC_i* are to be revealed *during* design. If the co-evolving *LCC knowledge* is made explicit and utilized *during* synthesis decision-making, it provides *additional* knowledge that can be employed for comparing alternatives in terms of their *wear LCCs*. If made aware of unintended LCCs, a designer may therefore retract a commitment to explore other alternatives.

3. A 'KC' APPROACH FRAMEWORK TO 'WEAR CONTROL'

This section introduces the developed *'Knowledge of life-cycle Consequences (KC)'* approach framework [3]. The concept of the 'KC' approach is to explicitly reveal the *wear LCCs* co-evolving with the design solution, so that the designers can take them into consideration *during* synthesis. In order to *timely* support the designer's thinking process, designers need principles disclosing how to describe evolving artefact life solutions from which *wear* LCCs can be inferred. Further, to adopt this approach, there is a need to describe *what* needs to be captured and modelled to generate a suitable *wear* LCC knowledge base. Thus, the LOD approach developed is based on three frames (Figure 6), an *Operational Frame,* a *wear LCC Knowledge Modelling Frame* and an *Artefact Life Modelling Frame*, these collectively forming the *'KC' approach framework.*

3.1 The Operational Frame

This concerns the operating principles of the approach. Basically, it assumes that many well-developed solution concepts (PDEs) and well-known life cycle phase elements (LCPEs) are reused during synthesis and encountered during the total life of an artefact. A fundamental operating principle is that a designer is engaged in concurrent synthesis - the *artefact life model* is *what* the designer generates and models *during* component life design synthesis. For this purpose, when a design sub-problem (Figure 6) is encountered (step 1), the designer interacts with a synthesis elements library to search (step 2) for a set of suitable elements (step 3). Based on designer intentions, preferences and circumstances, the designer commits (step 4) elements to evolve the artefact life model. This evolving model is monitored (step 5) by LCC inference knowledge which infers (step 6) any co-evolving wear LCCs. Relevant LCC action knowledge infers actions that need to be carried out, such as changes in performance measures of appropriate life-phase metrics to allow designers to monitor the artefact life behaviour. Collectively, this inferred knowledge is utilized (step 7) for exploring the avoidance/relaxation of the LCCs detected, including those related to wear.

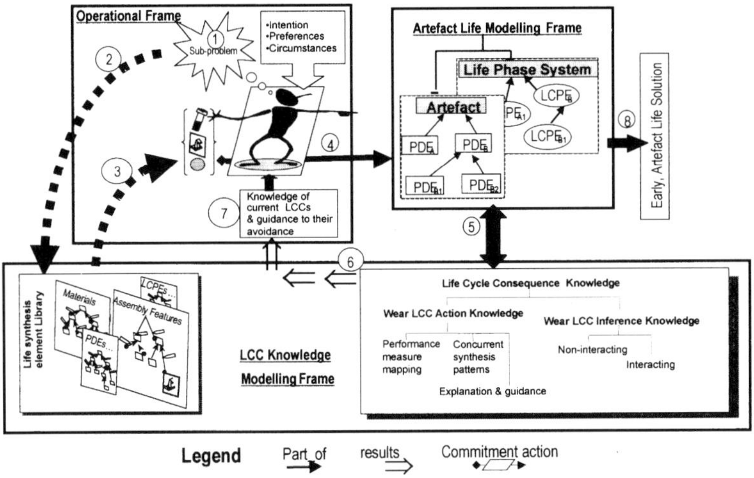

Figure 6 - 'KC' approach framework

3.2 The Wear LCC Knowledge Modelling Frame

Based on the LCC phenomena model generated, wear LCC knowledge can be modelled and compiled. The LCC knowledge modelling frame (Figure 6) discloses *what* to *acquire*, *model* and *relate* for an application domain, together with *how* to transform and structure the established relationships into meaningful *wear LCC inference knowledge* to support the inference of both *wear* LCC_{ni} and *wear* LCC_i relevant to the application domain synthesis elements. It also concerns modelling *LCC action knowledge.* The latter describes actions that arise from the detected LCCs.

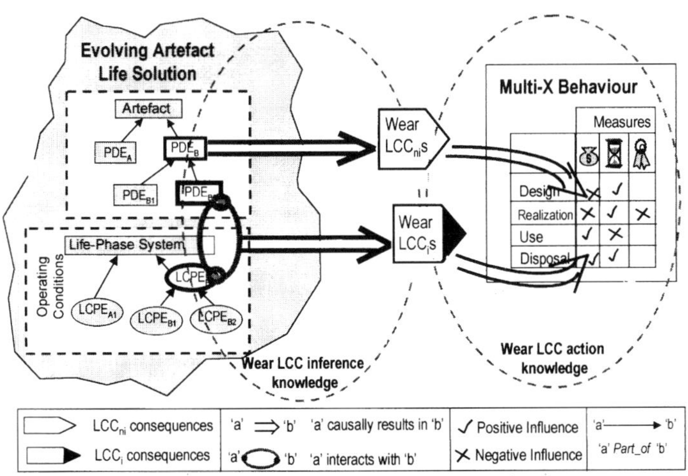

Figure 7 – Relationships embodying wear LCCs knowledge

Further, for *relevant* LCCs to be revealed and utilized at the *right time* with both *least* and *specific* synthesis decision commitments, this frame describes how LCC knowledge can be structured, the latter also necessary for knowledge management and scaling issues. Thus, this frame, provides a *formalism* for:

- *a life synthesis element library:* this consists of various PDE (eg. form features, assembly features, materials) and LCPE (e.g. fabrication system);
- *relationships* (Figure 7) between PDEs/LCPEs and LCCs describing how to model:

a) wear LCC inference knowledge: these are descriptions *logically* relating artefact life PDEs/LCPEs to LCC_{ni} and/or LCC_is e.g.

IF [PDE] is a *kind_of* Snap-fit
AND <Use phase>$_{\text{Operating Conditions}}$ = Very Frequent Dis-assembly
THEN $(LCC)_i$ = Abrasive Wear

As structured above, it means that if a *cantilever* or *annular type* of snap-fit is used during design, the same rule is applicable, this reducing LCC knowledge duplication.

b) wear LCC action knowledge this describing:

i. *mappings between LCCs and performance measures* - a LCC (e.g. abrasive wear) can cause a change (δ) to a *performance measure (PM)* e.g. cost, of maintenance process [P], forming part of the servicing phase; Changes in performance measure values are modelled *relative* to values resulting from other possibilities on a range –10 to +10 e.g.

IF $(LCC)_i$ = Abrasive Wear
THEN <Use phase>$_{\text{quality}}$ = -9
AND <Use phase>$_{\text{time}}$ = -5
AND <Servicing phase>$_{\text{cost}}$ = +5

ii. *concurrent synthesis patterns:* in some cases, a LCC is a decision space 'S' with a set of decision proposals D{O} having a set of options {O}. Sometimes, the set {O} has an associated default commitments (e.g. use a lubricant). Describing such patterns allows commitments to be automatically made to evolve a specific <model> e.g.

IF $(LCC)_i$ = Abrasive Wear
THEN <Design phase> = Add [lubricant] to <solution model>
OR <Design phase>= Re-design contact form
OR <Design phase>= Use material with a better hardness value
OR <Use phase>$_{\text{Operating Conditions}}$ = Reduce operating force

iii. *explanations of specific wear LCC and providing guidance to their* avoidance/ relaxation: having inferred a LCC (e.g. abrasive wear), designers need explanations of what the wear LCC means, what it effects and guidance to its avoidance/relaxation.

3.3 The Artefact Life-Modelling Frame

This frame provides a formalism for describing an *'artefact life' model* that supports *wear LCC* inference. For this research, the characteristics used in defining PDE models are those that support the inference of *wear LCCs*. PDEs which have been found relevant are, form features, assembly features, material, surface finish, dimensions and tolerances. The generic LCPE model developed in this research is based on Hubka & Eder's [7] model of a transformation process. LCPE characteristics that have been considered sufficient for supporting the inference of LCCs from the interaction of a conceptual mechanical component solution and a conceptual life-phase system solution are: process technology, technological properties, process parameters, process parameter values, process minimum economic quantity, process technical (physical) system, and technical system parameters. Also, for infering *wear LCCs*, characteristics defining the operating conditions of technical systems is essential if life specific *wear LCCs* are to be inferred.

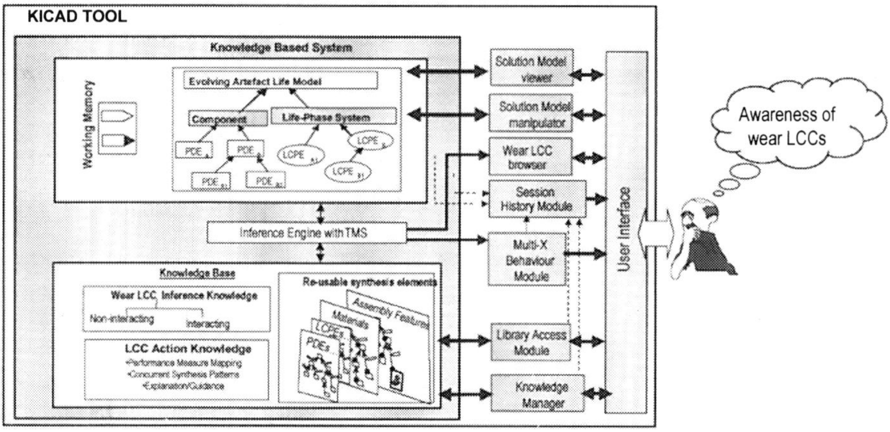

Figure 8 – KICAD prototype system architecture

4. PROTOTYPE 'CAD FOR WEAR CONTROL'

The 'KC' framework can be realized as a *KICAD* tool. Such a prototype for LOD of thermoplastic components, nicknamed 'FORESEE' is detailed in [4]. FORESEE allows designers to generate an early mechanical component compositional model and a number of early life-phase system compositional models. It also provides a list of LCCs including those related to wear, associated with the *current solution state*, an associated set of *multi-X* relative performance measure values, explanations of detected LCCs and guidance to their avoidance, a history of decision commitments made and their associated impact on performance measures. FORESEE's system architecture (Figure 8) consists of a *library access module* which allows the designer to search for synthesis elements (e.g. assembly features) that result in intended consequences (eg. easy dis-assembly). The *knowledge manager* provides utilities to allow designers to dynamically add/modify classes/instances of synthesis elements stored in the library. The *solution model manipulator* provides functions that allow the designer to add/detail/refine elements to the *solution models* being concurrently synthesized, these represented internally in the *working memory* of the *knowledge based system*. The *inference engine* employs *wear LCC inference knowledge* to reveal wear LCCs co-evolving with the solution and *LCC action knowledge* to enable the KICAD tool to take a pro-active role in component LOD. The wear LCCs detected are displayed in the hypermedia (Html) based *consequence browser* this allowing a LCC to be individually browsed to reveal details about its meaning, its source commitment(s) and guidance to its avoidance. A compositional hierarchy of an evolving artefact life solution model is displayed via the *solution model viewer*. The *multi-X behaviour module* employs *performance mapping knowledge* to estimate and display fluctuations in performance measures as a result of LCCs inferred. A *session history module* is used to automatically maintain the design history.

5. CONCLUSION

Experimentation [3] established that the LCC knowledge structuring concept provides an incremental step towards the systematization of LCC knowledge. This has been achieved by organizing the different synthesis elements (PDE/LCPE) into *kind_of* taxonomies, with known LCC_{ni} associated to the relevant abstraction levels in the taxonomy [3]. This allows more specific elements lower down in the taxonomy, to inherit the LCC_{ni} knowledge associated with more general elements, thus reducing the duplication of common knowledge. Similarly, wear LCC_i inference knowledge can be structured by modelling *interacting relationships* between classes of synthesis elements, allowing it to be inherited by PDE/LCPEs lower down in the taxonomies. However, since LCC knowledge is distributed, its acquisition is practically difficult. Nevertheless, only by knowledge captured can valuable LCC knowledge be codified, shared and explicitly reused *during* solution synthesis in new design situations. Employing the current LCC knowledge structure for practical applications still requires research concerning the scaling up and maintenance of the integrity of a very large LCC knowledge base. Research is also required to *establish* and *model* LCC certainty factors. Research to *benchmark* the relative fluctuations in performance measures is required. For example on a scale of 10, should the formation of abrasive wear result in a servicing cost increase of say 3 units or 8 units? To conclude, experimentation highlighted that the 'KC' approach framework contributes a step towards the development of KICAD tools aimed at pro-actively supporting designers in controlling artefact wear *during* design. However, further work [3] is required in order to practically exploit its utilization .

Acknowledgements
This paper evolves from work previously carried out by Jonathan C.Borg at the University of Strathclyde, UK, in collaboration with Dr. Xiu-Tian Yan and Prof.Neal.P.Juster.

REFERENCES

[1] Andreasen, M. M., "Design methodology", Seminar on Design Theory and its Application, Trondheim, Norway, 1991, The Norwegian Academy of Technical Sciences,.
[2] Bhushan, B., "Principles and Applications of Tribology", John Wiley & Sons Inc. New York, 1999.
[3] Borg, J. C., "Design Synthesis for Multi-X - A 'Life-Cycle Consequence Knowledge' Approach", Ph.D., University of Strathclyde, Glasgow 275, 1999.
[4] Borg, J. C., X. T. Yan and N. P. Juster, "A KICAD Tool For Pro-Active Exploration Support To 'Design Synthesis for Multi-X'" in Knowledge Intensive Computer Aided Design edited by S. Finger, T. Tomiyama and M. Mantyla. Tokyo, Japan, Kluwer Academic Publishers: 1998, pp 274-295.
[5] Duffy, A. H. B. and M. M. Andreasen, "Enhancing the Evolution of Design Science", Proceedings, International Conference on Engineering Design, Praha, 1995, 29-35.
[6] Duffy, A. H. B., M. M. Andreasen, K. J. MacCallum, et al., "Design Co-ordination for Concurrent Engineering", Journal of Engineering Design, Vol.4, No.4, 1993, 251-265.
[7] Hubka, V. and E. Eder., "Theory of technical systems: A total concept theory for engineering design", Springer-Verlag Berlin/Heidelberg, 1988.
[8] Krause, F.-L. and K. Martini, "Simulation Tools for the Development of disassembly-friendly products", 6th International Seminar on Life-Cycle Engineering, Queen's University, Kingston, Ontario, Canada, 1999, 286-294.
[9] Landsdown, A. R. and A. L. Price., "Materials to Resist Wear", Pergamon Press 1986.
[10] Ludema, K., "Friction, Wear, Lubrication, a Textbook in Tribology", CRC Press 1996.
[11] March, J. G., "A Primer on Decision Making - How Decisions Happen", The Free Press New York, 1994.
[12] Rabinowicz, E., "Friction and Wear of Materials", John Wiley & Sons Inc. New York, 1995.
[13] Tjalve, E., "A Short Course in Industrial Design", Newnes-Butterworths London, 1979.

The use of wear information for the development of disassembly-friendly products

F-L KRAUSE and **K MARTINI**
Institute of Machine Tools and Factory Management, Technical University Berlin, Germany

ABSTRACT

Consideration of disassembleability of products during their development today is one of many aspects a product designer has to think of. Since the use of CAx Systems was becoming state of the art in product development many activities to integrate all information the designer needs to a computer-supported development process were implemented. The problem of integrating recycleability and disassembleability aspects is the period of the product use, in which the product changes, among other things by tribological processes. These often difficult predictable processes can have influence on maintenance as well as the disassembleability at the end of the product life. The paper introduces possibilities to integrate tribological information into a feature-based CAD system, their visualisation with the help of a voxel modelling system and their processing into a disassembly simulation system.

1 INTRODUCTION

Today the increasing shortage of natural resources as well as the arising amount of waste force the companies to consider environmental aspects already during the development of new products. One of these aspects is the disassembleability at the end of the product life. That is why new demands are made on the designer to develop disassembly-friendly products without ignoring other design goals. Computer-based tools – possibly integrated in a CAD system – can help him to better fulfil this demands by providing necessary information about the product life cycle.

One of the main problems during the integration of disassembly aspects into CAD is the long time period between manufacturing and disassembly, in which the product normally is used. In this period the product is subject to several changes, among them tribological changes and corrosion. If the product changes during use, the disassembleability of the product at the end of its life also changes. That means, that the consideration of tribological changes is among other things necessary to assess the disassembleability of the product, and this should be done already during the product development process.

After a short conclusion of research related to this paper, section 3 shows one possibility to integrate wear information into CAD with the help of the feature technology. To achieve a powerful visualisation, the feature-based CAD system is combined with a voxel modeller.

Section 4 shows how the wear information included in a feature-based model can be processed within a disassembly simulation system. The wear degree here can be used, for instance, for the adaptation of forces and moments necessary to perform the disassembly process. So the simulation is becoming more realistic. An outlook over future work is given in section 6.

2 RELATED WORK

At the moment, there is a number of simulation system in the market, which can be used for modelling and planning assembly and disassembly systems [1, 2]. In contrast to the system presented in this paper they are not able to represent and utilise wear at the product. Another disadvantage is their inner structure (often based on Gantt diagrams) which brings losses of flexibility and extendibility of the system. Disassembly processes are often considered as reversal of assembly processes, whereas the concept presented here from the beginning is oriented towards disassembly. In this way other disassembly alternatives like destructive methods can be considered.

The computer-internal representation of wear with reference to its relevance for disassembly is not yet a topic for other research activities. However, DUTTA represents different materials and their distributions within parts computer-internally, though his aim is to manufacture heterogeneous parts by Layered Manufacturing Technology [3, 4]. Here the model of the part is split into regions that have the same distribution of material. These regions are stored additionally to the part geometry. Within these regions a homogeneous material distribution is assumed. The disadvantages of this approach are that double the space is needed for the additional storage of the material region and the assumption of homogeneous distribution within the regions.

3 INTEGRATION OF TRIBOLOGICAL INFORMATION INTO CAD

3.1 Computer-based Representation of tribological Information

There is a standardised classification of wear types documented in [5] (table 1). For every type another combination of the four mechanisms surface destruction, abrasion, adhesion and tribochemical reactions is effective.

Generally, the wear volume can be estimated to

$$V_w = k \cdot F_N \cdot s$$

where V_w: wear volume,
k: wear coefficient,
F_N: force in normal direction,
s: wear distance.

For every mechanism and material, the wear coefficient is different. Problematic are superpositions of the mechanisms because the influences of the mechanisms on each other are not completely discovered yet and cannot be calculated and represented computer-internally. An exception is the gliding wear, here some conditions for the occurrence of the mechanisms can be given, so that the wear volume keeps calculable. Furthermore the adhesion mechanism is not completely discovered yet. Table 1 shows which wear types and mechanisms can be represented computer-internally.

Table 1: Wear types and mechanisms [5, 6]

Type \ Mechanism	Surface Destruction	Abrasion	Adhesion	Tribochem. Reactions	**Assessment**
Gliding	x	x	x	X	+
Rolling	x				++
Pushing	x	x	x	X	-
Oscillating	x	x	x	X	-
Furrowing		x			++
Blasting	x	x			+
Erosion	x	x		X	+
Assessment	++	++	-	+	

++: representable
+: partly representable
-: not representable

3.2 Feature-based Approach

The calculation within the feature modeller is done with the help of the formula mentioned in section 3.1. By mouse interaction, the user selects the two wear partners, which feature description contains the material information necessary for the calculation in form of feature parameters. The contact surface between both partners is calculated, and with the help of the wear distance (also a feature parameter) the worn surface can be determined. The worn surface is divided into sub-surfaces according to its size. For every sub-surface the wear degree can be calculated. Figure 1 shows the calculation of wear between a fixed block, on which a sphere rolls as a simple example.

The reason for the arising shape is the occurrence of stresses according to the theory of Hertz and the wear distance. The calculated wear degrees are stored in an ASCII data file and together with original geometry, the position of the worn surface and the span of the calculated points moved to the voxel modeller [7].

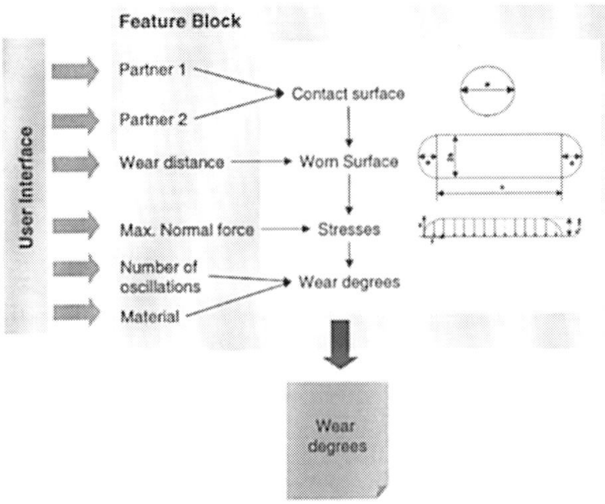

Fig 1 : Wear feature

3.3 Combination of Feature and Voxel Approach

As mentioned above the calculation of the wear degree takes place within the feature modeller. The calculation results are stored in a data file (ASCII Format), which is the input for the voxel modeller (figure 2). Within the voxel modeller the visualisation of the wear is made by removing voxels from the original geometry corresponding to the wear degree calculated by the feature modeller. Therefore a function for the preparation of the wear data was implemented. This function combines the wear degree, the position of the worn surface and the chosen distance to coordinate points. These points are combined to a tool path in the direction of the wear distance. A point-like invisible tool removes voxels removed along the path. The result of this procedure is a three-dimensional representation of wear, which additionally is supported by colour scales [8].

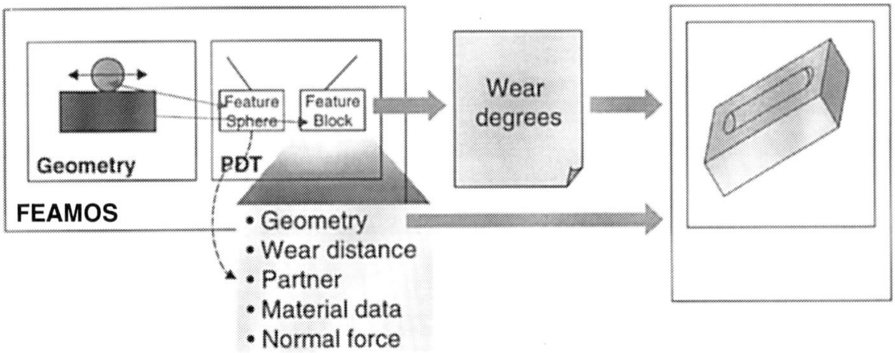

Fig 2 : Combination of feature and voxel modeller

4 DISASSEMBLY SIMULATION

A disassembly simulation system was implemented to process the wear information from the feature perspective to the voxel modeller. The base of the system is a simulation model that consists of the product structure and the process structure. The product structure, which can be read from the feature modeller, contains all feature information including wear information. Optionally the voxel model of the part can be loaded (figure 3). The process structure is derived from an AND/OR-Graph and contains all disassembly process alternatives for the corresponding assembly. The process structure can be loaded from an external program or can be developed from a product structure and a joint structure.

The simulation model is the base for some modules of the system. At the moment, the virtual teach-in module and the disassembleability check module are realised and tool editor and system editor are still under development. Within the virtual teach-in module three-dimensional tool paths for every part can be recorded interactively by space mouse. The player of the module can simulate the disassembly process. Within the disassembleability check module the wear information contained in the product structure is used to generate forces and moments necessary to remove the parts from the disassembly. These forces and moments are then compared with the maximum force or moment of the tool the user has selected for the actual disassembly operation. If the part is disassembleable, the other alternatives are disabled. The result of this procedure is a disassembly sequence that can be performed with the selected tools.

To extend the functionality of the simulation system, two new editors are actually implemented. With the help of the tool editor, points at the part and the corresponding tool can be selected interactively so that part and tool go together during the simulation (figure 4).

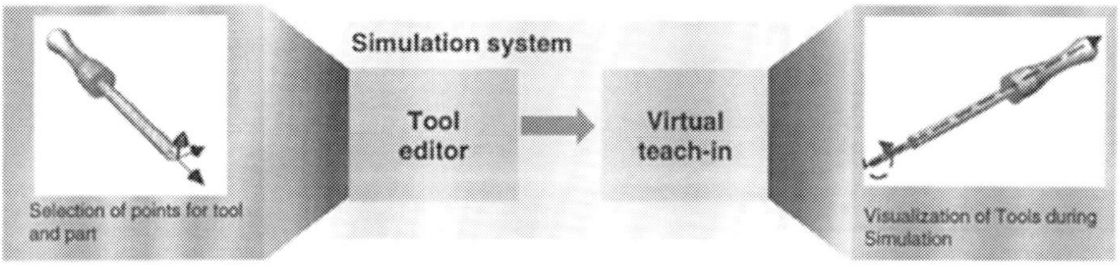

Fig 3: Tool editor and virtual teach-in module

The system editor allows the user to create the whole environment for the assembly or product. Geometry can be loaded from different formats and positioned on the floor so that a cell or system is growing. Movements can be added to every geometry, for that the editor includes some pre-defined movement patterns. The player of the virtual teach-in module can process all movements. Parts of the cell or system can be grouped to form a work-place. Every work-place can be connected to one assembly or part in the product structure, which is disassembled at this work-place. [9]

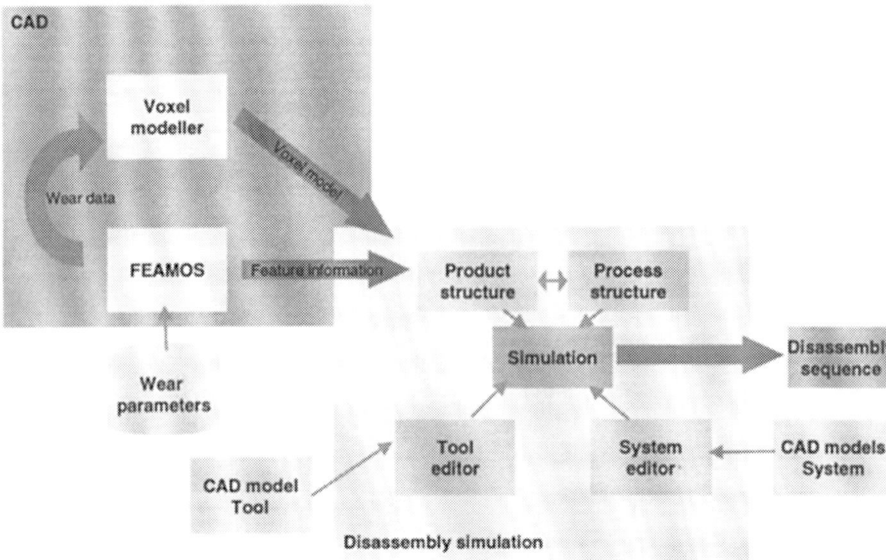

Fig 4: System architecture

5 OUTLOOK

The paper shows one possibility to represent changes of the product shape coming from wear with the help of a voxel model. This model is also appropriate for the representation of wear that changes the material structure. The implementation of these representations is planned for the next period of the project, for which some extensions of the voxel modeller have to be realised. On one hand, possibilities for the allocation of indices for several materials and their properties have to be made, on the other hand the space has to be minimised for very inhomogeneous parts.

Another topic during the next project period will be the representation of flexible parts. Especially for washing machines, the main focus is on such parts like cables and tubes which often become a problem. Due to their complex behaviour, disassembly processes of flexible parts are hard to automate. A disassembly-specific problem comes from the occurrence of wear like, for instance, the brittleness as a result of water containing calcium carbonate. This wear influences the behaviour of the part in such a way that forces and moments necessary for the removal of the part are changing and that the flexibility and handling of the part are getting worse. The investigation and representation of these connections as well as their integration in the simulation system will be a main focus for the coming project period, where a force feedback system for the interactions will be used.

With the help of appropriate sensors, wear-relevant parameters can be determined directly during the product use. A life cycle unit is developed in another partial project of the collaborative research area, which have to do this task. The data measured by this unit can be used for balancing calculated wear models with real values as well as for determination of parameters relevant for the calculation. That is why for the next period connection between wear model and life cycle unit is planned.

One application of the simulation system is the direct use at a disassembly work-place. To guide the worker during the disassembly process and to give him/her hints about worn parts, an Augmented Reality system can be used. Therefore the simulation results (tool paths, disassembly sequences) have to be prepared appropriately [10].

6 ACKNOWLEGDEMENTS

The work and results shown in this paper are part of the partial project D3 "Simulation Tools for the Design of Disassembly" of the Collaborative Research Area 281 "Disassembly Factories for the Recovery of Resources in Product and Material Cycles" funded by the German Research Association (DFG).

7 REFERENCES

[1] N.N.: http://www.deneb.com
[2] N.N.: http://www.technomatix.com
[3] Kumar, V.; Burns, D.; Dutta, D.; Hoffmann, C.: "A framework for object modelling". Computer-Aided Design, vol. 31, no. 9, Elsevier Science Ltd. 1999, pp. 541-556.
[4] Kumar, V.; Dutta, D.: "An Approach to Modeling and Representation of Heterogeneous Objects". Journal of Mechanical Design, vol. 120, no. 12, ASME 1998, pp. 659-667.
[5] Czichos, H.; Habig, K.-H., "Tribologie Handbuch, Vieweg Verlag Braunschweig, 1992.
[6] Bergmann, W.: "Werkstofftechnik. Teil 2: Anwendung", Carl Hanser Verlag München Wien, 1991.
[7] Krause, F.-L.; Stiel, C.; Martini, K.: "New Applications for Feature Modeling". Proceedings of the Israel-Korea Bi-National Conference on New Themes in Computerised Geometrical Modeling, Tel Aviv University, Ramat Aviv, Israel, February 18-19 1998, pp. 1-9.
[8] N.N.: "Sonderforschungsbereich 281: Demontagefabriken zur Rückgewinnung von Ressourcen in Produkt- und Materialkreisläufen. Arbeits- und Ergebnisbericht 1998-2000", Technische Universität Berlin, 2000.
[9] Krause, F.-L.; Martini, K.: " Simulation Tools for the Development of disassembly-friendly Products". Proceedings of „Life Cycle Engineering – The Next Millennium", 6th CIRP International Seminar on Life Cycle Engineering, Eds.: J. Jeswiet, T.N. Moore, P.H. Oosthuizen, Queens University, Kingston, Ontario, Canada, June 21-23 1999, pp. 286-294.
[10] N.N.: "Sonderforschungsbereich 281: Demontagefabriken zur Rückgewinnung von Ressourcen in Produkt- und Materialkreisläufen, Finanzierungsantrag 2001-2003", Technische Universität Berlin, 2000.

Optimum design concept of sliding bearing

A MARINKOVIC and **B ROSIC**
Department of Mechanical Engineering, University of Belgrade, Yugoslavia

SYNOPSIS

This paper presents an analytical concept and computer aided procedure for the multicriteria design optimisation of sliding bearing. According to hydrodynamic lubrication theory some bearing parameters and tribological performances are defined and used to determine the objective function(s), variables and the functional constraints in optimum design process. For the defined multicriteria optimisation model of the sliding bearing, a computer program based on an interactive dialogue is developed. The results of the program are presented in appropriate tables with Pareto - optimal solutions. A graphical illustration of the above problem is presented in terms of objective functions.

NOTATION

F	*bearing load*
d	*bearing diameter*
B	*bearing length*
p	*specific bearing load*
v	*sliding velocity*
n	*rotational shaft speed*
μ	*coefficient of friction*
k	*heat transfer coefficient (convection)*
A	*bearing surface area for heat transfer by convection*
t_b , t_o	*operational bearing temperature and air temperature*
ρ	*density of oil for lubrication*
η	*dynamic viscosity of oil for lubrication*
c, ψ	*clearance and relative clearance*
h_o , h_{olim}	*minimum of oil film thickness and it is critical value*
ε	*relative eccentricity*
S_o	*Somerfeld's number*

1 INTRODUCTION

Sliding bearings are in wide use today, they are applied to many of the machines we need and meet in life. This can be understood because of some of the advantages which this sort of bearing has compared with rolling bearings. Generally their production is not so complicated which makes for lower price; for simple mounting they can be made in separate parts and in operation they produce less noise and vibration. When correctly lubricated they are very practical for maintenance and have long operating lives.

In today's world it is necessary to either design a new "best system" or to optimise some existing systems to maximise efficiency and cost effectiveness. Optimum design concepts and methods can help to design such a system consisting of a sliding bearing.

This paper presents one attempt at applying the optimum design concept to a journal bearing. The optimisation model is based on well known hydrodynamic lubrication theory, which requires a whole range of tribological, geometrical and other parameters.

2 FORMULATION OF THE PROBLEM

According to hydrodynamic lubrication theory of journal bearings, we have a complex system of shaft and bearing, their clearance and bearing housing filled with oil. It is practically impossible to present in this paper a complete overview of all aspects of that theory. But in formulating this problem it is necessary to define some bearing parameters including the clearance value as well as other measures of tribological performance. These will be used to determine objective function(s), variables and functional constraints in the optimum design process.

2.1 Bearing parameters

To define all relevant parameters of the journal bearing, consider some basic relations in bearing calculation [2]. The specific bearing load and sliding velocity can be calculated using:

$$p = \frac{F}{d \cdot B} \quad \text{and} \quad v = d \cdot \pi \cdot n$$

If it is supposed that the heat produced in the bearing by friction is transferred by convection on shaft, housing and environment air:

$$P_f = F \cdot v \cdot \mu = k \cdot A(t_b - t_o)$$

2.2 Hydrodynamic lubrication theory of journal bearing

Some well known suppositions are made in the analysis of hydrodynamic journal bearing and they are used to define functional constraints in optimization model. All lubricating fluid (oil) flows are supposed to be laminar and this can be checked by [3]:

$$R_e = \frac{\rho \cdot v \cdot c}{2\eta} \leq 41{,}3\sqrt{\frac{d}{c}}$$

The minimum of oil film thickness must be greater than its critical value:

$$h_o = 0{,}5d \cdot \psi\,(1-\varepsilon) > h_{o\lim} \qquad \text{where is:} \qquad \psi = \frac{c}{d}$$

The dimensionless parameter named Somerfeld's number, relates operating and tribology aspects of the performance of a hydrodynamic journal bearing; it is given by:

$$S_o = \frac{p \cdot \psi^2}{\eta \cdot \omega}$$

The Somerfeld's number depends on relative eccentricity (ε) and relative bearing length (B/d) and its values are numerically calculated. Similarly it can be used to define value of relative friction coefficient ($\mu\ /\psi$). For engineering applications numerical solution both of these relations are presented in the literature in form of tables [3] or diagrams.

3 FORMULATION OF THE OPTIMISATION MODEL

In developing an optimisation model, one must start with very strict engineering requirements, which a sliding bearing should fulfil regarding the operating temperature, coefficient of friction, bearing dimensions, pv operating range etc. From the outset, the operating temperature has been used as a main criterion.
Based upon the requirements defined, it follows that it is practically impossible to describe the desirable performances of a sliding bearing with one criterion only. It is necessary to introduce a larger number of criteria considering the desirable performances, even the conflicting ones, represents a significant step towards the reality of a bearing model solved by the multicriteria optimisation methods. So in this paper, we add two more criterion: the coefficient of friction to represent the tribological aspect of the problem and also a parameter to represent the bearing dimensions, for example, the magnitude outside the bearing housing surface. Criteria regarding the desired performance are expressed by the criteria functions which, for the best journal bearing design, should reach the extreme:

$$\underset{x \in D}{extr}\ f(x)$$

The functional criteria for a sliding bearing can be written in form of the following relations:

$$f_1(x) = t_b = \frac{F \cdot v}{k \cdot A}\mu + t_o$$ - operating temperature of bearing

$$f_2(x) = A$$ - outside bearing (housing) surface area

$$f_3(x) = \mu$$ - coefficient of friction

In addition to that, it is also necessary to include functional constraints in the form of the inequalities:

$$g_1(x) = p_{\lim} - p > 0$$ - specific bearing load

$$g_2(x) = v_{\lim} - v > 0$$ - sliding velocity

$$g_3(x) = 41{,}3\sqrt{\frac{d}{c}} - \frac{\rho \cdot v \cdot c}{2\eta} > 0$$ - laminar oil flow

$$g_4(x) = 0{,}5d \cdot \psi\,(1-\varepsilon) - h_{o\lim} > 0$$ - oil film thickness

The following relation defines the "feasible" domain:

$$D = \left\{ x \in R^n \middle| \; g(x) > 0 \right\}$$

Based upon the objective functions given and upon the functional constraints, all the relevant values of the sliding bearing have also been identified, so that the vector of the variable values can be written in the form of the following relations:

$$x = x(\,d, \psi, \eta\,)$$

4 OPTIMISATION PROCEDURE

The optimisation model is characterised by the existence of mutually conflicting criteria and thus the optimum solution in relation to one objective function is not at the same time optimum in relation to another objective function. This means that the acceptable solution can be only that permissible solution which is not dominated by another permissible solution. Finding the set of nondominant solutions $\boldsymbol{X}^{\boldsymbol{p}}$ of the Pareto set is the first step in solving the multicriteria optimisation problem.

In solving multicriteria optimisation problems it is almost impossible to define exactly in advance the preference regarding the selection of "the best" design. In such circumstances, the most suitable procedure is to apply an interactive method, in which the decision maker (the designer) gradually sets out his/her preferences regarding the proposed set of Pareto solutions. The interactive method is based upon a dialogue between the designer and the computer in the process of finding "the best" compromising solution. The Pareto solution set gets narrower and narrower because in each iteration the designer gives his/her own preference regarding the desired performance, i.e., he places the upper confining values of the objective function vector. In the process, the calculation phase (determination of the Pareto solution set) is alternately replaced by the decision phase, in which the designer analyses the Pareto solution set, i.e., gives suggestions for reducing the observed set. This means that in each iteration the ideal point changes its position in the criterion space of the objective solution. Fig.1 shows a conceptual flow-chart for the interactive design optimisation process.

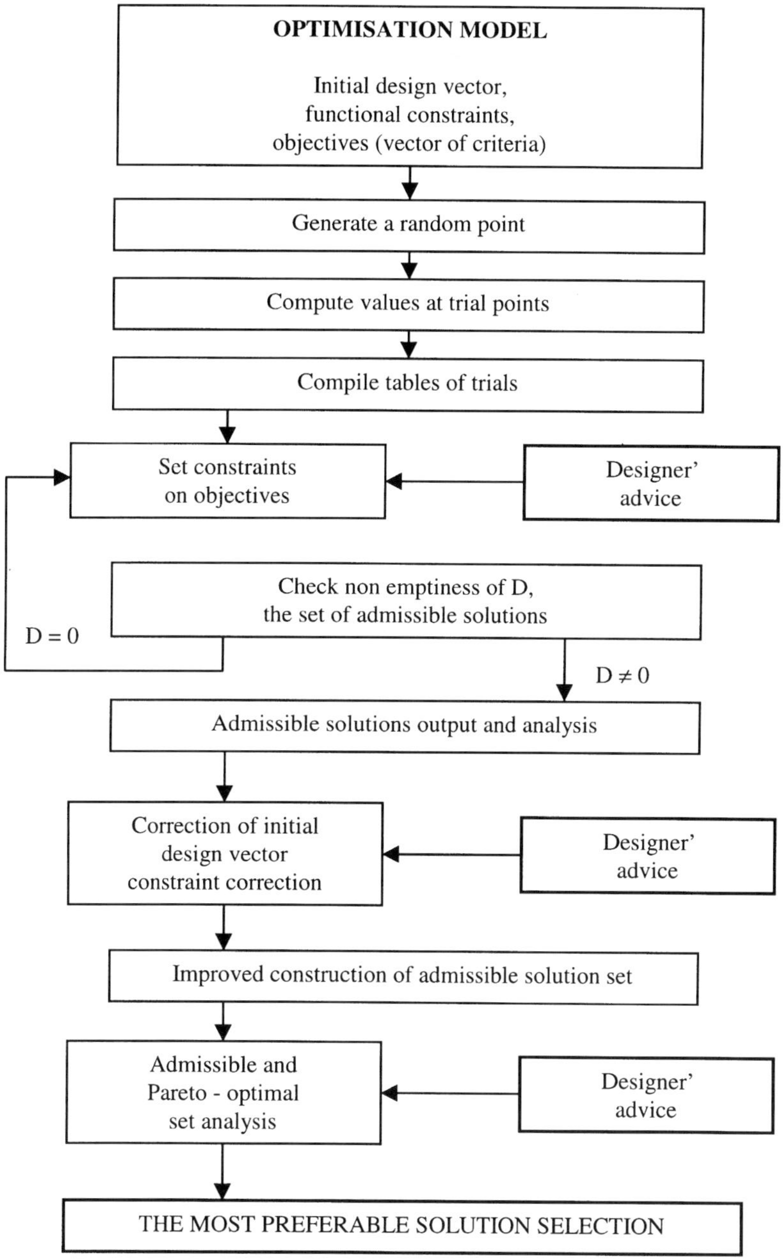

Fig. 1 Flow - chart for optimum design procedure

5 THE RESULTS

For the proposed model a computer program with an interactive approach has been devised. The approach is based upon generating random numbers according to the Monte Carlo method. The interactive approach ensures an efficient way of finding "the best" compromise solution. In this paper the most favourable design of journal bearing has been determined for the following input parameters:

$F = 2500N$	- bearing load
$n = 1200\ min^{-1} = 20\ s^{-1}$	- rotational shaft speed
$t_o = 20\ ^oC$	- external air temperature
$\rho = 900\ kg/m^3$	- oil density
$k = 20\ W/m^2\ K$	- heat transfer coefficient (convection)

For this numerical example was taken bearing dimensions d and B have been set equal, but note that the method and the model is applicable for any set of bearing dimensions. The bearing material is chosen from group of CuSn alloys. It could be said that it is only an example and this optimisation concept is applicable to any material of bearing. The results of computer program can be shown in the Table 1, where Pareto solutions for the given task are shown.

Table 1 Pareto - optimal solutions

MODEL	DESIGN VARIABLES		
	d	ψ	η
	mm	‰	mPa s
4774	34,7	0,577	10,1
6286	38,1	0,504	7,2
8998	34,5	0,679	14,9
9274	32,0	0,522	15,2

In Table 2 are shown the ordered sets of the objective function values for the corresponding models of the sliding bearing.

Table 2 Objective vector function

MODEL	OBJECTIVE FUNCTIONS		
	$f_1 = t_b$	$f_2 = A$	$f_3 = \mu$
	^oC	mm^2	-
4774	61,2	21038	0,003169
6286	54,9	25447	0,002969
8998	69,9	20855	0,003840
9274	73,0	17914	0,003776

The criterion space can be shown as a diagram illustrating the interrelations between certain objective functions. Fig. 2 presents the criterion space for the operating temperature - coefficient of friction and Fig. 3 for the operating temperature - bearing surface.

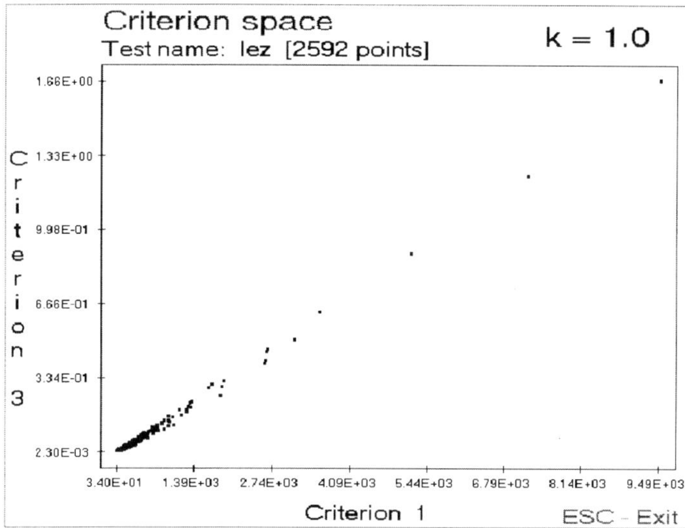

Fig. 2 The criterion space for the operating temperature - coefficient of friction

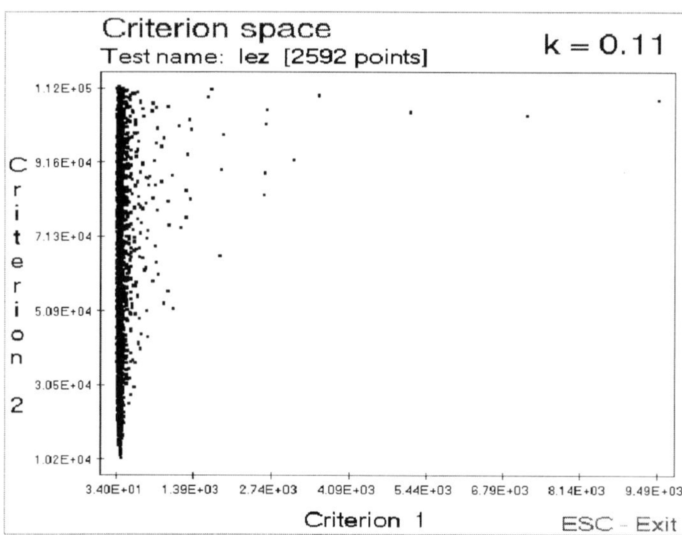

Fig. 3 The criterion space for the operating temperature - bearing surface

Based upon a geometrical interpretation of the results in these criterion space, the conclusion can be drawn that there exists very strong correlation between the criteria f_1 and f_3 (operating temperature - coefficient of friction) and also that criteria f_1 and f_2 (operating temperature - bearing surface) are almost in conflict.

Finally, in order to generate a practical design of a sliding bearing, it is necessary to adopt "the best" solution from the suggested set of Pareto solutions. By analysing the models obtained, it has been decided to adopt the model under number 6286 as "the best" sliding bearing design. The values of design variables and criteria f_1 value (operating temperature) in the chosen model determine the type of oil and also the clearance value for this "best" solution viz.:

$d = 38\text{mm}$
$c = 0{,}504 \cdot 38 = 19{,}15\mu\text{m}$
oil with $\eta \cong 7$ mPa s at $t_b = 55\ ^{o}\text{C}$

6 CONCLUSION

The present paper is a brief review of wider investigations carried out with the aim of applying powerful optimisation methods in solving the problem of the selection of optimum parameters in sliding bearings. Multicriteria modelling very well reflect the design process in which usually several conflicting objectives have to be satisfied. All relevant parameters that effect the sensitivity of the constraints are identified. The procedure developed in this paper is advantageous because the results are practical for engineering application. This basic concept of a sliding bearing optimum design could be extended with the adoption of some new parameters and objective functions which will make a higher quality of optimisation model. The concepts presented here can also be applied to the optimum design of other types of bearings.

REFERENCES

[1] Pramod J., Alice M., "Theory of Design: An Optimisation Perspective", Mech. Mach. Theory, Vol.25, No.3, 1990, pp.287-303.
[2] Trbojevic M., Jankovic M., "The analysis of contemporary methods of design of journal bearings", in proceedings Judeko, Zagreb 1976, pp.333-352.
[3] Decker K. H., "Maschinen elemente, Gestaltung und berechnung", Hanser Verlag, Muenchen; Wien, 1995.
[4] Eschenauer H., Koski J., Osyczka A., "Multicriteria design optimisation, Procedures and Applications", Springer Verlag, 1990.

Life-cycle design – economical and ecological decision making

J HESSELBACH and **C HERRMANN**
Institute of Machine Tools and Production Technology, Technical University of Braunschweig, Germany
O RICHTER and **M FLAKE**
Institute of Geo-Ecology, Technical University of Braunschweig, Germany

ABSTRACT

Intensifying environment legislation as well as increasing environment related costs lead to the consideration of new aspects as a part of the product development. This paper presents a method to support the decision making process with the objective to identify fields of impact on the economical and ecological level. On base of respective break-down structures, the effect of the application of design rules can be highlighted. This supports the recognition of relations between effects of decisions on the subsequent life stages.

1 INTRODUCTION

Over the last decade, the importance of protecting the environment has come more and more into focus. Along with this growing understanding of the need to protect the environment has come the desire to move towards sustainable development. As a consequence, more and more, restrictions and environmental related legislation come into force. In Europe, for example, the self-obligation for old car recycling underlines that the amount of non-recyclable parts have to be reduced to a maximum of 15 % until 2002 and further decreased to a maximum of 5 % until 2015. At the same time, it has to be considered that the recycling rate for old cars is about 75 % nowadays. Considering that the amount of electric and electronic assemblies still increases, as well as the amount of plastic and new materials, further effort will be necessary to reach such defined objectives.

Along with the establishment of environmental objectives at legislation level is the awareness of increasing related costs as well as the effect on other aspects such as manufacturing or assembly. These costs will be part of the pricing of the manufactures or will influence the decision for or against buying a product in the case the consumers have to pay for it when it comes to end-of-life. As a result, companies integrate the environmental responsibility into their company strategy and create an important prediction for the internal acceptance of the extended product requirements. Due to the fact that environmental impacts, as well as costs, are mainly determined at the design stage, approaches are necessary to support designers and decision makers during the design process.

2 LIFE CYCLE DESIGN

Life cycle design (LCD) can be defined as the systematic consideration of design issues associated with entire life cycle during the development of products and processes. LCD covers all design disciplines such as Design for Manufacturing (DFM), Design for Assembly (DFA), Design for Service (DFS), Design for Recycling (DFR) and Design for Environment (DFE). These are often referred to collectively as Design for X (DfX), Figure 1. Whereas Design for Environment aims at the minimisation of the environmental impact, the remaining elements of DfX mainly focus on the reduction of costs.

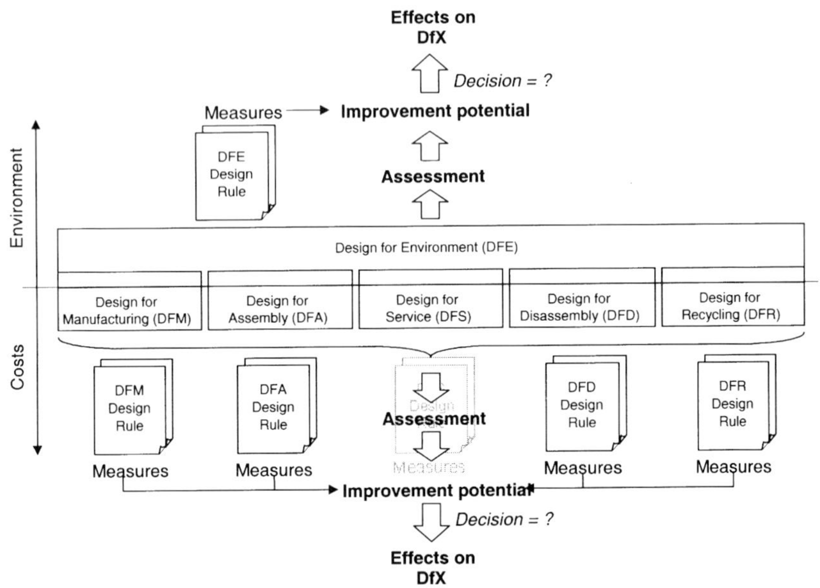

Figure 1: Design disciplines of LCD

If all elements of LCD are integrated into a single continuous feedback-driven design process, this is defined as concurrent engineering. Here, the main objective of LCD is to support design teams to create products taking into account the entire product life-cycle from conception through manufacturing and assembly to disassembly, recycling and disposal. To support the related design disciplines, various design approaches and tools are available. However, it has to be considered that this process includes all companies, suppliers or organisations may be concerned and, therefore, interacts with multiple life-cycles.

Design activities can be performed at various levels. In the case of DFE, for example, the environmental improvement of existing products can take place, a radical redesign based on existing concepts is possible, or more, a green function or system innovation can be achieved [1]. Nevertheless, starting point of design processes often is a more or less complete product model on the basis of an existing or planned product. With the aid of the available design tools, the respective process and/or product parameters can be determined. The corresponding assessment results are generated in this way. In the following step, improvement potentials must be identified based on these results and be assigned to the corresponding product properties [2]. Suitable improvement measures and design alternatives must be determined subsequently. Therefore, selection tables as well as design rules and guidelines are available, those are often summarised in the form of design guides or check lists, Figure 2.

Design Rule: *Avoid non-compatible labels on plastic parts*

Labels used for identification on plastic parts can prevent a material recycling because pigments, films and adhesives used for the fabrication of labels are often not compatible with plastic fixed on. In particular, paper labels can make a material recycling impossible. In the case of disassembly, removing these labels usually is very time-consuming and therefore extremely costly. Furthermore, hardly removable remains are often left on the part contaminating the recycling process and result in a low-grade plastic being unsuitable for use as a raw material. For this reason, non-compatible labels on plastic parts should be avoided.

Design Alternative: *3M™ Recycling Compatible Label*

3M™ Recycling Compatible Label Materials are compatible with housings made of ABS, PC, PC/ABS and HIPS. This material offers significant improvements for the electronic and computer industry's plastics recycling process.
By creating this new label material, 3M allows manufacturers and recyclers to eliminate the removal process and still obtain a high-grade "raw" recyclable material for reuse directly as new computer shells, allowing for a "closed-loop" life-cycle [3].

Figure 2: Example of a guideline for DFD/DFR

Different conceptual or design measures for product improvement with regard to a specific target can often be selected [4, 5, 6]. The problematic nature of this situation is that a chosen measure can have positive, no or a negative effect on other design targets, Figure 1. In particular, this applies to the DFE which covers all product life stages. Increasing the environmental compatibility can often go along with higher costs in another life stage. Under the aspect of the huge number of requirements design engineers already have to consider, it has to be mentioned, that, due to the complexity of DFE as well as DFD and DFR, design engineers can not and do not want to be experts in these fields. For this reason, models are required which support design engineers in the decision making process, Figure 3.

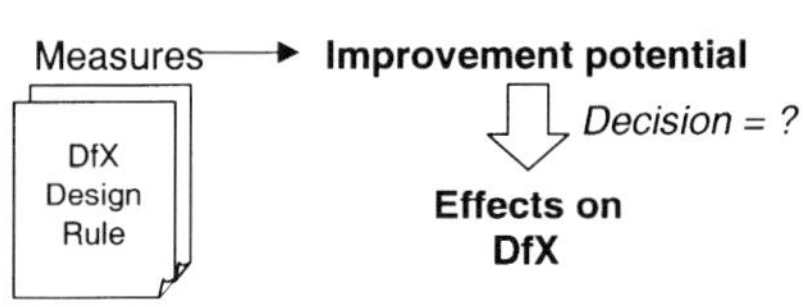

Figure 3: Decision problem

3 DECISION SUPPORT MODEL

In general, two objectives can be pursued using a cost and environmental impact model. As for costs, on the one hand, the aim can represent the determination and evaluation of the costs via the entire life-cycle (Life Cycle Costing). On the other hand, an optimisation and/or minimisation of the costs via the entire life-cycle can be pursued [7]. This process is generally based on a Activity-Based-Costing (ABC) and/or on the determination of the relations between costs and design features. However, particularised knowledge about costs and profits as well as about the existing system interrelations is required [8].

The environmental impact can be determined using a Life Cycle Assessment (LCA) method. Here, the collection of the inventory data is already linked with a high effort. On account of the resulting complexity and of the joined expenditure, the permanent applicability of these approaches in the design process is limited.

3.1 Objectives and Demands

The objective of the proposed model it is the determination of influences which are linked with an application of a measure to improve product design. Therefore, it is no question of a quantitative calculation of the life-cycle costs and of the environmental effects. Rather, the understanding of the system interconnections should be sensitised by showing possible fields of influence, Figure 4.

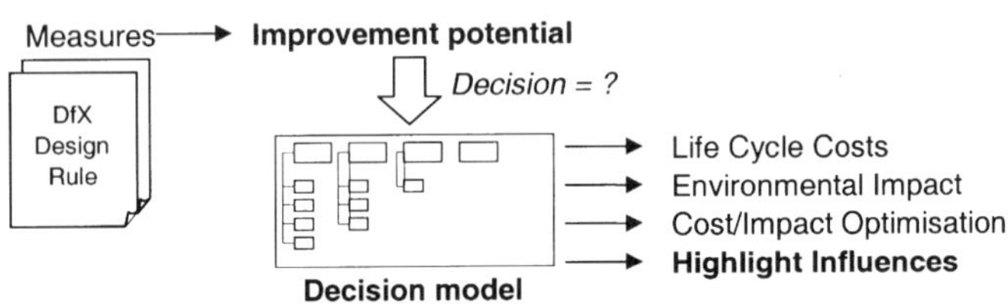

Figure 4: Decision models

In this way, the decision process should be supported. The demands can be summarised by the following factors:

- clear structuring,
- application oriented,
- focus on cost drivers (as far as possible),
- extendability.

In addition to these points, the basic concept should not only be tooled to an enterprise but show an universal character. In order to make adaptations to the enterprise-specific situation as simple as possible, a relatively high level of detail is striven. By this, positions can be removed or summarised, instead of carrying out extensive enlargements.

3.2 Life Cycle Stages

Starting point for the development of the decision model was the approach of Blanchard [9]. However, the horizontal (life-cycle stages) as well as the vertical (cost positions and environmental fields of impact) level of detailed description were increased. Here, the question about the clarity of the total model always has to be considered. The proposed model is subdivided into eight life-cycle stages:

- Research and development,
- Manufacturing (incl. material sourcing),
- Assembly,
- Distribution logistic,
- Use and service,

- Re-distribution logistic,
- Disassembly,
- Recycling and disposal.

3.3 Cost-Breakdown-Structure (CBS)

In addition to the horizontal structuring, the vertical structure has to be determined. Here, a two-stage division was chosen that can often be found: General terms describing the costs were assigned to the individual life-cycle stages summarising several cost positions, Figure 5.

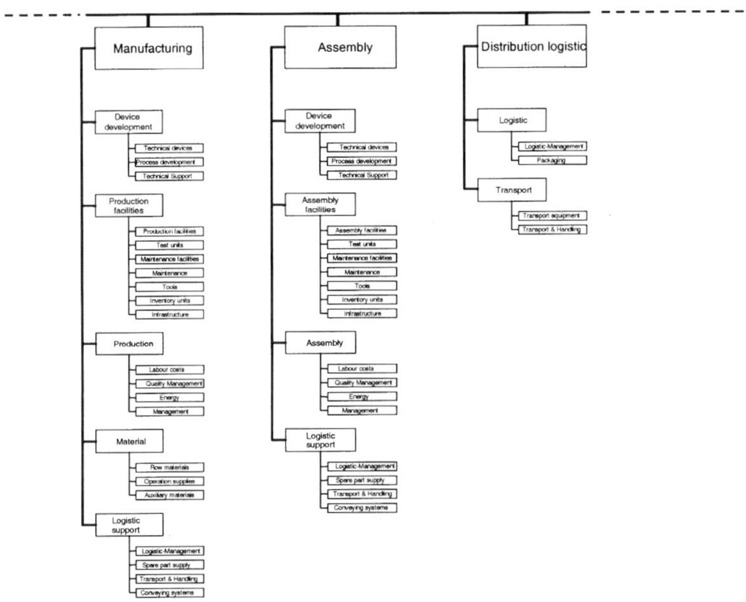

Figure 5: Example for the Cost Break-Down Structure

3.4 Environmental-Impact-Structure (EIS)

The association of categories of environmental impact to the life cycle stages of electronic appliances bases on data mainly originating in investigations of television sets. Television sets are very complex products, which probably represent an enormous part share of the burdens caused by products of the electronic industry for this reason. However, we have to presume that other appliances can have specific effects (for example: water consumption of a washer during its service life).

According to the data, a possible weighted relation of the burdens to the life cycle stages is not admissible. Especially for the energy consumption, the environmental impacts often depend on the different lengths of the use phase of the products (for example: use phase of a television set 10-20 years - of a PC 3 years on an average).

For these reasons, the categories of impacts are assigned referring to the principle of negative choice, i.e. that, in the first instance, all categories are considered on principle, apart from obviously not applying categories, which are excluded then. Hence it follows that not all sectors are involved evidently. On account of the available data sets, it is difficult to decide, whether a category is concerned. That is why no „yes/no" decisions are made for the association, but a hierarchy is established.

Category	Symbol
Category surely concerned	✓
Category eventually concerned	?
Category rather not concerned	~
Category surely not concerned	-

Table 1: Hierarchy of the Environmental Impact Assessment.

4 DESIGN CHANGES AND IMPACT RELATIONSHIPS

Generally, there are two different kinds of relationship in a decision model. On the one hand, these are the horizontal and vertical relationships between the different cost positions and environmental impact fields themselves. On the other hand, relationships between measures (modification of design parameters) and the different positions of the model exist. It can easily be imagined that the so resulting relationship net is joined with a high degree of complexity. The related expenditure of determination and presentation can be assumed as very high. Moreover, there is the problem that individual relations remain simply unconsidered with a corresponding effect on the results. The modification and enlargement expenditure of the model structure using a relationship based model can also be seen as very high and as a further restriction due to the circumstance that every existing position and all possible measures have to be checked concerning their relation to the additional position.

Following a more universal approach, the identification of such relationships is hardly or not possible. For example, the relationships among specific cost positions themselves or among specific design attributes and individual cost positions recognised in a system, can not be transferred to another system without further investigations.

4.1 Keyword Based Classification System (KBCS)

For the proposed method, the measures are the interface to the positions of the CBS as well as to the impact fields on the environment. The effect on different aspects of the positions of the decision model can be derived from the measures. As an assumption for this, the measures have to be defined clearly. In order to systematise this process, it was attempted to define general terms (Keywords) summarising similar aspects, Table 2.

Keyword	Definition	Example
Monetary	For all monetary quantities in the field of physical and immaterial goods and services.	- Material costs - Energy costs
Structure	For construction of assemblies and products and arrangement of components.	- Accessibility - Priority restrictions
Joining Techniques	Covers the joining of assemblies and parts and also includes the influence of the corresponding part geometry.	- Screw connections - Snap fits
Material	Related with the kind of materials used.	- Row material - Operating supplies

Table 2: Keyword definition

Starting from the defined keys, all available and documented measures have to be checked. The keys affected by a measure are assigned to. On the level of the CBS, the cost positions receive the counterpart, the lock. The assignment occurs with regard to the question: By which keys can the position be influenced? On what the position depends? This course of events must be carried out for every lock referring to every position of the CBS. Therefore, it has to be mentioned that identifying the effect of a design rule on the cost, it is essential to define the life-cycle fields and stage positions in an enough detailed way that impact drivers (cost drivers) can be directly assigned to.

4.2 Stage restriction and degree of similarities

In addition to the definition of keys, it is also suitable to determine the area (stages of the life-cycle) a measure applies to. This process has to be performed only once, as in the case of the determination of the keys. Such a restriction of the effect of the keys on some stages is reasonable, since, with application of a measure, no effect on each life-cycle stage is expected. With the aid of this stage restriction, the informative capability of the KWBC-system can be further increased, since an identical lock combination is assigned to many cost positions and fields of impact in the individual stages.

Basing the decision on the number of keys and of the corresponding locks, the degree of similarities on the decision model can also be considered. On the one hand, the possibility to activate positions presents itself, if a specific degree of similarities is achieved. On the other hand, the similarity degree can be displayed via a corresponding representation of the positions. Here, it has to be considered that both, the case "more keys than locks" and the case "more locks than keys", can occur.

5 REALISATION AND APPLICATION FOR THE CBS

Starting from a list of measures available as a database, the described model is ideal for a software implementation. For an identified improvement potential, suitable measures can be selected for product improvement. By means of the keys assigned to the measures, an assignment to the locks of the affected stages can be computed. As a result, the affected positions of the CBS are highlighted. The degree of similarities can preferably be displayed by means of a colour coding (e.g. red: high, orange middle, yellow: less influence).

Figure 6 shows an example of the function of the model for the introduced exemplary design guideline concerning plastic compatible labels (Figure 2). The assigned keys are: monetary, process, time and marking. The affected stages are assembly and disassembly as well as recycling and disposal.

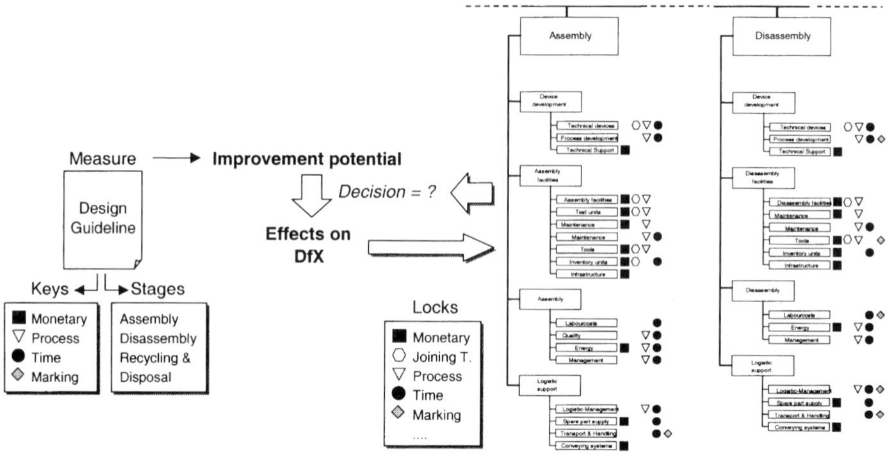

Figure 6: Example using the model

The positions affected by the measure are highlighted. For the individual positions, it can now be checked whether a modification is desired and/or whether a modification is within existing limits.

6 CONCLUSIONS AND OUTLOOK

The decision model is characterised by a relative small complexity. The interface for the relationships between costs positions and environmental impact fields is not on level of the CBS or of the impact structure but on the lesser complex level of the measures.

A inter-stage link-up between the positions does not occur. This leads to a raised clarity during realisation of the model, since only the logic correctness of the individual keys and locks has to be understood. In this way, the risk of forgetting individual relationships does not exist. However, the essential advantage of the model is the easy extendability and/or adaptability. In such a way, cost positions can be added to the CBS at any time just by assigning the related locks. The enlargement of the list of measures is also an independent process. Here, the correct keys and stages concerned with have to be assigned to the measures added. So, the system can simply be adapted so to individual needs of the user.

However, some difficulties which become clear during the realisation of the concept can also be derived from the advantages: On the one hand, a precise content(s)-related knowledge and definition of the cost positions is necessary in order to be able to assign the locks correctly.

For example, it must be clear that the maintenance costs of the product do not result from the personnel costs during the product maintenance only, but are also a function of the product quality. On the other hand, an exact definition of the keys is also presupposed in order to allow a logic and for all measures applicable assignment. Keys formulated or defined vaguely can easily result in difficult understandable outputs, that means an incorrect representation of the effects on the life-cycle. A further disadvantage represents the missing integration of a direction of the relationships. In the model, no output is given whether a corresponding position is positively or negatively influenced. This has to be decided in the respective case. Future work will focus on overcoming the mentioned disadvantages and increasing the information quality of the presented decision model.

Acknowledgements
The authors wish to thank Andreas Frey, Stefan Gelhausen und Kerstin Gottschalk for their work on this model.

REFERENCES

[1] Stevels, A., „Integration of EcoDesign into Business, a New Challenge", in Proceedings of the First International Symposium on Environmentally Conscious Design and Inverse Manufacturing (EcoDesign'99), IEEE Computer Society, Tokyo, 1999, pp. 27-32

[2] Kang, H. Y.; Lee, C.; Jung, J.W., "EcoDesign Application to Electronic Products Using the new Software Tool", in Proceddings of the International Symposium on Electronics and the Environment, IEEE Computer Society, San Francisco, 2000

[3] Süoss, B., „Wirtschaftliches Kunststoff-Recycling – Kompatibles Etiketten-Material unterstützt die Verwertung von Elektrogeräten", Umwelt, Bd. 29/Nr. 3, 1999, pp. 12-13

[4] Hesselbach, J.; Herrmann, C.; Kim, Y.-K., „Recycling Oriented Design – Weak-Point Identification and Product Improvement", in Proceedings of International Symposium on Sustainable Manufacturing, Jiaotong University, Shanghai, 1999

[5] Amezquita, T.; Hammond, R.; Salazar, M.; Bras, B., „Characterizing the Manufacturability of Engineering Systems", in ASME Design Engineering Technical Conference, Volume 1, DE-Vol.82, 1995

[6] Clark, T.; Charter, M., „Eco-Design Checklists – For Electronic Manufactures, 'Systems Integrators', and Suppliers of Components and Sub-assemblies", The Centre for Sustainable Design, 1999

[7] Dhillon, B., „Life cycle costing", Gordon and Breach Science Publishers, New York, London, Paris, 1988

[8] Wübenhorst, K., „LCC for Construction Project", in Long Range Planning, Volume 4, 1986, pp. 87-97

[9] Blanchard, B., „Design and Manage to Life Cycle Cost", M/A Press, Portland/Orego, 1978

Life-cycle Assessment for Optimized Products

Environmental consequences of tribological issues in mechanical systems using replacement refrigerants

C CIANTAR and **M HADFIELD**
Sustainable Product Engineering University Research Centre, Bournemouth University, UK

SYNOPSIS

In 1965, one of the main conclusions highlighted by Professor Jost was that designers are unaware of the consequences tribology has on product energy consumption, efficiency and life extension. This paper will highlight the importance this statement maintains and how these characteristics also influence environmental considerations. The domestic refrigerating compressor will be used as a case study to investigate how the removal of chlorine from refrigerants affects tribological issues and how these may result in adverse indirect environmental consequences. An environmental assessment using LCA as well as experimental bench tests on actual hermetic compressors were carried out to test this research hypothesis.

1 INTRODUCTION

The importance of lubrication is highlighted in what is today commonly referred to as the Jost Report [1]. Cost savings obtained from a reduction in friction of plant and machinery are identified to have been overlooked by industrial designers. The report concludes that energy consumption of products as well as efficiency and life extension all contribute towards a potential economic saving.

More than thirty years on, it is still evident how compartmentalisation in engineering, tribology being one good example, have been maintained and progressed [2]. Unfortunately, the deeper these compartments develop the more difficult the applicable domain becomes making it harder to produce universal solutions to global issues. Similar to the industrial designers addressed by the Jost Report, product developers are today still being faced with the task of having to comprehend scientific information ranging over many disciplines to be made aware of the consequences of their design decisions.

Until the mid-sixties, most industrial decisions were based on economics as the above mentioned report highlights. Environmental considerations were usually limited to compliance with local and national pollution regulation. Over the years, with an increasing demand from customers and international legislation, companies realised their commitment towards global environmental issues and how little they knew about the wider implications of the requirement that products need to satisfy environmental criteria. This paper attempts to point out how complex environmental issues in product development may become due to the different engineering fields that influence them. The identification of energy consumption, product efficiency and life extension as a measure of cost savings are also a measure of environmental savings that need to be addressed to ensure the development of sustainable products [3]. It is hoped that this research will assist designers understand the problem structure and present the various routes available to them to be able to make decisions with the least environmental side effects.

2 THE HERMETIC COMPRESSOR IN DOMESTIC REFIGERATORS

Due to international legislation, the chlorofluorocarbon (CFC-12) refrigerant has been phased out and a new hydrofluorocarbon (HFC-134a) refrigerant has been developed as a transitional, environmentally friendly alternative in domestic refrigerators. Compared to CFC-12, HFC-134a has removed the ozone depletion potential (ODP) and lowered the global warming potential (GWP) [4]. Although it is argued that better alternatives, such as hydrocarbon (HC) refrigerants already exist [5], an assurance on the availability of HFC-134a is required if the CFC-12 phase out is to be completed successfully [6]. The high initial investment in developing this transitional refrigerant [5] and due to the higher life cycle costs of HC products to overcome safety issues [7], HFC-134a production for the domestic refrigeration industry is on the increase [6].

This change in refrigerant required that the mineral oils originally used be replaced with synthetic oils to improve lubricity as well as miscibility with the new HFC-134a refrigerant. This is essential to limit any lubricant deposits on the heat exchangers as it is transported as mist with the refrigerant. This not only starves the compressor from an adequate lubricant supply but also reduces the thermal conductivity of these heat exchangers. As a result, any change in the working fluid is likely to influence the cooling capacity of the hermetic compressor altering the overall system efficiency. Furthermore, this research is based on the fact that surface films formed by gas-solid reactions can change the frictional behaviour in suitable systems. Chlorine, for example, reacts strongly with many metals and the resultant chloride films can reduce friction [8]. Research work carried out so far [9, 10] identifies that tribological characteristics on high pressure contacts within a hermetic compressor operating in an HFC-134a environment are a cause for concern. The ensuing enhanced heat of friction may also influence system efficiency as well as the overall product durability.

It is still debatable whether refrigerator lifetime should be increased or reduced but any unforeseen problems, as a result of the new lubricants, may not necessarily mean complete product failure. An implication could be that during the first few years of operation system performance deteriorates due to issues of friction and wear. The arising conflicts and trade-offs required to address the above constraints are likely to have been overlooked when seeking a CFC-substitute. With a phenomenal increase in the rate of domestic refrigerator production, the energy efficiency as well as the extension of the useful life of this product, or its components, are a prerequisite to ensure that the ensuing environmental effects do not overshadow the direct effects the change in refrigerants attained [11].

In view of the above, this study will focus on a case study (the domestic refrigerating appliance) to address the environmental implications resulting from the manufacture, operation and disposal of this product. The direct effect resulting from the inadvertent release of the refrigerant and foam blowing compounds to the atmosphere will not be considered since the CFCs would overshadow our results. One implication of this work could be that ensuring a 100% reclamation of the CFCs rather than banning its use could be a more sustainable solution from an engineering viewpoint.

3 A PRODUCT ENVIRONMENTAL EVALUATION

The aim of this study was to determine the influence of the change in working fluid on the overall environmental performance of the domestic refrigerating appliance. It would be misleading to look at the environmental impacts of the refrigerant itself if, for example, the energy consumption in the use phase of the refrigerator is altered by the new working fluid. Alternatively, it would be a misunderstanding to focus on the energy consumption only if this new change will alter the life span of the compressor. A life cycle assessment (LCA) on the components making up the domestic refrigerator would help characterise this further [12].

Although much research efforts have been made to improve the overall performance of a domestic refrigerator, the highly inefficient mechanical system has received less consideration. This LCA study identified how the mechanical system compares with the rest of the components within the overall product system from an environmental viewpoint. This was essential since this research work focuses on the hermetic compressor to improve system performance. The impact categories (global warming, acidification, human toxicity, nutrification and summer smog) focused upon for the product components extended from the production of materials, component manufacture, distribution, assembly, use and disposal. Production and disposal of the CFC-12 and the HFC-134a refrigerants were also taken into consideration to study the differences on the above mentioned impact categories resulting from this substitution. Lubricant production and disposal was ignored throughout due to a lack of information available on the public domain. Nonetheless, its possible significance on the outcome of the results needs to be addressed more fully.

3.1 The system studied

The product under investigation was a domestic refrigerator incorporating a hermetic compressor identical to the one being laboratory bench-tested. It consisted of a fridge freezer having a total gross volume of 289 litres and a total net volume of 267 litres. On a mid-range thermostat setting and at an ambient temperature of 20°C, the unit can maintain an internal refrigerator temperature of 5°C. Other secondary qualities included self-defrosting, evaporation of the water from defrosting, shelves and boxes as well as a refrigerant charge of 115 grams. The average electrical power input to the refrigerating unit, including that for the internal light, is 110 watts.

Throughout the LCA study carried out a comparison between the refrigerant and the foam blower used in the insulation of the refrigerator was also performed. CFC-12 was compared to HFC-134a by assuming either of these compounds to be the foam blower and the refrigerant. This is of importance since Papasavva et al. [4] argue that the manufacture of HFC-134a is more energy intense compared to CFC-12. Such assumptions are unrealistic since neither CFC-12 nor HFC-134a can be used as a blowing agent. Instead CFC-11 was used in the former whilst other HFC blends were developed for the latter. Nonetheless, this was assumed here to ensure simplicity of the analysis and account for the actual mass of compound used per appliance. However, it is worth mentioning that for CFC-11 the energy of production is less than that for CFC-12 [4] with a consequent rise in environmental impact in this study. Experimental observation identified approximately 20% (by weight) reduction in HFC-134a per refrigerator as compared to CFC-12 to attain the required operating pressures. This difference in weight has been accounted for throughout this study. The procedure and boundary conditions of this study are disclosed in detail elsewhere [13].

3.2 Outcome of this environmental assessment

This LCA study identified that the compressor contributed significantly towards the overall product environmental impact [13]. Figure 1 highlights this by identifying which impacts are the most significant and by how much the compressor contributes towards each of these impacts. Allocation of impacts between the compressor and the remaining items within the refrigerator was done by evaluating the compressor independently. For clarity, all the remaining items (excluding the insulation and the refrigerant) within a domestic refrigerator were grouped together as Total (Figure 1). The impacts shown in Figure 1 were obtained from the choice of materials making up the individual components using the CML Problem Oriented Valuation technique [14]. The energy required for the production of the components was ignored in this comparison as this was not accurately known and any assumptions could have influenced the results. It is right to mention though that the manufacturing process of the hermetic compressor must be very energy intense due to its robust design. It is likely that the environmental impacts will be drastically increased compared to the other components. The importance of such an observation was that tribological characteristics in the mechanical system resulting from a change in the refrigerant/lubricant combination had to be addressed more rigorously.

Another observation (Figure 2) was obtained from the energy required to manufacture and recover the CFC-12 and HFC-134a compounds. This energy, required for the synthesis and recovery of the two refrigerants, is the most detrimental since other life cycle stages are relatively identical for either of the two compounds [4]. In this study, a worse case scenario was assumed in which the electricity was generated from coal. It is clear that even though the refrigerating system requires 20% more CFC compound than HFC, the environmental impact resulting from the production of HFC-134a is greater than that of CFC-12, assuming that no leaks occur as these would influence the ODP. The global warming impact, in particular, is just below 50% higher for HFC-134a, this contributing to the direct GWP. Other impact categories, like acidification and human toxicity, are also increased by around the same percentage.

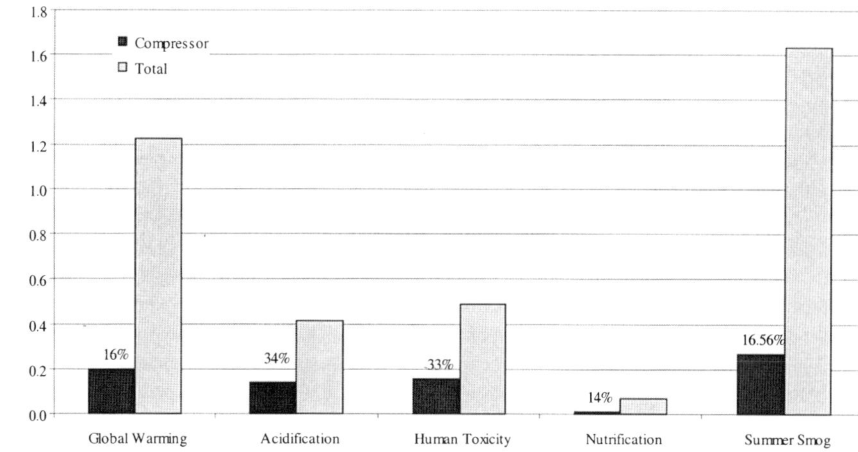

Fig. 1 A comparison of the most significant impacts between the compressor and the remaining items in a domestic refrigerator, assuming no energy of production

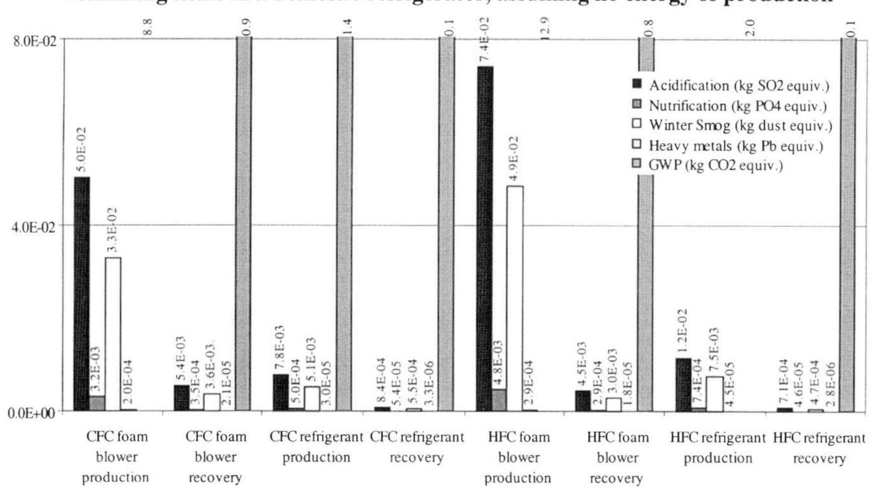

Fig. 2 The environmental contribution of the refrigerant compounds

4 AN EXPERIMENTAL VALIDATION

4.1 Experimental analyses

The environmental evaluation outlined above identified the hermetic compressor as the focal point of this research work whose technical life had to be assessed in view of the change in the working fluid. Experiments were devised to perform accelerated functional compressor testing and to collect test data to be able to relate to the physical lifespan of the product. As the wear and friction in compressor components are a function of a number of parameters, the use of actual refrigerating compressors presents a more near-to-reality scenario and a better evaluation. The use of actual compressors, as opposed to the standard benchtest equipment, addressed issues relating to contact geometries, relative motion, impurities, surface finish and treatment as well as actual material combinations. With close control of temperatures and pressures, as well as refrigerant concentrations, any side effects, such as the detrimental behaviour of additives in the lubricant, were considered. Also, since in actual compressors very little oxygen is present at the rubbing surfaces to form beneficial oxide films, wear in an HFC-134a environment is more pronounced [15].

Figure 3 shows a schematic representation of the experimental liquid/vapour refrigerating compressor test rig designed and developed by the authors. This experimental rig allows for continuous and intermittent operation tests required to monitor the effect of operating scenarios on the energy consumption in the product use phase as well as the ensuing wear and friction effects. Pressure and temperature data as well as the energy consumed are electronically inputted and stored on computer. A more detailed explanation of the experimental setup is given in [12].

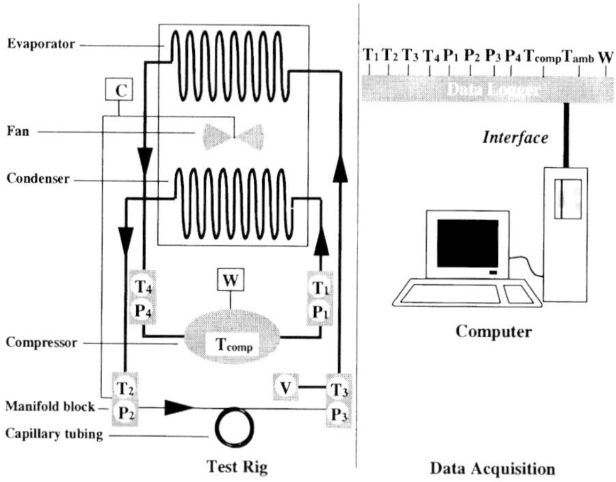

C	Fan controller
T_1 and P_1	Discharge temperature and pressure
T_2 and P_2	Discharge temperature and pressure prior to refrigerant expansion
T_3 and P_3	Suction temperature and pressure after refrigerant expansion
T_4 and P_4	Suction temperature and pressure
W	Power meter
T_{comp}	Compressor temperature
T_{amb}	Ambient temperature
V	Vacuum meter

Fig. 3 Refrigerating compressor test rig [12]

4.2 Experimental conditions

A number of experiments have been performed for two types of synthetic lubricants. One set of experiments was a 30 bar (double that experienced in a domestic refrigerating system) high pressure test in which the polyol ester (POE) synthetic lubricant, characterised by three different lubricant viscosities, was tested [9]. In another experiment, the POE of ISO viscosity grade (VG) 32 was compared to a polyvinylether (PVE) synthetic lubricant of the same viscosity. With this setup, the oils were tested under normal pressure conditions and under high pressure conditions. Table 1 gives average values for pressures and temperatures obtained over the time the experiments were running continuously for the number of hours shown. From laboratory experience it is known that the bulk oil temperature is approximately 15°C higher than the compressor sump shell temperature. Hence, the normal pressure tests were operating at a higher bulk oil temperature than the high pressure tests. Further details of the test conditions may be obtained from [12].

Table 1 Operating and environmental test conditions [10]

Test Description	Lubricant Type	Discharge Pressure (bar)	Condenser Temperature (°C)	Suction Pressure (bar)	Evaporator Temperature (°C)	Shell Temperature (°C)	Ambient Temperature (°C)	Hours Run
High Pressure Wear Test	POE of ISO VG 10	26.6	-	0.8	-	74.6	21.4	360
	POE of ISO VG 22	27	-	1.1	-	74.6	21.4	366
	POE of ISO VG 32	24.9	-	0.9	-	73.1	21.4	332
Normal Pressure Wear Test	POE of ISO VG 32	14.9	50.1	0.7	-31.3	100.95	22.1	972
	PVE of ISO VG 32	14.8	50.1	0.8	-26.5	100.7	22.2	900
High Pressure Wear Test	POE of ISO VG 32	26.9	94.8	1.2	25.9	71.5	22.2	300
	PVE of ISO VG 32	24.7	94.4	1.2	30.6	71.2	22.2	300.1

4.3 The sliding surfaces investigated

During refrigerant compression, maximum pressure is applied between the small end of the connecting rod and the gudgeon pin of the piston/connecting rod assembly. During this compression cycle, the connecting rod oscillates through an arc relative to the pin. The high pressure contact occurring at this conforming interface is of a major concern due to the difference in material hardnesses, the continuous change in sliding velocities and due to the forced-feed lubrication technique adopted. The presence of a lubricant is also critical at the interface between the large end of the connecting rod and the driveshaft. This, although assessed microscopically, was not given much attention in this study since the contact pressures are likely to be reduced due to the larger contact area the forces are acting upon.

4.4 Experimental results

4.4.1 The POE viscosity test

The objective of this accelerated test was to determine the influence of the oil viscosity on the high pressure contacts. Although low viscosity oils may be thought of to be more detrimental to components these may improve system performance due to reduced work done by the compressor and enhanced thermal conductivity at the heat exchangers. Furthermore, since domestic compressors are approximately running for 20% of the product lifetime, the operating temperature of the machine is low (less than 70°C), possibly making the lower viscosity oils more beneficial to lubricate contacting parts at start up.

During this 500 hour high pressure continuous test two significant observations were made. Operating temperatures for the lower viscosity oils were difficult to maintain so much so that the compressors had to be cooled externally. Microscope observations [9] revealed that this was due to the high heat of friction which caused a significant rise in the operating temperature due to a lack of lubricating film at the conforming contact between the gudgeon pin and the connecting rod. This resulted in severe wear as compared to the high viscosity oil.

Another observation throughout this set of experiments was the amount of material transfer, by the process known as *adhesion*, from the aluminium alloy connecting rod to the steel gudgeon pin [9]. This wear process, evident significantly on the lower viscosity oils, resulted in debris that might have circulated around the refrigerating system. Such aluminium transfer, deposited on the gudgeon pin, could also have lent itself to the increased operating temperatures mentioned above since the sliding of similar materials causes friction. With the high viscosity oils the amount of material transfer was drastically reduced but burnishing marks on the steel were observed. This implied that more lubricant was present at the conforming contact but was decomposed due to the high temperatures involved. Further details of these tests have been published elsewhere [9].

4.4.1. The POE versus the PVE tests

All lubricants, even the type studied here, include anti-wear additives that are known to generate a surface coating over the contacting surface [16] lowering the friction of aluminium alloys sliding against steel. Although such additives are generally unknown, due to a commercial interest, these set of tests compare two types of lubricant blends of an identical viscosity and hence assessing different anti-wear additives.

Results obtained from this study and accepted for publication [10] identified a series of wear mechanisms occurring at the high pressure contacts for either of the lubricants tested. In Figure 4, it is evident how the PVE has performed better than the POE lubricant. The machining marks on the gudgeon pin were still evident for the PVE type (Figure 4b) but have been worn out for the POE combination (Figure 4a).

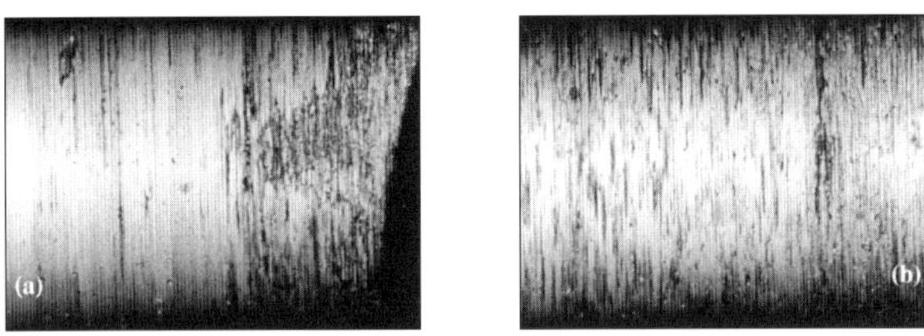

Fig. 4 Comparison of wear characteristics; (a) POE gudgeon pin at 30 bar (x52.5); (b) PVE gudgeon pin at 30 bar (x52.5) [10]

However, the PVE seemed to be more chemically active compared to its counterpart. In Figure 5 it is evident how the destruction of the aluminium connecting rod, possibly due to sub-surface reactions, resulted in the formation of debris.

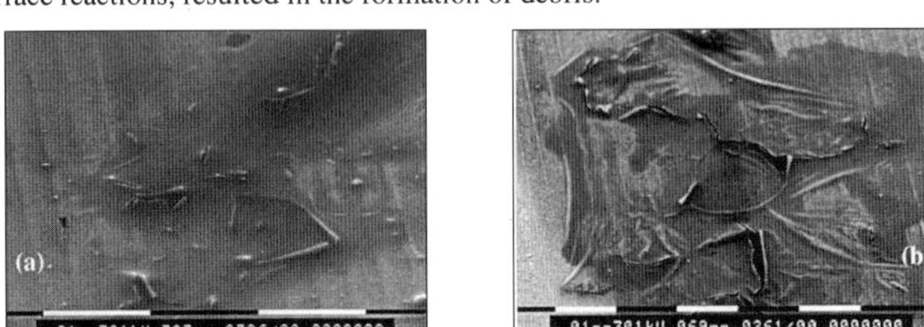

Fig. 5 Sub-surface reactions; (a) POE connecting rod at 30 bar; (b) PVE connecting rod at 30 bar (x170) [10]

Other comparisons were carried out to study the extent of wear, as well as the different wear mechanisms occurring with the new HFC-134a/synthetic lubricant combinations. Although only wear characteristics for the high pressure tests are shown it can be said that the normal pressure conditions showed similar traces of material degradation. There were instances were the wear characteristics for the normal pressure tests were more evident as compared to the higher pressure tests. It still needs to be verified whether this observation is due to the increased bulk temperature of the oil in the normal pressure tests or whether the high pressure tests have concealed vital run-in information on the high pressure contacts considered here. It is hoped that the start/stop operating setup will shed more light on this.

5 CONCLUSIONS AND OUTLOOK

This paper highlights the importance that issues of tribology have on the environmental costs throughout the life cycle of the product. It has been outlined that energy consumption, efficiency and life extension of the hermetic compressor in a domestic refrigerator need to be re-assessed in view of the efforts being carried out to identify an environmentally friendly alternative refrigerant. It was the intention in this paper that conspicuous environmental considerations be brought to light in an attempt to ensure the sustainable development of mechanical systems using replacement refrigerants.

Research work carried out so far indicates that the indirect environmental consequences pertaining to the shift in refrigerants may in fact increase. Friction and wear have been preliminarily investigated using two types of lubricant that were characterised by viscosity and blend. With either of the lubricant blends tested the wear characteristics were improved with the higher viscosity oils. However, this needs to be further addressed to study the influence viscosity has on the energy consumption. Different compressor operating scenarios will also be investigated since on start up the wear effect could be enhanced with the high viscosity oils. Such work will all be addressed in relation to experiments being carried out with the CFC/mineral oil combination.

Finally, extreme pressure tests on used (contaminated with refrigerant) and unused lubricants are also being carried out in an attempt to understand the influence of the refrigerant on lubricant performance. Although at a premature stage of the investigation, it may be said that there is considerable evidence that the presence of HFC-134a in the lubricant has a potential to reduce the performance of the lubricant. How this is likely to influence product durability is not yet understood although it is most likely that the overall influence will not be beneficial.

Further project information may be obtained by visiting our website at www.designforlifecycle.org.

REFERENCES

1. Jost, H.P., "Lubrication (Tribology)", Department of Education and Science, London, 1966.
2. Tomiyama, T., "A manufacturing paradigm toward the 21st Century", Integrated Computer Aided Engineering, 4, 1997, pp. 159-178.

3. Allenby, B.R., "Industrial Ecology; Policy Framework and Implementation", Prantice Hall, New Jersey, 1999.
4. Papasavva, S. and Moomaw, W.R., "Life-cycle global warming impact of CFCs and CFC-substitutes for refrigeration", Journal of Industrial Ecology, 1 (4), 1998, pp. 71-91.
5. Papasavva, S. and Moomaw, W. R., "Adverse implications of the Montreal Protocol Grace Period for developing countries", International Environmental Affairs, 9 (3), 1997, pp. 219-231.
6. DOE, "Report on the UK Use and Emissions of Selected Halocarbons", March Consulting Group, Department of the Environment, UK, 1996.
7. Kramer, D.E., "Why not mineral oil?" in ASHRAE Practical Guide, Vol. 41, ASHRAE, 1999, pp. 55-61.
8. Hutchings, I.M., "Tribology; Friction and Wear of Engineering Materials", Edward Arnold, London, 1995.
9. Ciantar, C., Hadfield, M., Smith, A.M. and Swallow, A., "The influence of lubricant viscosity on the wear of hermetic compressor components in HFC-134a environments", Wear, 236, 1999, pp. 1-8.
10. Ciantar, C., Hadfield, M., Swallow, A. and Smith, A., "The influence of POE and PVE lubricant blends within hermetic refrigerating compressors operating with HFC-134a refrigerant", Wear, 2000, accepted for publication.
11. Ciantar, C. and Hadfield, M., "Addressing tribology as a design opportunity for sustainable domestic refrigerators" Journal of Industrial Ecology, 1999, submitted for publication.
12. Ciantar, C. and Hadfield, M., "Investigating the sustainable development of domestic refrigerating systems working in an HFC-134a environment" in Proceedings of the Recent Developments in Refrigeration and Heat Pump Technologies, Professional Engineering Publishing Ltd., London, 1999, pp. 47-59.
13. Ciantar, C. and Hadfield, M., "An environmental evaluation of mechanical systems using environmentally acceptable refrigerants", International Journal of Life Cycle Assessment, 2000, accepted for publication.
14. Pira, "PEMS Version 4 - User Manual", Pira International, UK, 1996.
15. Yoon, H.K., Poppe, C.H. and Cusano, C., "Predicting lubricant performance in refrigerant compressors: A comparison between component testing and benchtesters", ASHRAE Transactions, 102 (1), 1996, pp. 86-95.
16. Dizdar, S., "Formation and failure of chemireacted boundary layers in lubricated steel contacts", PhD Thesis, KTH - Royal Institute of Technology, Stockholm, 1999.

ACKNOWLEDGEMENTS

The authors acknowledge the Engineering and Physical Science Research Council (EPSRC), UK for funding this three year research project as part of the Design for Whole Life Cycle Programme.

Life-cycle studies on different natural fibre reinforced components and thermoplastics for automotive parts

M FLAKE and **A HANSEN**
Institute of Geo-Ecology, Technical University of Braunschweig, Germany
T FLEIßNER
Senior Consultant, München, Germany
G FISCHHABER
Audi AG, Ingolstadt, Germany

Abstract
An ecological assessment for different natural fibre reinforced components is introduced. The journey of life regarded contains the agricultural cultivation of fibre plants, which is specific for different regions (global scope), the method of harvesting and the processing of the harvested crops. Finally, the analysis also includes the further processing of the fibre, from the manufacturing of the fibre composite matrix, on which the production of form press components for the automotive industry bases. The differences of energy demand and resulting emission amount during the use phase of a passenger car and different scenarios for the end of the life cycle (recycling possibilities, deposition or incineration) are modelled and assessed.

1 INTRODUCTION

In the past decade the potential applications of renewable resources have become a focus of public attention for several reasons. They are potentially valuable as industrial materials and regenerative fuels in place of non-renewable resources and they are generally considered to be environmentally friendly because they are biodegradable. In addition, they also contribute to reducing the effect of anthropogenic global warming since their use has largely no impact on the amount of CO_2 in the atmosphere. The potential uses are in this context as varied as the species of plants and types of animal materials (e.g. wool, animal fats, etc.) available for use. In the field of biomass, by-products (e.g. unused timber) can be used just as well as plant oils or grains as part of whole plant exploitation [1]. There are also many applications as industrial raw materials in the field of chemical engineering. Cosmetic products as well as laundry detergents and cleaning agents benefit from the use of natural contents (e.g. plant oils) in terms of additional flexibility in their performance and in product compatibility. Chemical products made of organic raw materials used, as special materials for technical applications are frequently equivalent to or even better than those substances synthesised from petrochemical ingredients in terms of their price/performance ratio [2].

Aside from economic aspects and questions concerning the technical suitability of organic raw materials, ecological considerations must be increasingly included in the decision-making processes, which are an integral part of product development. Rape oil methylester is a powerful example of how one should not automatically assume there is a comparative environmental advantage from using biomass instead of fossil fuels [3], [4]. Despite CO_2 neutrality and recycling potential, an analysis of the overall life cycle of a product should show that the expenditure of material and energy as well as the resulting emissions are in fact greater than an alternative product based on non-renewable resources. However, reliable information on the overall product life cycle is frequently not available for the purpose of considering the ecological aspects of a question. Especially in agriculture a definitive ecological evaluation is frequently very difficult since the material flows involved in farming are integrated into the natural ecosystem material cycles. Solid evidence of ecological advantages and disadvantages from the use of organic materials therefore require complex life cycle analyses and observations of various environmental aspects (raw material use, energy expenditures, surface availability, etc.).

One method, which has proven itself for the analysis and quantification of environmental impact of industrial production processes and their products, is the life cycle assessment (LCA). Besides the analysis of ecological effects measuring the consumption of materials and energy as well as the resulting emissions in the life of the product and their overall evaluation, this method also offers the possibility of identifying weak points and optimising potentials and is thus useful as an instrument to support continuous product improvement [5]. During studies of alternative or substitute products, a yardstick on which to base a comparison is always necessary, the equivalent benefit being in the foreground.

2 CASE STUDY

The use of natural fibrous materials in the automotive industry is now widespread. Environmental protection, weight reduction and improved recycling characteristics are arguments which car manufacturers frequently put forward to familiarise their customers with the ecological benefits of such materials. Whether organic materials really have so many advantages in comparison to their petrochemical opposites is the subject of a study funded by the carmaker Audi AG. Based on various trim mouldings, this study quantitatively examines the comparative material and energy assessments as well as the resulting environmental impact throughout the whole life cycle of the product, from the raw material, the cultivation of fibre plants, to component manufacture and the useful life phase of a car, all the way to the existing options of recycling and disposal.

The results of this study are meant to help decide whether ecological considerations support more intensive use of organic materials in future. The ecological advantages of cultivated raw materials ought to be quantitatively evidenced by such a study both by reference to the material and energy expenditures as well as any environmentally relevant emissions. This report presents the results of the comparative life cycle analysis of the interior side trim moulding of the Audi A3. The ecological assessment focused on comparing the interior trim supports. The manufacturing expenditures are assessed starting from the raw materials in storage to the functional equivalence of the two supports made of flax/jute epoxy resin and the injection-moulded ABS (acrylnitrile-butadiene-styrene copolymer). The trim moulding including natural fibrous content is made using flax and jute fibre and a bonding matrix made of epoxy resin and hardeners. The flax is planted and processed in Belgium, the jute is grown in Bangladesh. The manufacturing expenditures for the supports made of injection-moulded ABS as a reference component are derived from the manufacture of the side interior trim moulding of the Audi A3.

The following figures illustrate the assessment period for the manufacture of supports made of different materials up to component equivalence. The analysis of the energy and material expenditures for the ABS granulate and the epoxy resin hardener system is based on the results of various studies carried out by the APME (Association of Plastics Manufactures in Europe) [6], [7].

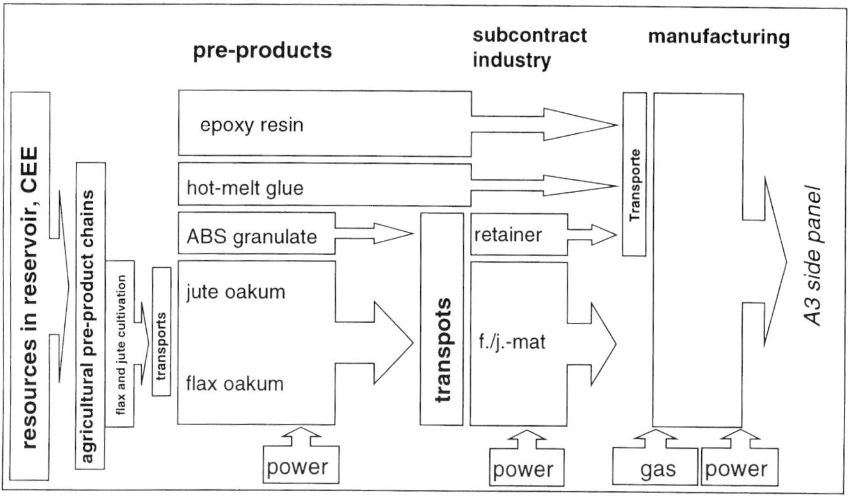

Fig. 1: System boundaries of the life cycle of the natural fibre reinforced composite.

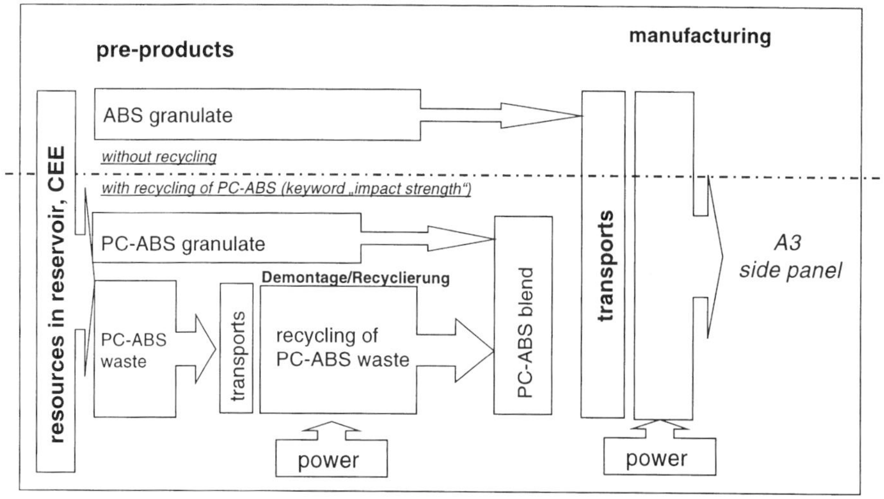

Fig. 2: System boundaries of the life cycle of injection-moulded ABS composite.

3 METHODOLOGY

The project presented here involves a comparative ecological assessment. The procedural methods are based, for one, on ISO 14041 for performing product ecology assessment [8] and secondly the VDI guidelines for the establishment of cumulated energy expenditure [9].

The underlying structure of product ecology assessments is structured as follows:

- goal and scope definition, determination of the system boundaries and depth of the assessment,
- life cycle inventory (LCI),
- impact assessment (LCIA),
- uncertainty and sensitivity analysis,
- weak-point and optimisation analysis.

The goal definition of the product ecological assessment is largely outlined by describing the subject of the investigation. The focus of the work is the quantitative recording of material and energy flows throughout the whole life cycle and their comparison in the sense of a product comparison. The method is illustrated in Fig. 3: the production chain is divided into individual steps and for each manufacturing step, taking into account the allocation rules for linked products, an input-output analysis is made of the material and energy flows. The merging of the individual processes into an overall account regarding the functional unit, in this case the functionally equivalent trim moulding, makes up the quantitative core of an ecological assessment. The life cycle inventory (LCI) provides the basis for an ecological assessment, but the results are not aggregated using a special appraisal method.

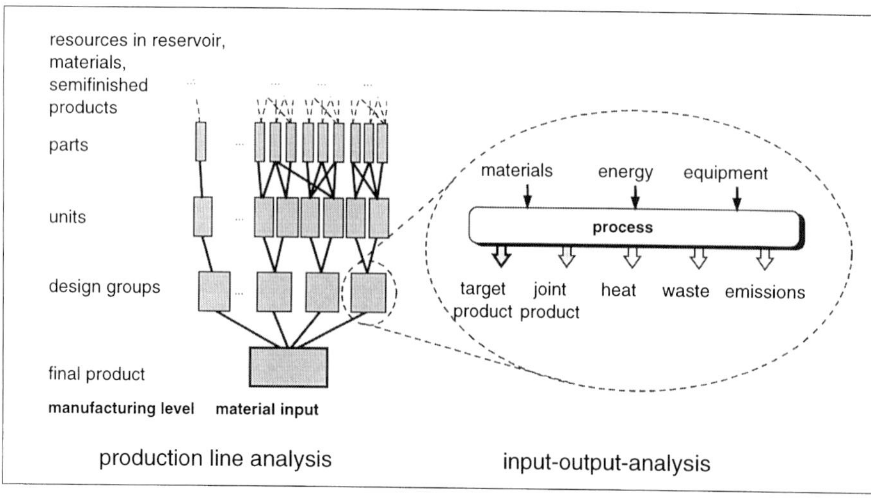

Fig. 3: Structure of the input-output-analysis.

In the concluding impact assessment the data cumulated in the LCI are aggregated into impact categories regarding their environmental impact by means of weighing factors. The following table provides an overview of the impact categories registered as part of a standard list [10]. The environmental impact statement focuses on the demonstration of the cumulated energy expenditure (CEE) and resource consumption, as well as global warming potential as the result of energy-related emissions and the potential contribution to acidulation and eutrophication, which is especially crucial in the agricultural process chain.

Table 1: Impact categories according to the Umweltbundesamt.

impact categories	**balanced materials (including weighting of their effects)**
resource consumption	Cumulative Energy Demand mineral raw materials
global warming potential	$CO_2 = 1$ / $N_2O = 320$ $CH_4 = 24,5$
ozone depletion potential	CCl_3F (FCKW 11) = 1
human toxicity	see acidification + NMHC, particles, dust, benzene, formaldehyde, benzopyrene und dioxines
ecological toxicity	see human toxicity
formation of photooxidants	NMHC, NO_x
acidification potential	$SO_2 = 1$ / $NO_x = 0,7$ $HCL = 0,88$ / $NH_3 = 1,88$
eutrophication	phosphate equivalents $PO_4^{3-} = 1$
space consumption	–
noise pollution	–

From the use of different materials, as a result of weight differences, there are also effects on the use phase of a car. The principal consumption of energy is in the use phase for fuel consumption, which represents 80 % of the total energy for manufacture, use and disposal of a car [11].

- The data regarding savings potential by reducing weight cover a wide range of values (0.15 – 1.0 l / (100 kg * 100 km)).
- The quotient for the reduction in consumption is 0.34 – 0.48 l / (100 kg * 100 km) for petrol in the New European Driving Cycle (NEDC) and at 0.29 – 0.33 l / (100 kg * 100 km) is somewhat lower for diesel-powered vehicles [12].

The reduction in fuel consumption as a result of weight differences in the trim moulding compared is taken into account in this study for two different motorization types. The corresponding fuel economy is calculated in terms of primary sources of energy.

The disposal phase as observed through the different processes and assessment methods is shown in Fig. 4. Beside the expenditures for disassembly and disposal, credit entries resulting from material recycling and incineration must be taken into account. Since the trim moulding parts consist of a mixture of materials (flax/jute supports, foils, decorative fleece, metal add-on parts, etc.) and a separation by material is not possible at present because of the unacceptably high cost of labour and energy, incineration is appraised as the disposal option in the product ecological assessment. The energy credit entry is calculated as the heat value of the materials used.

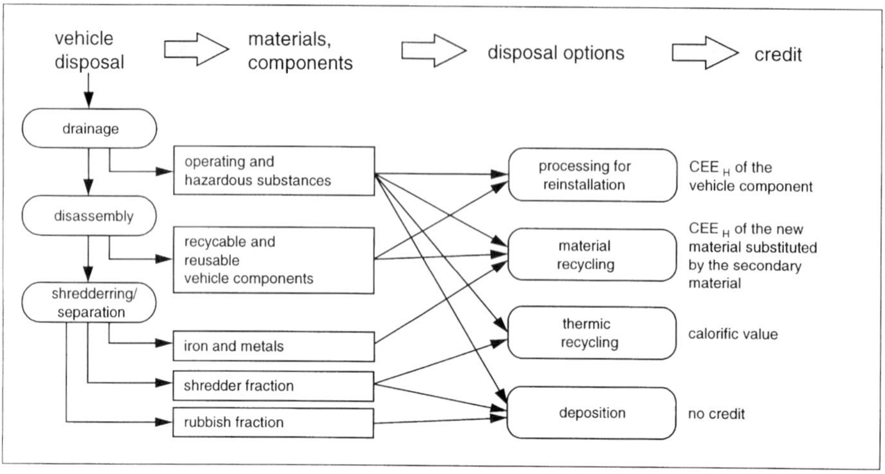

Fig. 4: Different processes and assessment methods in the disposal phase.

4 RESULTS

First, in Fig. 5, the results of the life cycle section agricultural production" are shown for the CEE (columns) and the global warming potential (line). The energy expenditures and the emissions are linked to a reference surface area necessary to produce the fibre for the manufacture of one support, in the case investigated this area amounts to approx. 11 m². The central European fibre cultivation (flax & hemp) clearly requires more energy than jute cultivation in Bangladesh, where work is manual and farming usually does not include fertilisers, and sisal cultivation on the plantations of East Africa. The comparison to central European food production shows, however, that there the intensity of machine use, and especially the quantities of fertilisers (mainly nitrogen fertiliser) and the chemical plant protection expenditures are clearly greater. The differences between the winter barley and winter wheat are affected directly by the level of nitrogen fertilising, whereas for sugar beets energy intensive harvesting must also be taken into account. The cultures fallow land, grass/clover seeding and flax (oil) stand for variations of the EU's rules regarding taking land out of production. Regarding this subject, it should be stated that flax, as a renewable raw material grown for oil, is comparable with fibrous plants, while the energy expenditure per surface area planted purely to hold the soil may be slight, no product is grown which can then be processed. The relevant emissions of the GWP are proportional to the CEE and can be largely traced back to the use of primary energy.

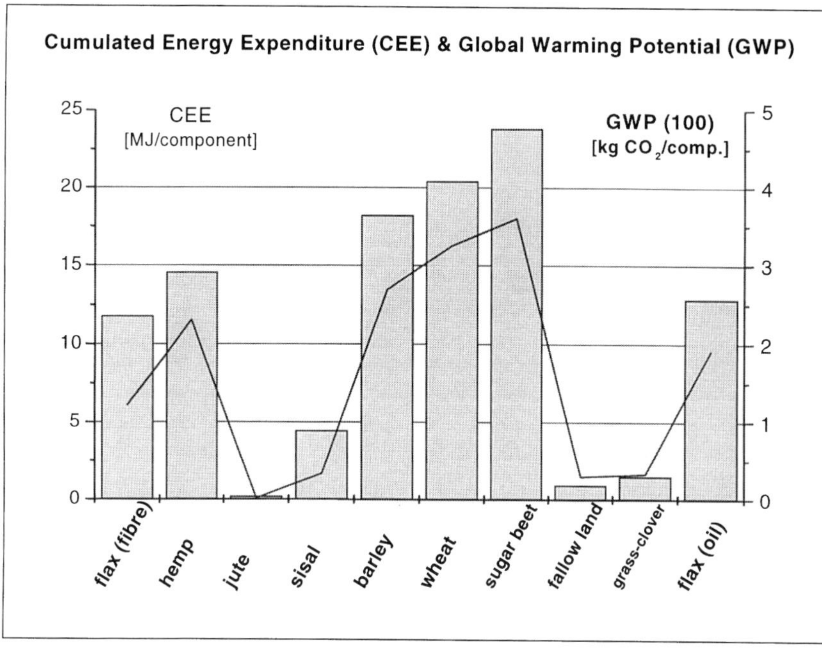

Fig. 5: Results of different agricultural production schemes.

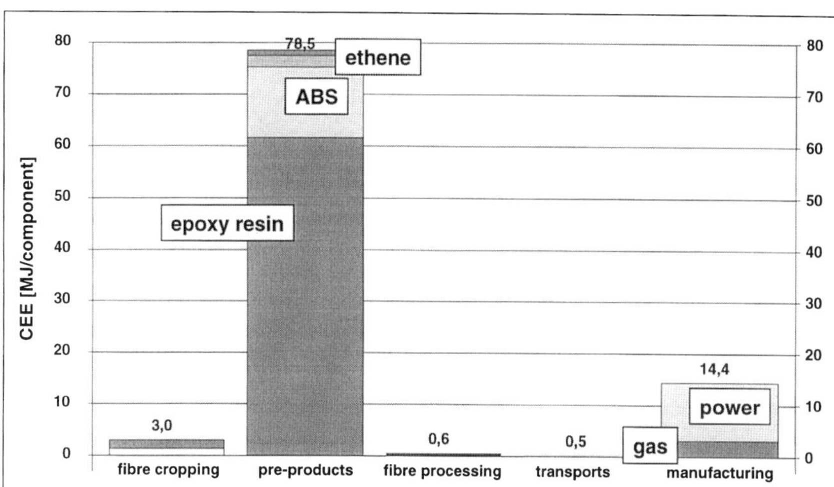

Fig. 6: Shares of the cumulated energy expenditure of the life cycle phases for the manufacture of the side panel made of natural fibre reinforced composite.

Fig. 6 shows the cumulated energy expenditure (total of roughly 98 MJ) divided into the life cycle phases for the manufacture of supports made of a natural fibre + epoxy resin composite. The expenditures for cultivating fibrous plants are still clearly reduced in comparison to Fig. 5 due to the division of the fibre gross proceeds into the different product lines and, for the whole assessment, only of minor importance. The energy assessment is clearly dominated by the epoxy-resin-hardener system, significant are also the expenditures related to hardware (ABS retainer & hot-melt-adhesive based on ethylene) and natural gas and power consumption during the actual manufacture of the supports.

Fig. 7 shows a compilation of the material expenditures and the energy consumption for manufacturing and installation of the complete interior side moulding of the Audi A3. The difference in the CEE by the time the component is installed amounts to roughly 72 MJ in favour of natural fibre components. For the component weight there is also a difference of approx. 0.3-kg in favour of the natural fibre support. This lower weight is then calculated with the consumption reduction coefficients (cf. Fig. 8) corresponding to the motorization related to the use phase of the car.

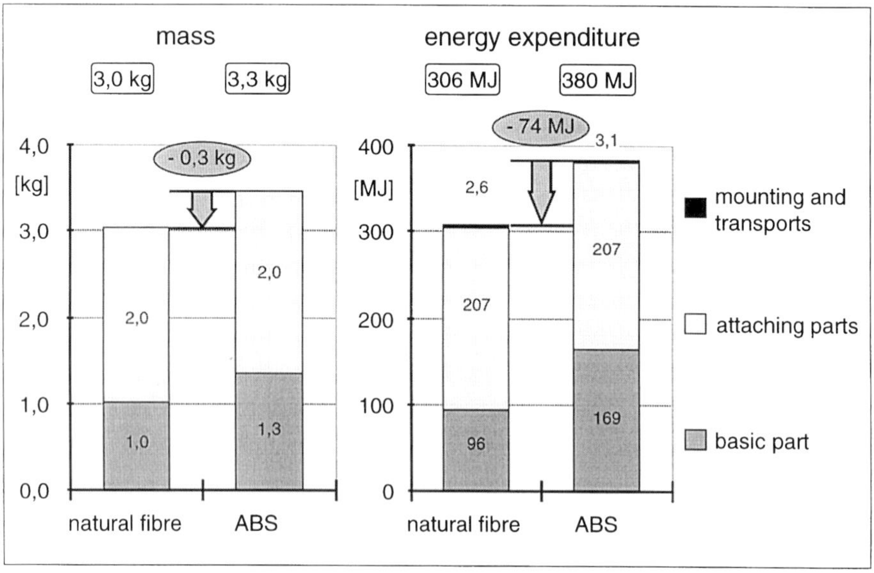

Fig. 7: Material expenditures and the energy consumption for manufacturing and installation of the complete interior side moulding of the Audi A3.

The primary energy fuel economy, depending on driving performance, can be found in Fig. 8. The energy credit for the natural fibre component is roughly in the same range of 64 – 100 MJ (for a driving distance of 200000 km) as the difference in the manufacturing expenditures.

The acumulation of the whole life cycle for the CEE can be found in Fig. 9. The disposal option suggested in this case is incineration. The energy credit for the ABS support, based on the greater heat value and the larger material expenditure, is approximately calculated 23 MJ greater than for flax/jute epoxy resin supports. Overall, based on the energy analysis the resulting ecological advantage ranges 115 – 155 MJ depending on motorization and with a driving life of 200000 km.

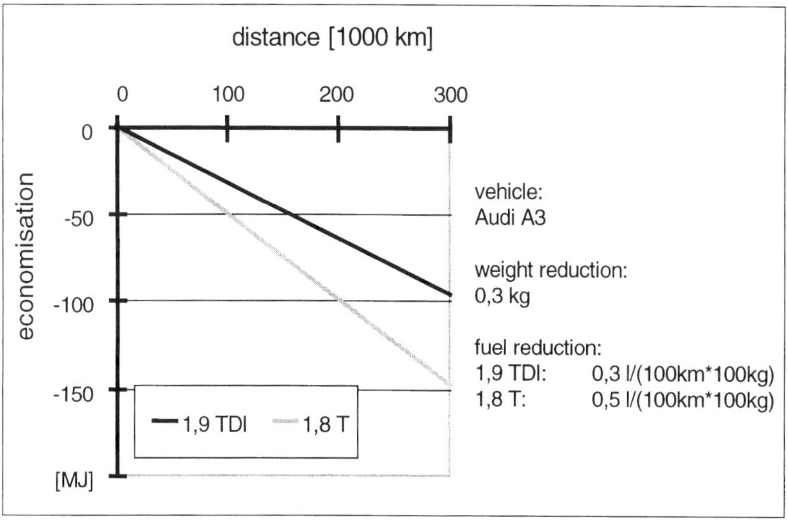

Fig. 8: Primary energy fuel economy, depending on driving performance.

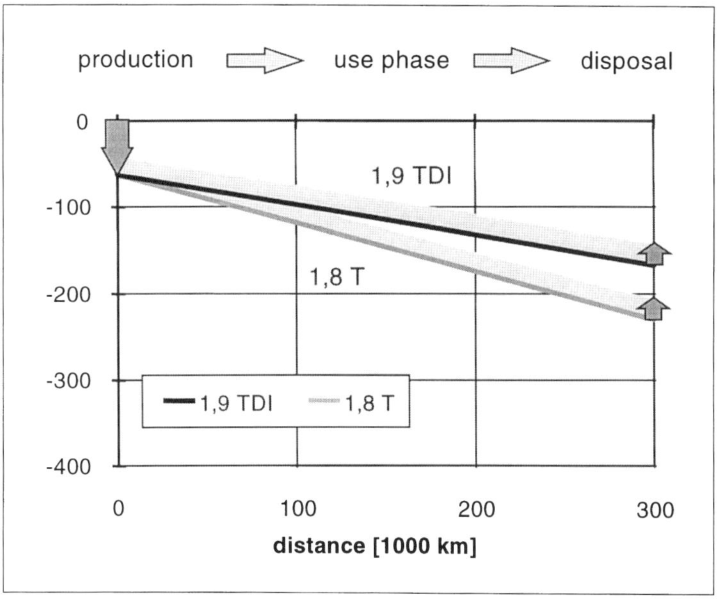

Fig. 9: CEE of the whole life cycle.

5 CONCLUSIONS & OUTLOCK

The current results of this study can be summarised as follows:

- The support made of natural fibre composite material, in comparison to a reference component made of ABS die cast, in terms of material and energy deployed, as well as environmentally relevant emissions is more advantageous ecologically speaking.
- The potential for ecological improvement lies in using the epoxy-resin-hardener system as a substitute.
- Cultivating and processing of fibre plants as understood by the assessment limits are ecologically harmless.
- Transport expenditures are insignificant for the life cycle.
- The use phase of the car reveals substantial ecological advantages for the natural fibre composite materials due to the weight difference and the resulting lower fuel consumption.
- Material recycling is not possible for natural fibre resin composite materials. The best disposal option is incineration.

For the optimisation of the life cycle of automotive parts, only sensitivity analyses are available until now. They are used to compare several scenarios, e.g. effects of the distances of transports, usability of regrind thermoplastics, extensive crop rotation systems or further innovations (matrix of short fibre and bio-polymers or resins based on renewable materials).

6 REFERENCES

[1] Kaltschmitt, M. & G. A. Reinhardt: Nachwachsende Energieträger: Grundlagen, Verfahren, ökologische Bilanzierung. Vieweg & Sohn Verlagsgesellschaft, Braunschweig Wiesbaden, 1997.

[2] Eierdanz, H. (ed.): Perspektiven nachwachsender Rohstoffe in der Chemie. VCH, Weinheim, New York, Basel, 1996.

[3] Umweltbundesamt: Ökologische Bilanz von Rapsöl bzw. Rapsölmethylester als Ersatz von Dieselkraftstoff (Ökobilanz Rapsöl). UBA-Texte 4/93. Berlin, 1993.

[4] Umweltbundesamt: Aktuelle Bewertung des Einsatzes von Rapsöl/RME im Vergleich zu Dieselkraftstoff. UBA-Texte 79/99. Berlin, 1999.

[5] Schaltegger, S. (ed.): Life cycle assessment (LCA) - quo vadis? Synthesebücher SSP Umwelt. Birkhäuser Verlag, Basel, Boston, Berlin, 1996.

[6] Boustead, I.: Co-Polymers of Acrylonitrile-Butadiene-Styrene (ABS) and Styrene-Acrylonitrile (SAN). APME - Association of plastics manufacturers in Europe, Brussels, 1997.

[7] Boustead, I.: Liquid Epoxy Resins. Brüssel, Association of Plastics Manufacturers in Europe, Brussels, 1997.

[8] DIN ISO 14041: Umweltmanagement - Ökobilanzen - Festlegung des Ziels und des Untersuchungsrahmens sowie Sachbilanz (dt. und engl.). Beuth, Berlin, 1998.

[9] VDI (ed.): Richtlinie VDI 4600 - Kumulierter Energieaufwand - Begriffe, Definitionen, Berechnungsmethoden. VDI-Verlag, Düsseldorf, 1997.

[10] Umweltbundesamt: Methodik der produktbezogenen Ökobilanzen – Wirkungsbilanz und Bewertung. UBA-Texte 23/95, Berlin, 1995.

[11] Schweimer, G. & M. Schuckert: Sachbilanz eines Golf. VDI-Berichte 1307, 1996, 235-256.

[12] Eberle, R. & H.A. Franze: Modelling the use phase of passenger cars in LCI. Proceedings, Total Life Cycle Conference, 139-148. Society of Automotive Engineers, Warrendale, 1998.

Surface Engineering

HVOF coatings to combat fatigue – the quest continues

R AHMED
Department of Mechanical and Chemical Engineering, Herriot-Watt University, Edinburgh, UK
M HADFIELD
School of Design, Engineering, and Computing, Bournemouth University, UK
S TOBE
Ashikaga Institute of Technology, Tochigiken, Japan

ABSTRACT

Thermal spray coatings exhibit excellent tribological properties in corrosive, erosive and sliding wear. Microstructural defects can however compromise the performance of these coatings in rolling contact fatigue applications. This study shows some preliminary results which indicate that by appropriate selection of the coating thickness and, the use of dense coating microstructure produced by some of the advanced HVOF systems e.g. liquid Fuel JP5000; it is possible to achieve an RCF life which is comparable to the conventional steel ball bearings. Results are discussed in term of the surface observations using the Scanning Electron Microscope (SEM), surface profilometry, microhardness tests and residual stress measurements.

1 INTRODUCTION

Innovative advancements in thermal spraying technology [1] have led to numerous applications enabling them to become an integral part of the aerospace [2-5], automotive [6-8], metal/paper rolling [9-10], oil [11], construction [12], marine [13] and biomedical [14] industry. The thrust behind the increasing number of applications is the improved microstructure offered by the advanced coating systems such as the HVOF (High Velocity Oxy-Fuel), LPPS (Low Pressure Plasma Spraying) and, emerging technologies such as the use of hybrid laser systems [15]. Excellent corrosive [4, 5], erosive [2, 3] and abrasive [16] wear resistance of these coatings e.g. WC-Co, WC-Co-Cr has resulted in environmental friendly solutions to the engineering problems. Improved efficiency, longer life and salvaging of worn/undersized/over-machined components to meet the dimensional requirement indicate only a short list of these improvements.

The improved microstructure of WC cermet coatings is a result of the better understanding of the influence of the starting powder and the process parameters on the thermodynamics/kinetics of the deposited coatings. This has helped optimise the harmony of the process and spraying conditions and resulted in a better control of the average dwell time and the process parameters. These investigations indicate that the coating microstructure can have a significant influence on the corrosive/erosive/abrasive wear resistance but scarce information is available on the fatigue behaviour of these coatings.

Recent studies by the authors on concentrated rolling/sliding contacts [17-19] have revealed that the RCF performance does not only depend upon the tribological conditions but also on the selection of the coating material, thickness, substrate material and the deposition conditions. This study shows that by the appropriate design of the coating substrate system and a suitable choice of the process and the tribological parameters, it is possible to obtain a fatigue life in excess of 70 million stress cycles under laboratory. Results presented in this paper although are qualitative nevertheless, are consistent with the previous findings under similar Hertzian contact conditions [20].

2 EXPERIMENTAL PROCEDURE

2.1 Coating Deposition

WC-12%Co coatings produced by the HVOF (JP5000) process were deposited on the surface of the 440-C steel rolling element cones. These cones had a diameter of 14.5 mm and an apex angle of 109.4 degree. The particle velocities for this thermal spray system are in the range of 1,005 to 1,118 ms^{-1}. These velocities are significantly higher than most of the other HVOF systems resulting in a better coating quality. In addition, this system also gives much more uniform distribution of particles within the spray stream. The spraying direction, which was perpendicular to the apex of the cone, was achieved by placing the rolling elements on a rotating disc. A schematic of the mechanism is shown in Figure 1. The spraying arrangement comprised of six spray guns at an approximate angle of 70 degree to the plane of the rotating plate. Although this was a complex arrangement than conventionally used techniques i.e. to spray individual cones along the direction of cone axis, this technique had the advantage to allow sufficient cooling whilst spraying almost normal to the cone apex to avoid shadowing effects. The HVOF process conditions, which are industrially optimised for this spray system to produce a dense WC-12%Co coating microstructure with minimum carbon loss, were used in this study. Prior to the spraying process, the substrate material was sand blasted and preheated to a temperature range of 100 to 150 °C. The average as-sprayed coating thickness of the rolling element cones were 300, 200 and 100 μm. The rolling elements were ground and polished to give an average coating thickness of 250±15 μm, 150±15 μm and 60±10 μm respectively. The root mean square (RMS) surface roughness (R_q) of the coated rolling elements after grinding and polishing was measured to be 0.065±0.015 μm using a stylus profilometer.

2.2 Rolling Contact Fatigue Tests

A modified four-ball machine was used to conduct the Rolling Contact Fatigue (RCF) tests under immersed lubrication conditions using B.P. Hitec-174 as the test lubricant. Hitec-174 is a high viscosity hydrocarbon lubricant having a kinematic viscosity of 200 mm^2s^{-1} at 40 °C. These tests were conducted at an ambient temperature of approximately 23 °C and at a spindle speed of 4000±5 rpm. The cup assembly was loaded at a contact load of 160 N in conventional steel ball bearing (steel planetary balls) and hybrid ceramic bearing (ceramic planetary balls) configurations. The ratio of the Elasto-Hydrodynamic Lubrication (EHL) film thickness to the average surface roughness (λ) for the test conditions were calculated to be greater then three, indicating a full film separating the surfaces.

3 RESULTS & DISCUSSION

Table 1 shows the RCF performance evaluation results of thermal spray coatings in terms of the contact stress and the number of stress cycles to failure. These results are not intended for statistical fatigue life prediction but for a comparative analysis of the tribological conditions on the fatigue failure. The first observation evident from these results was that these coatings can combat fatigue for a number of stress cycles in excess of 50~70 million without failure (test A and C). These tests were suspended to confirm via surface observations that no fatigue damage has occurred after a number of stress cycles comparable to the fatigue life of the conventional steel ball bearings. Figure 2 shows the surface observation of the rolling element of the suspended test (A). This Figure shows the overall view of the wear track and a higher magnification image. It can be appreciated from Figure 2(b) that no fatigue damage occurred during the tests. However, the tests results shown in Table 1 also indicate that the fatigue life can be seriously compromised as soon as the stress level is increased (Test B & D) or the coating thickness is considerably reduced (Test E & F). Figure 3 shows the surface observations of the failed coated rolling element D. Figure 3(a) shows the overall view of the wear track. Surface wear and micro-pitting led to the failure of this rolling element as shown in the higher magnification images in Figures 3(b) and 3(c). This is a typical failure mode with ceramic planetary balls. Previous studies by authors [17] have shown that this failure mode is triggered not only by a higher stress caused by the higher elastic modulus of the silicon nitride ceramic rolling element balls (E=310 GPa) but also, due to their higher hardness (HV_{300} = 1600).

An interesting result from these tests was that for the cases considered in this study no delamination was observed within the coating microstructure. This is indicative of the good quality of the coatings with high cohesive strength in comparison to similar coatings produced by the high velocity plasma spraying system. The difference in the cohesive strength of the coating was further investigated by the indentation fracture toughness tests [20]. Figure 4 shows the surface observations of the indentation fracture tests for the HVOF coatings considered in this study and the high velocity plasma spraying system from a previous study [17]. High velocity plasma spray coatings led to the micro-fracture at various locations under the action of the applied indentation load of 4.9 N (Figure 4(a)). These measurements were performed across the coating thickness to ensure that the results are representative of the through thickness variations. It can be appreciated that the HVOF coatings did not fracture up to an indentation load of 4.9 N (Figure 4(b)). This further confirmed the higher cohesive strength of these coatings. Although the absence of cracks meant that the fracture toughness could not be calculated at these loads, previous studies on coatings produced with similar coating powder and process parameters but with lower substrate preheat temperature have indicated a value of K_{Ic} and G_c as 1.8 $MNm^{-3/2}$

and 10.7 Jm^{-2} respectively [16]. This difference could be due to slightly higher degree of compressive residual stress expected at higher preheat temperature in the current study. However, no studies were performed to estimate the through thickness variation in the residual stress fields. Near surface residual stress observation of the pre and post RCF test conditions using the x-ray diffraction technique however, showed a high compressive stress in the coatings as shown in Figure 5. These studies were performed along and across the rolling direction. The details of the measurement procedure can be seen in Ahmed and Hadfield [21].

Delamination at the coating substrate interface was however observed in all the cases of the coating thickness less then 60 μm. This is indicative of the failure caused by the lower adhesive strength of the coating substrate system. Figure 6 shows the surface observations of the wear track of a typical failure in the Back Scattered Electron Image (BEI). The coating completely delaminated from the substrate in this case (Test F). This is typical of the exposure of the interface to the high shear stresses under the Hertzian load i.e. the coating thickness is less then or equal to the depth of the maximum shear stress. Appropriate selection of the coating thickness or other measures of mitigating the stress gradient at the interface can thus avoid such failures.

4 CONCLUSIONS

1) Tribological conditions such as contact stress, hardness and elastic modulus of the contacting surfaces can significantly influenced the performance of thermal spray coatings.

2) It is possible to achieve the RCF life in excess of 50~70 million stress cycles for HVOF coatings by appropriate selection of the coating thickness and tribological conditions.

3) Indentation fracture toughness tests can be used as a guide to access the coating resistance to the delamination within the coating microstructure.

Authors Closure

These results indicate a step forward in the use of thermal spray coatings to combat RCF. However, microcracks, poor metallurgical bonding and secondary phase particles within the microstructure can significantly influence the fatigue life, compromising the *reliability*, which is essential for stress critical components. It is therefore cardinal to find ways to produce a coating microstructure, which is uniform and free from such microstructural defects.

REFERENCES

[1] Pawlowski, L.,"The science and engineering of thermal spray coatings", ISBN 0-471-95253-2 (1995).
[2] Rhys-Jones, T. N., "The use of thermally sprayed coatings for compressor and turbine applications in aero-engines", Surface and coatings Technology, 42, 1-11 (1990)
[3] Donald, S. P., "Practical applications of HVOF thermal spray technology for navy jet engine overhaul & repair", Plating and surface finishing, 82(7), ISSN 03603164 (1995)

[4] Daniel, R. L., Sanders, H. L. Mendrek, M. J., "Replacement of environmentally hazardous corrosion protection paints on the space shuttle main engine using wire arc sprayed aluminium", Conference Proceedings, National Thermal Spray Conference, Boston, USA, ISBN 0871705095, 93-98, (1994).
[5] Sahoo, P., "High Performance Wear Coatings-the Quest Continues", Powder Metallurgy International, 25, 73-78 (1993).
[6] Holloman, L., & Levy, A. V., "Ceramic coated components for the combustion zone of natural gas engines", Journal of thermal spray technology, 1(1), 27-31 (1992)
[7] Bymes, L., Kramer, M., "Method and apparatus for the applications of thermal spray coatings onto aluminium engine cylinder bores", Conference Proceedings, National Thermal Spray Conference, Boston, USA, ISBN 0871705095, 39-42, (1994).
[8] Miyamoto, T., & Sugimoto, S., "Current conditions and future trends of thermal spray technology in the Japanese automobile industry", Conference Proceedings, International Thermal Spray Conference, Kobe, Japan, ISSN 12413074, 3-8 (1995).
[9] Mayor, V., "Applying traction coatings to steel mill rolls", Conference Proceedings, National thermal spray conference, Ohio, USA, ISBN 0871705834, 61-64, (1996)
[10] Pawlowski, L, "Technology of thermally sprayed anilox rolls: State of art, problems, and perspective", Journal of Thermal Spray Technology, 5(3), 317-334 (1996)
[11] Keshavan, M. K. and Kembaiyan. K. T. "Wear Characterization and Practical Applications of Thermal Spray Coatings in Drilling Applications", Conference Proceedings, National Thermal Spray Conference, Anaheim, CA, USA, ISBN0871704706, 635-641, (1993).
[12] Cramer, S. D., Covino, B. S., Holcomb, G. R., Bullard, S. J., Collins, W. K., Govier, R. D., Wilson, R. D., Laylor, H. M., "Thermal spray titanium anode for cathodic protection of reinforced concrete bridges", Journal of thermal spray technology, 8(1), 133-145 (1999)
[13] Rogers, F. S., "Thermal spray for commercial shipbuilding", Journal of thermal spray technology, 6(3), 291-293 (1997)
[14] Normand, B., Renaud, F., Coddet, C., & Tourenne, F., "The effect of spray conditions on the corrosion resistance of titanium coatings for surgical implants", Conference Proceedings, National Thermal Spray Conference, Ohio, USA, ISBN 0871705834, 73-78, (1996)
[15] Zhou, Z., Eguchi, N., Shirasawa, H., & Ohmori, A., "Microstructure and characterization of zirconia-yttria coatings formed in laser and hybrid spray processes", Journal of thermal spray technology, 8(3), 405-413 (1999)
[16] Slavin, T. P., & Nerz, J., "Material characteristics and performance of WC-Co wear-resistant coatings", Conference Proceedings, National Thermal Spray Conference, Long Beach, USA, ISBN 0871703920, 159-164 (1990)
[17] R. Ahmed & M. Hadfield, "Failure modes of plasma sprayed WC-Co coated rolling elements", Wear, 230, 39-55, 1999
[18] R. Ahmed & M. Hadfield, "Fatigue behaviour of HVOF coated M-50 steel rolling elements", Surface Engineering, 14(6), 1998
[19] R. Ahmed & M. Hadfield, "Rolling contact fatigue performance of detonation gun coated elements", Tribology International, 30(2), 129-137, 1997
[20] R. Ahmed & M. Hadfield, "Wear of HVOF coated cones in rolling contact", Wear, 203-204, 98-106, 1997.
[21] R. Ahmed & M. Hadfield, "Experimental measurement of the residual stress field within thermally sprayed rolling elements", Wear, 209, 84-95, 1997

[22] Beshish, G. K., Florey, C. W., Worzala, F. J., & Lenling, W. J. "Fracture toughness of thermal spray ceramic coatings determined by the indentation technique", Journal of thermal spray technology, 2(1), 35-38 (1993)

Table 1, RCF results for HVOF coated rolling element cones.

Test No	Coating thickness (μm)	Contact load (N)	Lower balls	Contact stress□ (GPa)	No: of stress cycles (10^6)
A	225	160	steel	2.7	68.58■
B	225	160	ceramic	3.1	30.78
C	150	160	steel	2.7	54.45■
D	150	160	ceramic	3.1	19.71
E	50	160	steel	2.7	1.242
F	50	160	ceramic	3.1	0.756

□ Un-coated case, ■ Suspended test

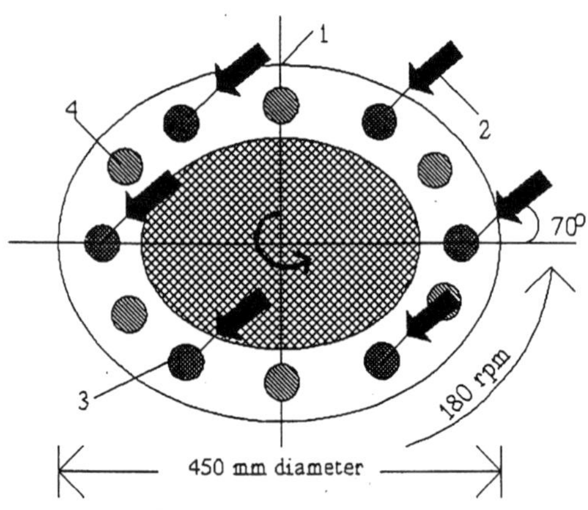

Figure 1, Schematic of the HVOF spraying arrangement.
(1; Rotating plate, 2; Spraying direction, 3; Sample, 4; Sample holder)

(a) Overall view (b) Wear track

Figure 2, Surface observations of the rolling element subject to the RCF test A.

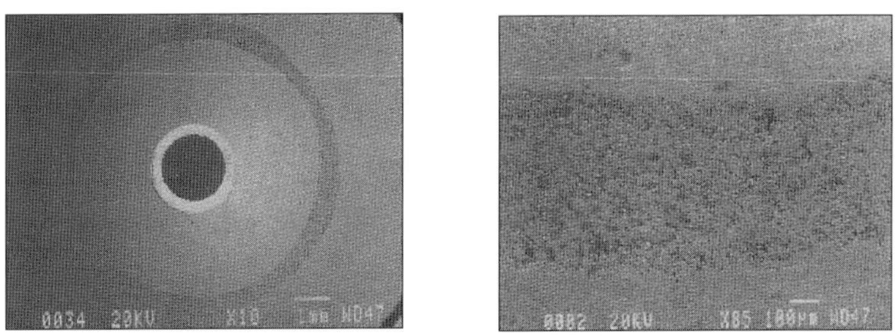

(a) Overall view (b) Wear track

(c) Higher magnification

Figure 3, Surface observations of the rolling element subject to the RCF test C.

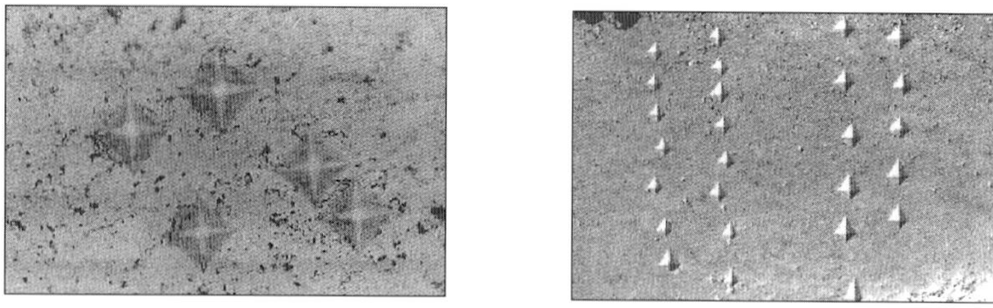

(a) High velocity plasma spray coating (4.9 N) (b) HVOF coating (2.9 and 4.9 N)
Figure 4, surface observations of the indentation fracture tests.

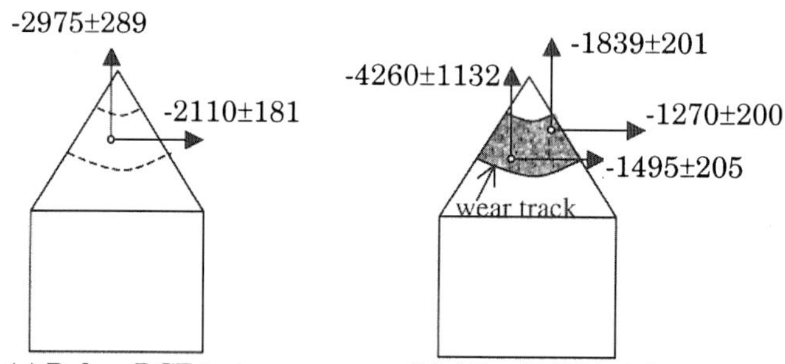

(a) Before RCF test (b) After RCF test A

Figure 5, X-ray residual test measurements of the pre and post test rolling element.

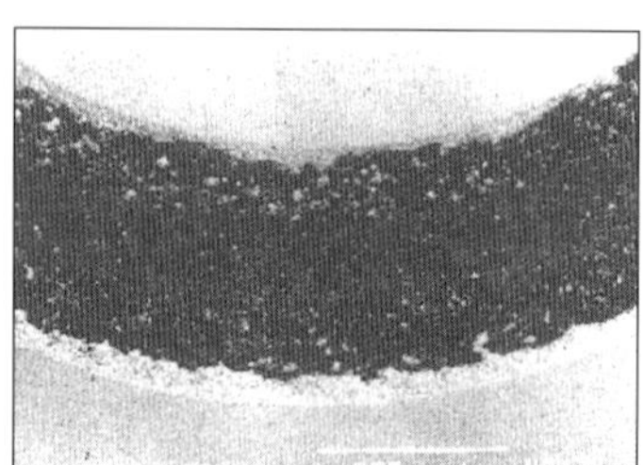

Figure 6, Wear track of the rolling element subject to the RCF test F.

Wear resistance of steels nitrided by nitriding using supersonic expanding plasma jets

Y ANDO and **S TOBE**
Department of Mechanical Science, Ashikaga Institute of Technology, Tochigiken, Japan
H TAHARA and **T YOSHIKAWA**
Department of Mechanical Science, Osaka University, Japan

ABSTRACT

Plasma jets have been successfully used as heat sources of thermal plasma spraying process. However, since the plasma jets is accelerated to supersonic under a low pressure environment, the plasma jets can be used low temperature plasmas with high chemical reactivity due to supersonic adiabatic expansion and frozen flow. From this viewpoint, nitriding of titanium plate using supersonic expanding nitrogen plasma jets at 30 Pa in chamber pressure was carried out in our previous study. As a result, it was proved that the plasma jets had enough reactivity to form a hard and thick titanium nitride layer on the surface of a titanium plate by only a few minutes nitriding. In this study in order to develop a practical low temperature and high rate nitriding process, nitriding of nitriding steel, carbon steel and stainless steel using this process were carried out and wear resistance of these nitrided samples were investigated. Consequently, though surface hardening was hardly occurred in the case of pure nitrogen plasma jets, surface hardening was obviously promoted by addition of hydrogen to nitrogen plasma jets. Especially, hard layers with over 1000 HV in hardness were formed on the surfaces of the samples by only 5 minutes operation in the cases of nitriding steel and stainless steel. Besides, according to the results of wear testing, wear mass loss of nitrided samples were much lower than those of non-nitrided samples in any cases. From these results, this process was found to have a high potential in the case of nitriding of steel.

1 INTRODUCTION

Steel have been used as engineering material for a long time owing to its mechanical strength, toughness and low price. Recently, surface modification of steel is increasingly demanded in a lot of applications, therefore, surface hardening process becomes more important. As for surface hardening techniques, nitriding has been successfully used as a high rate and low temperature surface modification process which replaces quenching and carburizing since this method was developed in the former part of the 1920's. Especially, since nitriding could be conducted at a significantly lower temperature with a much shorter operating time in comparison with the conventional gas nitriding method by development of ion nitriding in the latter part of the 1960's, application field of nitriding was widely expanded [1], [2]. Besides, quite recently, by the development of duplex surface tailoring which PVD film was deposited on the surface hardened substrate, the application field is still widely expanded even now. However, though ion nitriding is a high rate surface modification technique, it takes a few hours to accomplish the treatment. Hence, Surface coated steels by plating are widely used in comparison with nitrided steels in the industries where resistance to wear, corrosion are required.

While, though plasma jets have been used as heat sources of thermal plasma spraying process, the temperatures of the plasma jets fell down drastically due to adiabatic supersonic expansion but plasma jets can transport many radical particles on the substrate due to frozen flow under a low pressure environment. That is, this supersonic expanding plasma jets can be used as low temperature plasma with high radical particle density. In our previous study, in order to develop high rate and low temperature nitriding process, nitriding of titanium using supersonic expanding nitrogen plasma jets at 30 Pa in chamber pressure was carried out. Consequently, it was proved that the plasma jets had enough reactivity to form a hard and thick titanium nitride layer on the surface of a titanium plate by only a few minutes plasma jets irradiation.

In this study, in order to obtain useful information for the practical applications of the process using supersonic expanding plasma jets, nitriding of nitriding steel (SACM645), carbon steel (S45C), mould steel (SKD61) and austenitic stainless steel (SUS304) using supersonic expanding nitrogen plasma jets were carried out and tribological property of the nitrided samples were investigated.

2 EXPERIMENTAL PROCEDURE

Experimental apparatus for the surface nitriding consists of vacuum chamber, plasma torch, gas supply system, power supply system and vacuum pump shown as Fig. 1. Sample holder was placed in the vacuum chamber and plasma torch has the optimally designed supersonic expansion nozzle for use at 30 Pa chamber pressure. Three samples were horisontally set on the sample holder and the sample of the left side was placed as the axial center of the plasma jet irradiates this sample (This sample will be indicated as "center" in the following sentences). The discharge power was 6 kW (40 V, 150 A), and the mass flow rate of nitrogen was 9.6 SLM in the case of the nitriding using nitrogen plasma jets. In the case of the nitriding using hydrogen/ nitrogen mixture plasma jets, the discharge power was 7.5 kW (50 V, 150 A), and the mass flow ratio of hydrogen/ nitrogen was 1/3 (3.2 SLM/ 9.6 SLM). The nitriding time was 5 min. Irradiating distance, which is distance between nozzle outlet and surface of the sample, was 250 mm. Nitriding temperatures during operation were measured by CA thermocouple fixed on the other surface of the samples. The surface structures of nitrided samples are investigated by means of X-ray diffraction (CuK α, 40 kV, 30 mA). The surface hardness of the nitrided sample was measured by Vickers hardness testing. Wear resistance testing was conducted by so-called Block on Disk method using the apparatus shown schematically in Fig. 2 The disk covered with #2000 sandpaper was 200 mm in diameter, the rotational speed was 1450 rpm and the load was 500 g. The wear mass loss was measured every 3000 m wear distance by the time that total wear distance reached 15000 m.

Samples are ϕ 25 mm x 3 mm^t coin shaped SACM645, S45C, SKD61 and SUS304. The surfaces of these samples were polished by #1000 emery paper before operation. Components of these samples are shown in tables 1- 4.

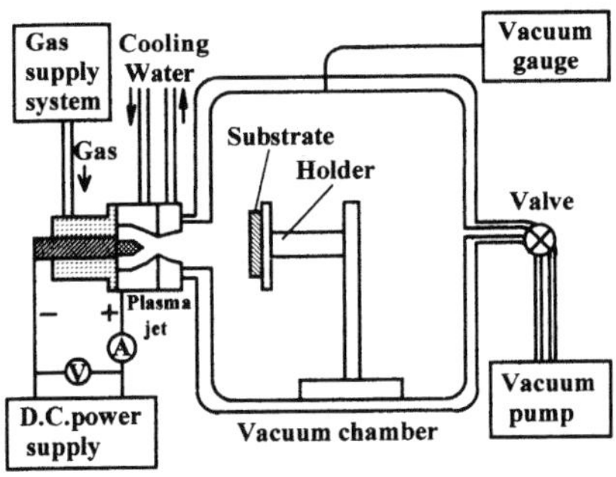

Fig.1 Schematic diagram of the apparatus for surface nitriding

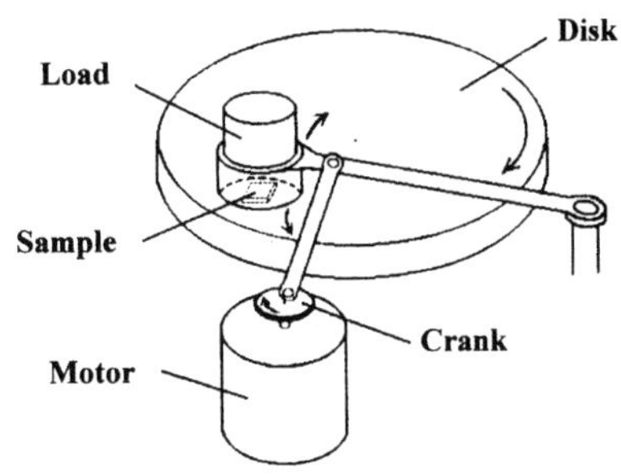

Fig.2 Schematic diagram of the wear testing apparatus.

Table 1 Component of SACM 645 steel

C	Si	Mn	Cr	Mo	Al	Fe
0.40	0.15	0.06	1.30	0.15	0.70	Bal.

Table 2 Component of S45C steel

C	Si	Mn	Cr	Ni	Cu	Fe
0.47	0.21	0.66	0.13	0.06	0.14	Bal.

Table 3 Component of SKD61 steel

C	Si	Mn	Cr	Mo	V	Fe
0.36	0.98	0.36	5.16	1.23	0.83	Bal.

Table 4 Composition of SUS304 steel

C	Si	Mn	Cr	Ni	P&S	Fe
0.08	1.0	2.0	18.0	8.0	0.07	Bal.

3 RESULTS AND DISCUSSION

3.1 Relation between irradiation distance and nitriding temperature

In our previous study[3], [4], plasma diagnostic study in a plasma torch in 30 Pa in chamber pressure was carried out by using optical emission spectroscopy. Consequently, though heavy particles temperatures at the nozzle inlet were apprximately 10000 K, the temperatures drastically fell down in the nozzle, and finally, vibration temperature and rotation temperature of the plasma jets became 4000 K and 2000 K respectively at the nozzle outlet. Therefore, this plasma is thought to be treated as low temperature plasma according to circumstances. However, since plasma jets still have enough heat energy to melt the substrate if irradiation distance is very short and axial variation of plasma jets' temperatures are drastical, it is important to know relation between irradiation distance and nitriding temperature so as to operate the low temperature nitriding process using supersonic expanding plasma jets. Then, first of all, the relation between irradiation distance and the nitriding temperature was confirmed. Concequently, nitriding temperatures at the center and the sample at the position which was 60 mm distant from the center fell down to 1173 K and 953 K respectively in the case of 250 mm in irradiation distance though these temperatures were 1273 K and 993 K respectively in the case of 180 mm. These results indicated that nitriding temperature decreased drastically with increasing irradiation distance, Especially, the nitriding temperature at the position 60 mm distant from the center in the cae of 250 mm in irradiation distance was almost the same as that of conventional nitriding process. Hence, only the data about the sample nitrided at the position 60 mm distant from the center in the case of 250 mm are explained in the following sentences as for wear resistance.

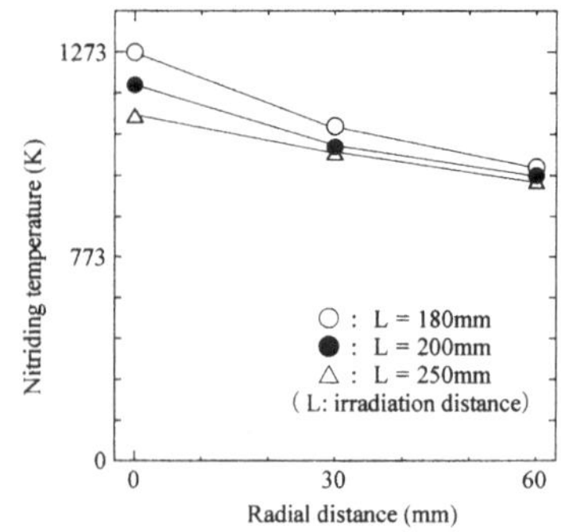

Fig. 3 Relation between radial distance and irradiation distance in the case of N_2/H_2 mixture plasma jets use

3.2 Nitriding of SACM645 and wear resistance of the nitrided sample

Supersonic expanding pure nitrogen plasma jets had enough reactivity to form thick nitride layer on the surface of the titanium plate by only a few minutes nitriding treatment [5]. Hence, in order to obtain more practical information, nitriding of SACM645 steel using pure nitrogen plasma jets were conducted. Besides, nitriding using hydrogen/ nitrogen mixture plasma jets were carried out. In conventional nitriding processes, nitriding rate was enhanced by hydrogen addition to nitrogen or nitrogen plasma. Generally, the effect of hydrogen addition on the nitriding rate was thought that chemically reactive species, i.e. molecular radicals, generated by hydrogen addition to nitrogen promote nitriding. In the previous study, one of the authors of this paper studied the N_2+$3H_2$ arc structure and the flow field in a plasma torch by means of optical emission spectroscopy [3], [4]. Consequently, since many kinds of molecular radical, such as not only N_2 but also NH, were observed near the outlet of the plasma torch, amonia and/or hydrogen/ nitrogen mixture plasma are expected to be much more reactive than pure nitrogen plasma also in nitriding using thermal plasma jets.

Fig. 4 shows radial profiles of surface hardness of the non-nitrided and nitrided samples on this condition. In the case of pure nitrogen plasma jets use, SACM645 surface was hardly hardened on the condition that nitride layer was formed on the titanium substrate. Even on the condition that sample was so close to plasma torch that the sample melted down during operation, surface hardening didn't occur. While, in the case of hydrogen/ nitrogen mixture plasma jets use, unlike the case of the nitriding using pure nitrogen plasma jets, the hard layer which was over 800 HV in Vickers hardness was formed on the surface of the sample by only a few minutes operation at each position. Fig. 5

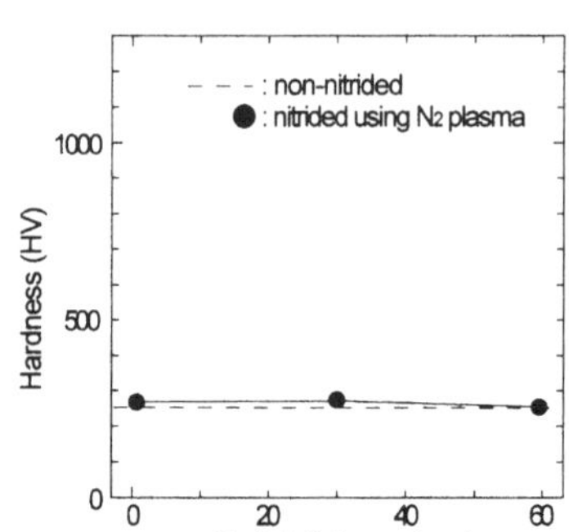

Fig. 4 Radial profiles of surface hardness of an SACM645 steel and the sample nitrided by pure nitrogen plasma jets

shows radial profiles of surface hardness of the samples nitrided by hydrogen/ nitrogen mixture plasma jets. According to optical micrographs of the surfaces of the nitrided samples, surface morphology of the sample was unchanged except the slight change of the surface color during operation though surface hardening occurred at each position. These phenomena were frequently investigated in the case that NH radicals generated in plasma jets contributed to the nitriding in the conventional nitriding processes. These results suggest that NH radicals contribute to surface hardening also in the case of nitriding using supersonic expanding plasma jets.

Fig. 6 shows wear mass loss of a non-nitrided sample and the samples nitrided at the position 60 mm distant from the center on the condition that hydrogen/ nitrogen mixture plasma jets were used. While the total wear mass loss of the non-nitrided sample and nitrided sample at 15000 m wear distamce were of mass 10.2 mg, the mass loss of the nitrided sample by hydrogen/ nitrogen plasma jets was of mass 2.3 mg. From these results, the improvement of wear resistance ability by nitriding could be confirmed. Though difference between nitride samples and non-nitrided sample couldn't be confirmed from the results of X-ray diffraction patterns even in this case, it considered that surface hardening occurred due to NH radical reaction because surface hardening didn't occur except the case that nitrogen and hydrogen were used simaltaniously.

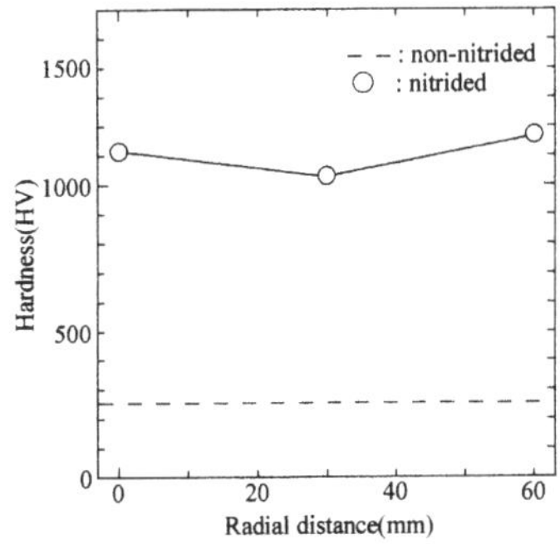

Fig. 5 Radial profiles of surface hardness of the nitrided samples in the case of H_2/N_2 mixture plasma jets (SACM645)

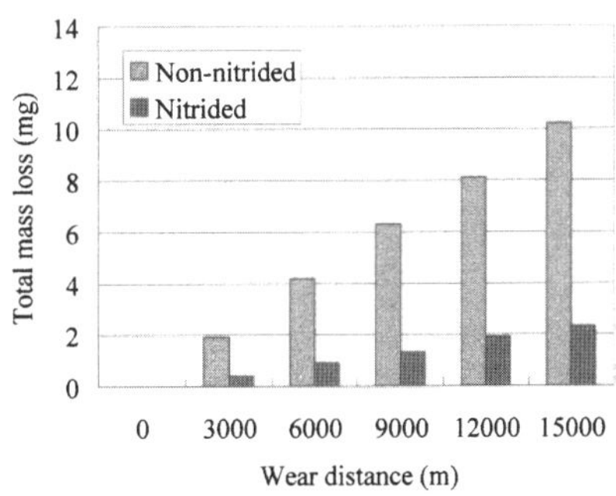

Fig.6 Wear test results of the non-nitrided sample and the sample nitrided by H_2/N_2 mixture plasma jets (SACM645)

3.3 Nitriding of S45C, SKD61 and SUS304 and wear resistance of the nitrided samples

Since SACM645 is a steel designed as surface hardening due to nitriding occur easily, SACM645 is very expensive compared to the other steels and mechanical strength of this steel isn't so high. Hence, it is very difficult to use this steel as engineering material. Then, in order to obtain more useful information for application this nitriding process to practical use, nitriding of S45C, SKD61 and SUS304, which were practical steel materials, were conducted. However, since surface hardening of the samples didn't occur and wear resistance improvement of them weren't observed in all cases of pure nitrogen plasma jets use, only the results in the cases of hydrogen/ nitrogen mixture plasma jets use are introdused in the following sentences.

Fig. 7 shows radial profiles of surface hardness of the S45C samples nitrided by pure nitrogen plasma jets and hydrogen /nitrogen plasma jets. Also in the case of S45C, though surface hardening didn't occur on the condition of pure nitrogen plasma use, the surface hardening was promoted by addition of hydrogen. However, maximum surface hardness of the nitrided S45C sample was about 500 HV and this value was very low in comparison with that of nitrided SACM645 samples. Fig. 8 shows radial profiles of surface hardness of the SKD61 samples nitrided by pure nitrogen plasma jets and hydrogen /nitrogen plasma jets. In the case of SKD 61, surface hardening was occurred only on the condition that hydrogen/ nitrogen mixture plasma jets were used and the surface was slightly harder than that of S45C. From the results of XRD, materials causes surface hardening couldn't be observed. Fig. 9 shows radial profiles of surface hardness of the SUS304 samples nitrided by pure nitrogen plasma jets and hydrogen /nitrogen plasma jets. In this case, unlike the case of S45C and SKD61, drastic surface hardening promotion occurred and surface hardness of the sample located at the position which was 60 mm distant from the center came to be over 1000 HV after nitriding. From the results of XRD, unlike the other cases, iron nitride was confirmed on the surface of each nitrided sample. By the way, the reason why surface hardening hardly occurred at the center in any cases was thought to be because the atmosphere at the axial center of the plasma jets came to be hydrogen poor as the plasma jets go downstream since hydrogen and hydride molecular had a tendency to diffuse radially from the axial center.

Figs. 10, 11, 12 show wear mass loss of a non-nitrided sample and nitrided S45C, SKD61

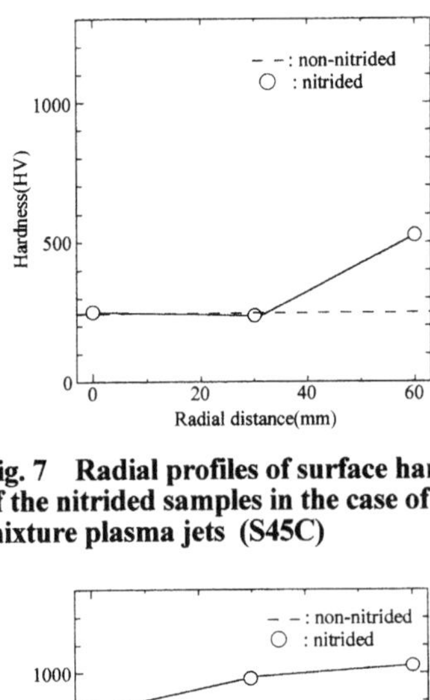

Fig. 7 Radial profiles of surface hardness of the nitrided samples in the case of H_2/N_2 mixture plasma jets (S45C)

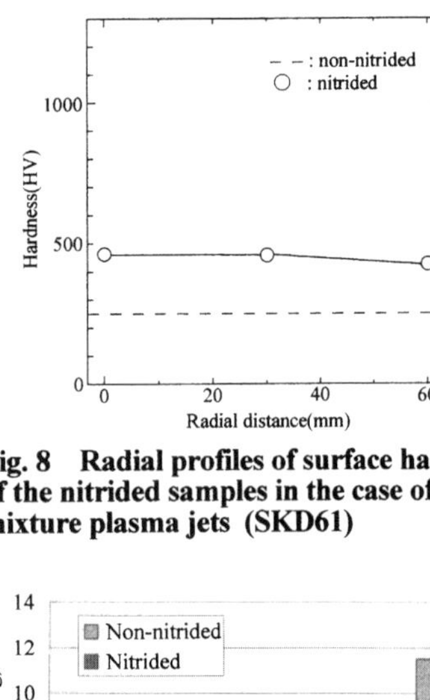

Fig. 8 Radial profiles of surface hardness of the nitrided samples in the case of H_2/N_2 mixture plasma jets (SKD61)

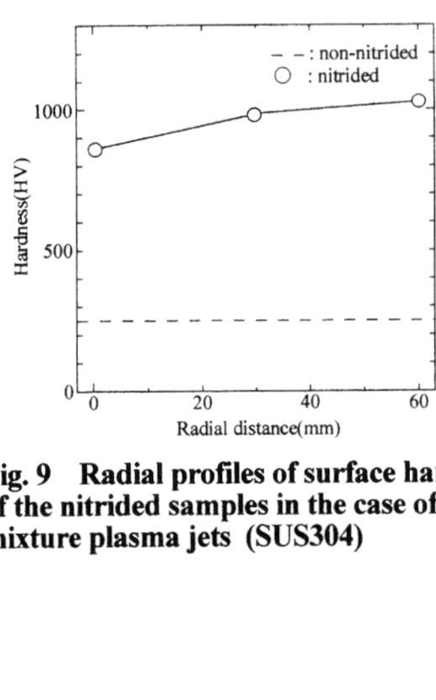

Fig. 9 Radial profiles of surface hardness of the nitrided samples in the case of H_2/N_2 mixture plasma jets (SUS304)

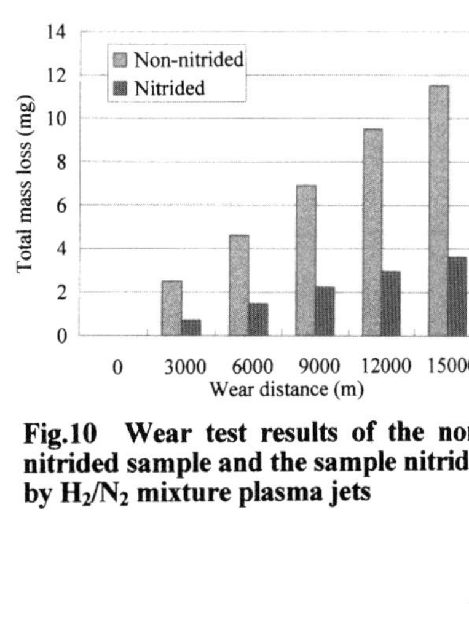

Fig.10 Wear test results of the non-nitrided sample and the sample nitrided by H_2/N_2 mixture plasma jets

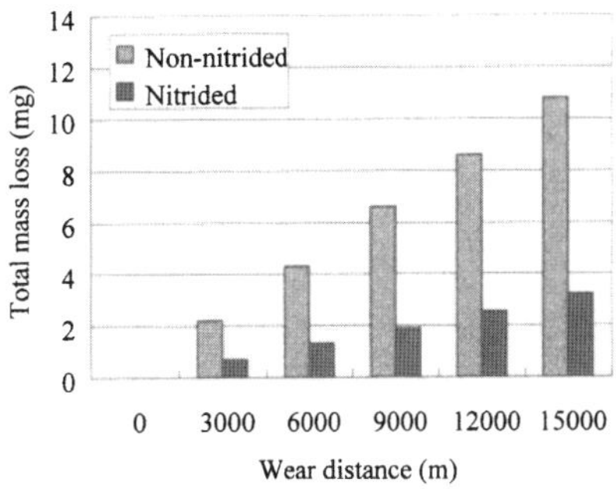

Fig.11 Wear test results of the non-nitrided sample and the sample nitrided by H_2/N_2 mixture plasma jets

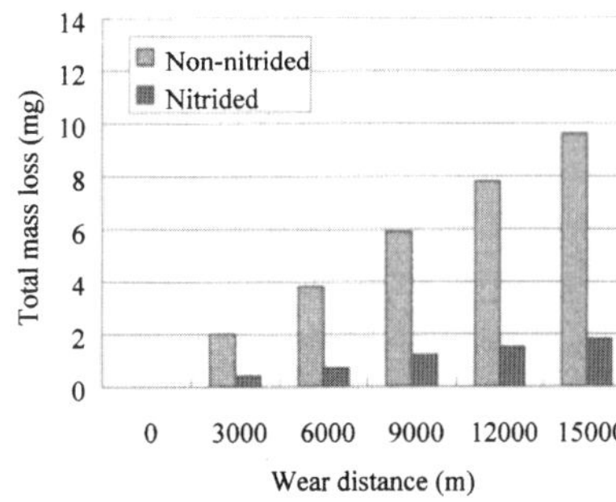

Fig.12 Wear test results of the non-nitrided sample and the sample nitrided by H_2/N_2 mixture plasma jets

and SUS304 samples, respectively. In the case of S45C and SKD61, wear resistance of the samples weren't improved so much on this condition. However, in the case of SUS 304, wear resistance of the sample was obviously improved by nitriding, total mass loss at 15000 m wear distamce was diminished from 9.6 mg to 1.8 mg. Incidentally, surface morphology were unchanged by the operations in spite of occurrence of surface hardening also in the case of S45C, SKD61 and SUS304.

From these results, this nitriding process was found to have a high potential for improvement of wear resistance of steels.

CONCLUSION

Nitriding of SACM645, S45C, SKD61 and SUS304 steel using supersonic expanding nitrogen plasma jets were carried out. Consequently, we could conclude as follows.

1) Though hard layer was not formed on the surface of the sample in the case of pure nitrogen plasma, surface hardening was promoted by hydrogen addition to pure nitrogen plasma.
2) In the case of ammonia and nitrogen-hydrogen mixture plasma jets, surface morphology of the samples retained the same condition as the non-nitrided sample had though the surface hardening occurred extremely.
   According to the experimental results by conventional nitriding processes, it was considered that the phenomena occurred because NH radicals generated in plasma jets contributed to the nitriding.
3) Nitriding using supersonic expanding plasma jets improve not only surface hardness but also wear resistance of steels.
4) From these results, this process was found to have a high potential for surface modification even in the case of not only SACM645 but also commercial S45C, SKD61 and SUS304 steels .

REFERENCES

[1] Hudis, M., "Study of ion-nitriding", J. Appl. Phys., 44, 4, 1973, pp.1489- 1496.

[2] Meletis, E. I. and Yan, S., "Low-pressure ion nitriding of AISI 304 austenitic stainless steel with an intensified glow discharge.", J. Vac. Sci. Technol. A, 11, 1, 1993, pp.25- 33.

[3] Tahara, H., Yonezawa, T., Andoh, Y. and Yoshikawa, T., "Emission spectroscopic measurement of ammonia or mixture of nitrogen and hydrogen plasma in a direct-current arc jet generator with an expansion nozzle", IEEE Trans. Plasma Sci., 26, 4, 1998, pp.1307-1313.

[4] Tahara, H., Shibata, T., Andoh, Y., Yasui, T., Onoe, K. and Yoshikawa, T., "Plasma characteristecs of supersonic ammonia and nitrogen/ hydrogen- mixture plasma jets under a low pressure environment", Proc. United Themal Spray Conf. (UTSC' 99), 1999, pp.720-725.

[5] Andoh, Y., Tobe, S., Tahara H. and Yoshikawa T., "Nitriding of titanium plate and atmospheric plasma sprayed titanium coating using nitrogen plasma jets under a low pressure environment", Proc. United Themal Spray Conf. (UTSC' 99), 1999, pp.234-239

Fretting behaviour of ion nitrided 34NiCrMo6 steel

A RAMALHO
Departamento de Engenharia Mecânica, Universidade de Coimbra, Portugal

SYNOPSIS

The main objective of this study was to investigate the influence of **ion nitriding** surface treatment on **fretting** behaviour of **34 NiCrMo6 steel**. Fretting tests were performed for this material, with treated and untreated surface, tested against AISI 52100 steel on an electromagnetic fretting machine. All experiments took place in air at controlled humidity and temperature using a plane-sphere contact geometry.

In order to evaluate the effect of ion nitriding treatment, the fretting tests were done for both materials using an amplitude displacement ranging from 20 to 80 μm and a normal load of 20 N. The ion nitrided specimens have been tested on different humidity levels in order to study the effect of humidity on fretting behaviour.

Friction force was acquired and measured during the tests. Fretting scars were observed by scanning electron microscopy and measured by profilometry for wear evaluation and to identify the mechanisms of surface failure.

1 INTRODUCTION

Fretting phenomena results from the reciprocating displacement with small amplitudes between surfaces in contact [1]. Fretting can lead to different types of damage, function of materials, environmental conditions and contact parameters. Usually three types of damage can be observed: fretting fatigue, fretting wear and fretting corrosion.

Ion nitriding is a method of surface hardening using glow discharge technology to introduce nascent nitrogen to the surface of a metal part for subsequent diffusion into the material. The typical structure of nitrided steels includes a diffusion zone with or without a compound surface layer. Ion nitriding becomes competitive relatively to other surface treatments, their main applications are mechanical components to improve mechanical properties and to achieve high wear resistance [2].

The aim of this work is the study of the fretting behaviour of ion nitrided steel against AISI 52100 steel. The effect of humidity on fretting has been also studied.

2 EXPERIMENTAL DETAILS

2.1 Specimens

The lower specimens were made from DIN 34CrNiMo6 steel quenched and tempered, with a disc form and dimensions of $\phi 30 \times 10\ mm^2$. Some specimens were surface treated by ion nitriding. The surface treatment increases the surface hardness on a depth of 350 μm. Untreated steel have an hardness of 360 HV30 and the steel with surface treatment reach a value of 750 HV0.2.

Prior the surface treatment discs were polished. The roughness measurements show that the nitriding treatment leads to a light increase on the surface roughness, as shown on the table 1. Commercial ball bearings, of 100Cr6 steel, with diameter of 10 mm were used as counterbodies.

Table 1 Roughness parameters of treated and untreated surfaces

	Untreated	Ion Nitriding
R_a (μm)	0.328	0.527
R_z (μm)	3.297	5.238
R_{max} (μm)	4.526	6.236
R_k (μm)	0.87	1.55
R_{pk} (μm)	0.27	0.47
R_{vk} (μm)	0.76	1.02

2.2 Fretting rig

Figure 1 shows a schematic outline of the computer controlled fretting wear rig used on this study. An electro-dynamic actuator was used to drive the upper specimen in horizontal harmonic movement. The normal load was applied using a spring. A piezoelectric force transducer attached to the lower stationary specimen allows to the measurement of the tangential force developed in the fretting contact due to friction. A displacement transducer branched to the mobile specimen, lead to the measurement of the fretting stroke and also drive the closed loop control system in order to have a real time displacement PID control.

An acrylic climate small chamber covers the specimens providing a selectable constant relative humidity. The automatic humidity system controlled by a PLC includes a circulating pump and desiccator or humidifier unites. A humidity sensor was used to measure the RH in the testing chamber and to drive the PLC.

2.3 Fretting tests

Fretting tests were carried out in the ball-plane configuration. The steel counterbody ball is forced onto the flat specimen at a normal force of 20 N. Prior to each test the specimens were ultrasonically cleaned in ethyl alcohol.

In order to evaluate the effect of displacement three stroke values of 20, 50 and 80 μm were selected.

On each test a harmonic oscillation with a frequency of 170 Hz was applied during 1.22×10^6 cycles.

The influence of relative humidity on the fretting behaviour was investigated for the treated material. Four different relative humidity levels were selected, namely 7%, 20%, 50% and 90%.

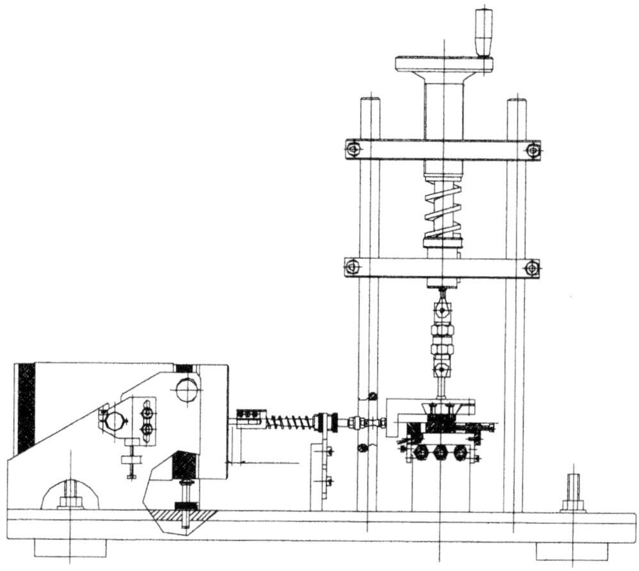

Fig. 1 Schematic outliner of the fretting rig

The extension of the wear damage was quantified using a laser profilometer, Rodenstock RM600-3D. The area of the transversal central profile was used as a measure of the wear of the flat specimen, whereas for the ball the transversal area was calculated by the expression:

$$A = \left[\frac{h}{a} \left(\frac{h^2}{4} + \frac{4a^2}{3} \right) \right]$$

were $h = r - \sqrt{r^2 - a^2}$

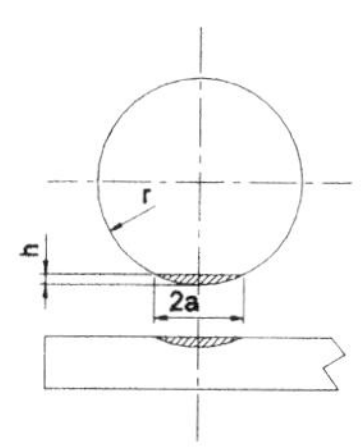

A Philips XL30 TMP Scanning electron microscopy (SEM) was used to analyse the fretting scars.

3 RESULTS AND DISCUSSION

The evolution of tangential force during the tests has been analysed. In spite of sliding is not always been conveniently established, the ratio between the amplitude of the tangential force and the applied normal load, named friction coefficient, is used to compare the results of the different tests. The amplitude of the friction force was calculated after applying a digital second order filter with zero-phase distortion to reject high frequency noise. Figure 2 shows the obtained results for ion nitriding discs tested under 7% of relative humidity.

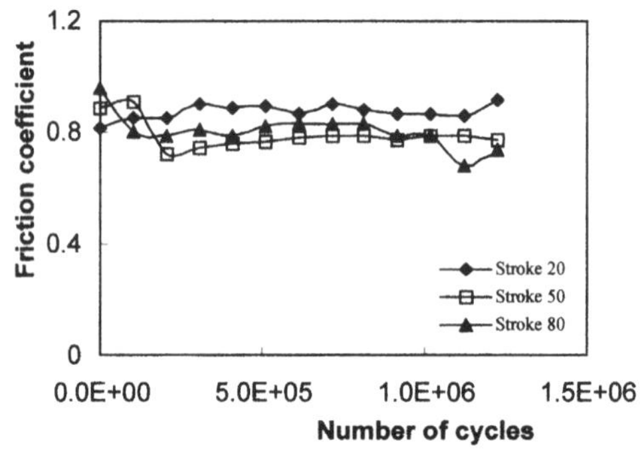

Fig. 2 Evolution of the tangential force during the tests (treated disc RH 7%)

Figure 3 shows the mean value of the friction coefficient calculated for the stabilised regime for all the tested pairs. The tangential force at the contacts of untreated material reveals a quite low value for all the studied stroke values, when compared with the treated pairs. Also, the evolution of the friction coefficient with the fretting stroke is different for treated and untreated pairs. For the untreated steel, tested under the small stroke of 20 μm, a quite lower tangential force have been obtained when compared to the higher stroke values, while the treated steel do not reveal a significant change of friction with the stroke.

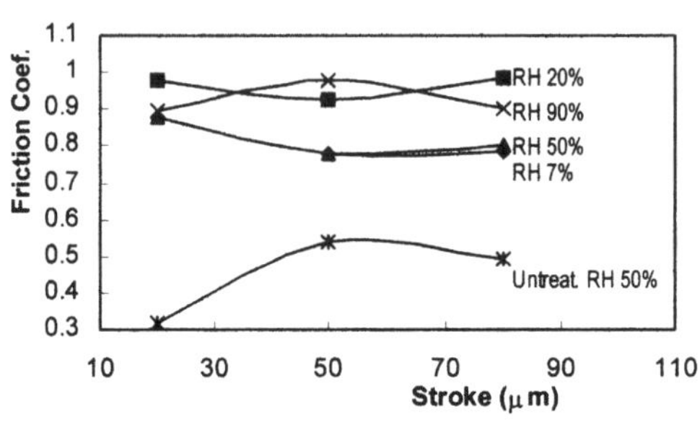

Fig. 3 Evolution of the tangential force during the tests.

At the end of each test the wear was measured on both specimens of the sliding pair, ball and disc. For the plane specimens the wear was evaluated by the middle profile. Figure 4 compares the measured profiles for nitrided specimens as a function of the humidity and fretting stroke.

The effect of the ion nitriding treatment on the wear decreasing can be observed in figure 5, where are compared the wear profiles of steel discs with and without surface hardening by ion nitriding tested under 50% RH.

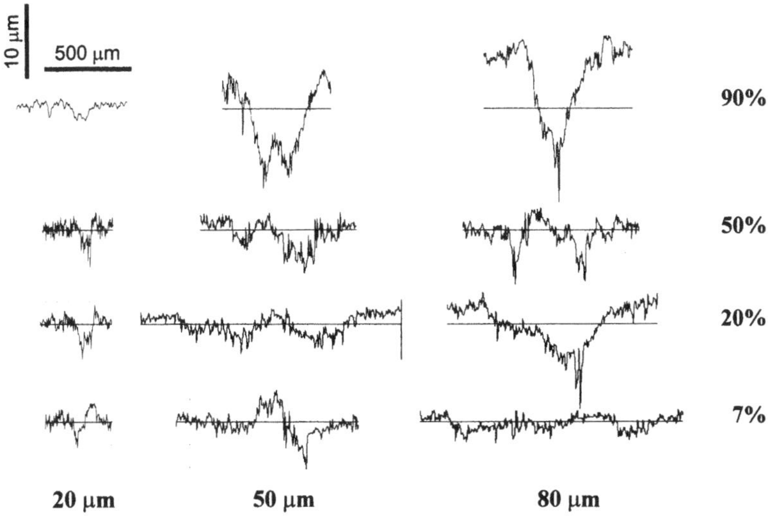

Fig. 4 Effect of humidity on the wear profiles of the nitrided discs.

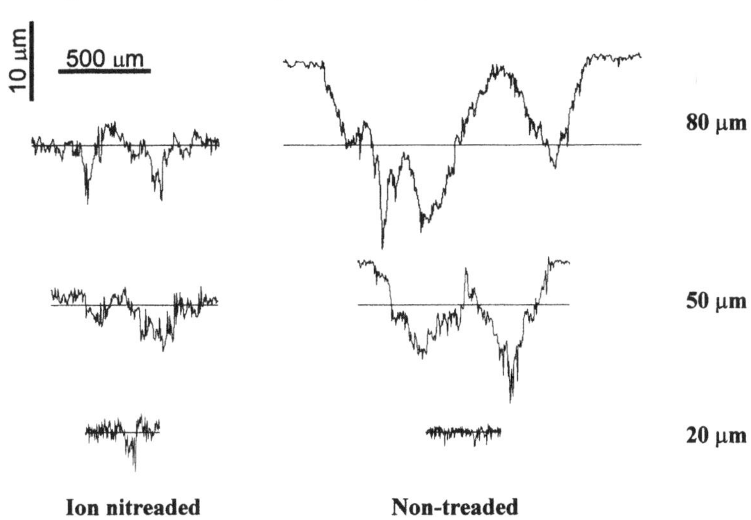

Fig. 5 Comparison of wear profiles of nitrided and non-treated discs.

The measured values allow to the determination of the transversal area of the scars, so a measure of the wear. Figures 6 and 7 show the obtained results for balls and discs respectively.

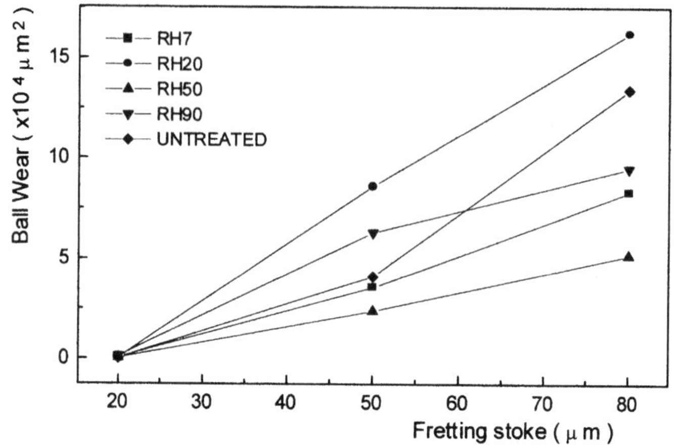

Fig. 6 Ball planimetric wear as a function of fretting stroke

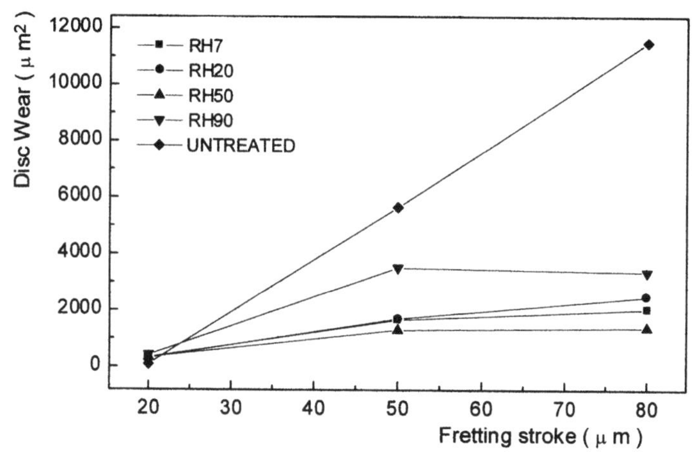

Fig. 7 Disc planimetric wear as a function of fretting stroke.

According to figure 4 to figure 7, for both ball and disc, for the smallest stroke value the wear is almost the same for all the tested conditions. Nevertheless, for displacement amplitudes of 50 and 80 μm the humidity has a strong effect on the amount of removed material.

Surface hardening resulting from ion nitriding has a strong effect on the decreasing of disc wear, figure 7. In spite of the increasing effect of surface hardness produced by ion nitriding, the ball tested against non-treated disc show also a higher wear than those tested against surface hardened disc at the same humidity value, RH 50%, figure 6.

It is not easy to conclude about the effect of humidity on the wear amount of ion nitrided steel, but, within the treated discs, the RH value of 50% seems to point out the best interaction conditions with the small wear and also the lowest friction values.

For the lowest tested value of stroke non-significantly difference was observed on wear comparing all of the tests with treated and untreated materials.

Using the acquired values of the friction force and of the relative displacement, a graphic of the friction force against displacement can be plotted. The figure 8 is a representation of those plots for the tests with the substrate and with the treated steel, tested under 50% of relative humidity. The area of those cycles corresponds to the realised work of the friction force, thus is a measure of the input energy in the system and can be correlated with the quantity of wearied material [3-5].

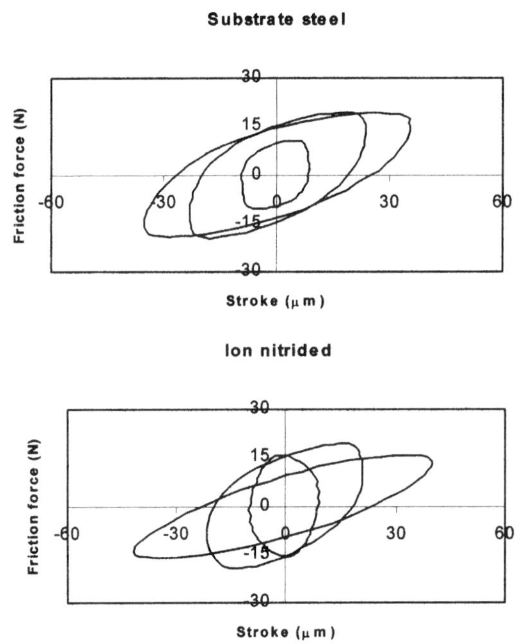

Fig. 8 Friction force vs. displacement for both tested materials.

For the same amount of dissipated energy the wear resulting on the substrate steel is greater than the wear of the surface treated specimens, as can be seen in figure 9.

At the end of each test the scars produced by fretting have been observed by SEM in both balls and discs. Figures 10 and 11 summarise the SEM observations.

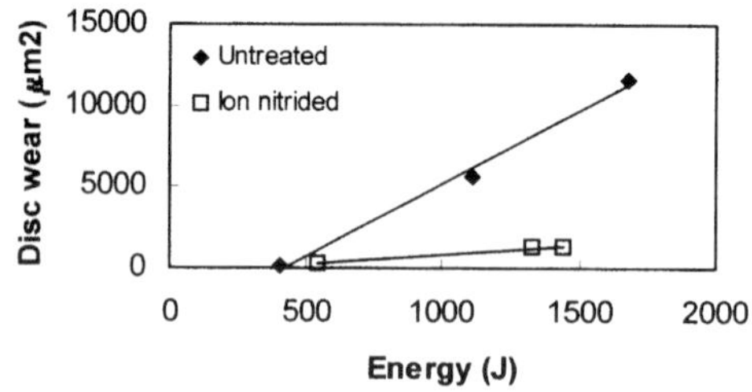

Fig. 9 Correlation between the disc wear and the cumulated dissipated energy.

Figure 10 shows the morphology of the wear scars for the tests with non-nitrided steel. For the

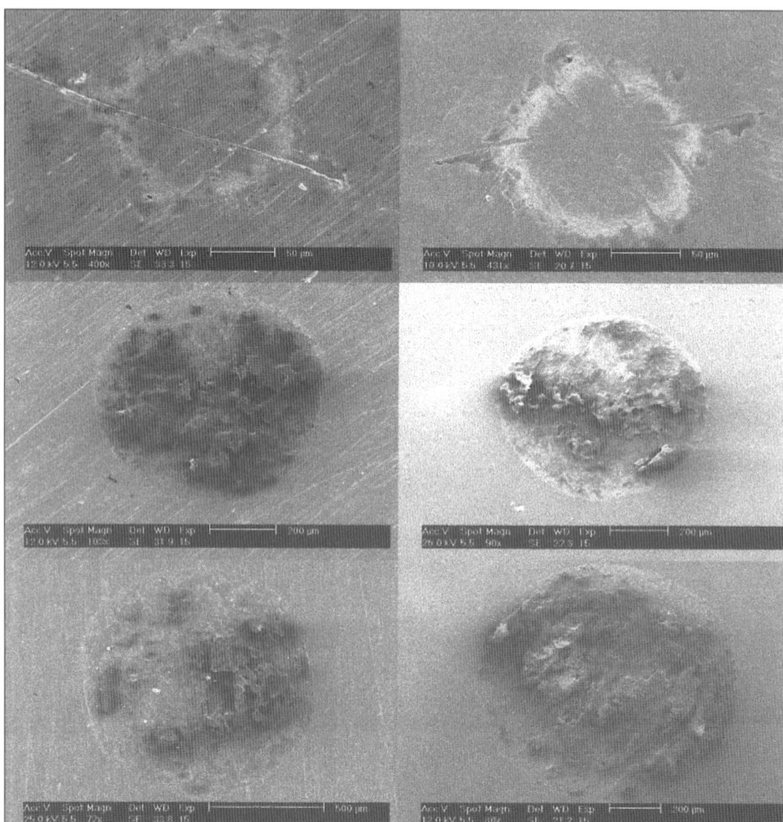

disc ball

Fig. 10 SEM observations of worn surfaces for non-nitrided discs tested under 50% of relative humidity. (Top from bottom stroke of 20 µm, 50 µm and 80 µm).

small tested stroke, 20 μm, the wear occurs only on an annular domain surrounding a central area without any damage features. This type of morphology characterises a partial slip regime well described by Mindlin [6] and verified in many studies [7]. The evidence of this interacting regime justifies the low value of the friction force obtained for this condition, figure 8. For the higher strokes the damage spread out to the entire contact region, and a large amount of oxide platelets can be observed adherent to the surfaces.
Figure 11 shows the morphology of wear scar of tests with nitrided discs. Great amounts of adherent platelets resulting from oxidation can be observed on the wear surfaces of ion nitrided discs. For these tests, even for the lowest stroke, the scar shows wear features on whole the contact area. So, gross slip took place at the contact area for all the tested values of the fretting stroke. This result can be expected considering that the maximum values of friction are almost the same for all the tested stroke values, figure 8.

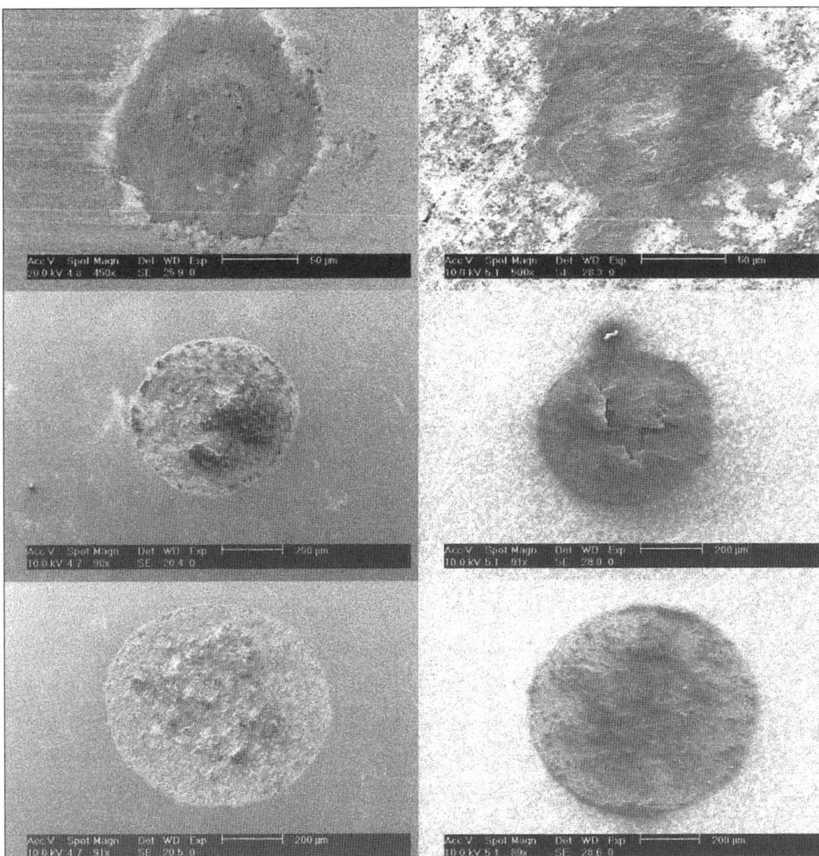

ball disc

Fig. 11 SEM observations of worn surfaces. Nitrided discs tested under 50% of relative humidity. (Top from bottom stroke of 20 μm, 50 μm and 80 μm).

4 CONCLUSION

(1) The ion nitriding can efficiently improve the fretting wear-resistance of DIN 34CrNiMo6 steel.

(2) Ion nitrided DIN 34CrNiMo6 steel fretting wear resistance depends on the relative humidity. The role of the relative humidity seems to be their influence on the surface oxidation and adherent layers.

(3) While the DIN 34CrNiMo6 steel without surface treatment shows a transition from mixed to gross slip fretting regime, when the stroke increase from 20 to 80 μm, the ion nitrided steel shows always a gross slip fretting regime.

ACKNOWLEDGEMENTS

The author would like to thank the Portuguese Foundation for Science and Technology for their financial support to this research (Project Praxis XXI 2/2.1/ TPAR 2041/95).

REFERENCES

[1] R.B. Waterhouse, Fretting Wear, ASM Handbook, Vol. 18, ASM International, (1982) 242 – 256.
[2] James M. O'Brien, Plasma (Ion) Nitriding, ASM Handbook, Vol. 4, ASM International, (1982) 420 - 424.
[3] S. Fouvry, Philipe Kapsa and Leo Vincent, Quantification of Fretting Damage, Wear 200(1996) 186 – 205.
[4] M.Z. Huq and J.P. Celis, Fretting Wear of Multilayer (Ti, Al)N / TiN Coatings in Air of Different relative Humidity, Wear 2225 – 229 (1999) 53 – 64.
[5] P.-Q. Wu, H. Mohrbacher and J.P. Celis, The Fretting Behaviour of PVD TiN Coatings in aqueous Solutions, Wear 201(1996) 171 – 177.
[6] K.L. Johnson, Contact Mechanics, Cambridge University Press, (1985).
[7] Olof Vingsbo and Staffan Soderberg, On Fretting Maps, Wear 126(1988) 131 – 147.

Assessing the ability of metal-on-coating pairs to modify frictional responses with application to the cutting process

W GRZESIK, Z ZALISZ, and **P NIESLONY**
Department of Manufacturing Engineering and Production Automation, Technical University of Opole, Poland

NOTATION

d	sliding distance
k	specific wear rate = $V/(d \times F_n)$
v_s	sliding speed
F_f	friction force
F_n	controlled normal load
V	volume of removed material
μ	coefficient of sliding friction = F_f/F_n

ABSTRACT

The principle aim of the paper is to investigate some cutting tool coatings that are frequently used in the machine building industry and to assess their frictional behaviour under the test conditions equivalent to those for the cutting of medium carbon steel when using sintered carbide and multilayered coatings. A modified pin-on-disc tester was used to record the friction coefficient curves, and the volumetric wear rate was proposed as a parameter for comparing the frictional behaviour of the tribo-pairs tested. The current research was mainly focused on the quantification of friction and wear as a result of variations of the sliding distance and the sliding speed. In particular, three characteristic stages in friction curves recorded during 1200s and the dominant wear mechanisms that developed at different sliding speeds were distinguished. It is concluded that the modification of friction occurring for metal-on-coating pairs can be successfully utilised for the optimum selection of coated cutting tools for the dry cutting of steels.

1 INTRODUCTION

The determination of friction is one of the most essential problems emerging in the design of cutting operations [1]. From a number of investigations it is clear that metal machining is characterised by extreme contact conditions at the tool-workpiece and the tool-chip interfaces including high contact pressure and temperature, and especially highly active and freshly generated surfaces. Usually it is not possible to reproduce in full-scale the contact conditions in machining using conventional pin-on-disc testing because the wear mechanisms involved are not relevant to that observed in machining. Modified pin-on-disc [2,3], pin-on ring [4] and ball-on-disc [5] test devices have been developed in order to perform sliding wear tests that simulate dry machining when using coated cutting tool materials. Hedenqvist and Olsson [4] have shown that the coating tested and the substrate should be considered as one element termed as *a coating/substrate system*. Recently, Lebrun et al. [3] have carried out extensive friction experiments using a special test device including the thermal and mechanical outputs of the plain-plain contact tribo-system. In both this and in [6,7] the vital role of the frictional heat flux (friction mechanical dissipation) in the frictional behaviour of the work (chip)-coating /substrate system was confirmed.

A new approach to the use of friction laboratory testing, which employs a special pin-on-disc arrangement, to compare the frictional behaviour of the selected metal-on-coating and metal-on-WC/Co carbide pairs that reproduce orthogonal cutting conditions at low and moderate sliding speeds is proposed.

2 EXPERIMENTATION

2.1 Test equipment

As mentioned above, uncoated and coated carbides sliding against the carbon steel disc were tested at ambient under dry friction conditions using a modified tribo-tester equipped with control and computer DAQ systems. The functional structure of the experimental set-up and instrumentation for process measurement and control are presented in Fig.1.

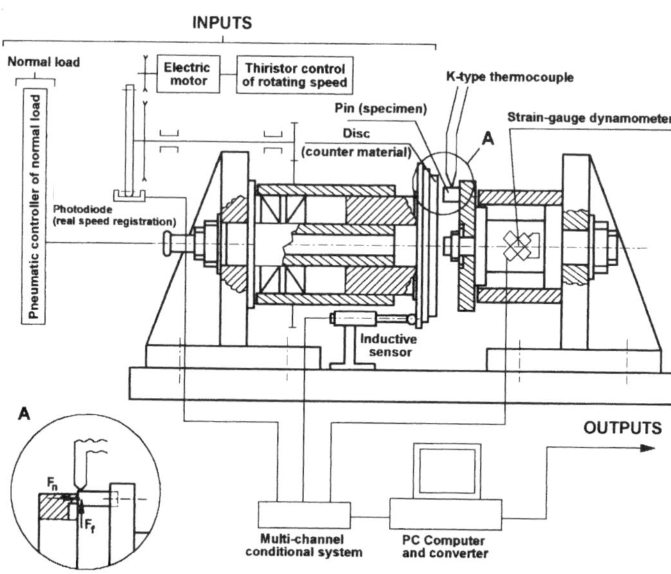

Fig. 1 A scheme of the experimental set-up based on pin-on-disc tribometer

In this investigation, the contact conditions were generated by letting the pin in the form of a rounded tool insert slide over the flat face surface of rotating steel disc. Thus, the real contact area was always irregular but approximately of an elliptic shape like for a Hertz contact [8].

The main features of the device used are as follows [9]:

a) the inputs: the normal load and the rotating speed can be changed continuously and precisely stabilised by means of a pneumatic load and steeples power regulation systems, respectively.

b) the outputs: the friction force, the linear wear of coupling materials and the contact temperature of the specimen (pin) are measured and analysed in *real-time* mode as functions of the sliding distance or equivalently of time. Finally, the coefficient of sliding friction, defined as the ratio of the variable friction force to the constant normal load, is continuously computed and recorded.

c) the tribo-tester operates under the control of specially designed computer software, which was installed in the memory of a PC computer equipped with a 16-channel A/D converter.

2.2 Materials

A normalized medium carbon steel AISI 1045 (UTS=670 MPa; 35 HRC), frequently used in comparative machinability tests, was selected as a countermaterial. Three different states of the disc face surfaces were examined: fresh grounded, runed-in (worn) after one trial and runed-in (worn) after two trials. The mean diameter of the disc-pin contact was assumed to be equal to 79.7 mm. The test pins were rounded inserts of 25.4 diameter and of RNMG 250900 ISO designation.

Three multilayered CVD coatings: two-fold TiC/TiN, three-fold TiC/Al_2O_3/TiN and four-fold TiC/Ti(C,N)/Al_2O_3/TiN were included in the friction tests performed. The total thicknesses of these coatings were equal to 8μm, 8μm and 10μm respectively. As a reference, P30 uncoated cemented carbide was used.

2.3 Test procedure

Sliding friction tests were carried out with four material combinations, keeping the normal force at F_n=50 N. The sliding speed was varied from 0.5 m/s to 1.5 m/s, which corresponds to the cutting speeds normally employed when cutting carbon steels using uncoated and carbide tools. For each tribo-pair, the sliding distance was selected to be equivalent to a test time (tool life) of 1200s. Additionally, prior to each test the working surfaces of specimens were ultrasonically cleaned with tetrachloromethane (carbon tetrachloride) and dried in a warm air stream. A minimum of three trials were conducted for each material/sliding speed combination. To quantify the friction behaviour of tool materials, the friction force was measured, and by using a DAQ system instantaneous values of the coefficient of friction were computed and recorded for certain intervals of time (ca. 4 s).

Light optical microscopy (*lom*) was used to examine mating surfaces and to characterise the dominant wear mechanisms of the tool materials. In this investigation, an Olympus PME3 optical microscope equipped with a JVC TK-1281colour video camera was employed. The selected wear scars produced on coated inserts (pins) were visualized and the video signals were transmitted to a PC computer. Subsequently, the raw images were processed and dimensioned by means of a computer image processing system developed earlier by the authors [7].

3 DETERMINATION OF FRICTIONAL RESPONSES OF THE TRIBO-SYSTEMS

In order to characterise the frictional interaction for selected tribo-pairs, the friction force was measured with a frequency of 0.1 s^{-1} and converted into the coefficient of friction, which was stored in the computer memory as a function of time (distance). The selected relationships obtained for 60 trials were plotted in Figs. 2-7 in order to show the characteristic tendencies in frictional behaviour of the four tribo-contacts. Variations of the friction coefficient obtained for three different sliding speeds v_s= 0.5, 1.0 and 1.5 m/s are presented in Figs. 2-5.

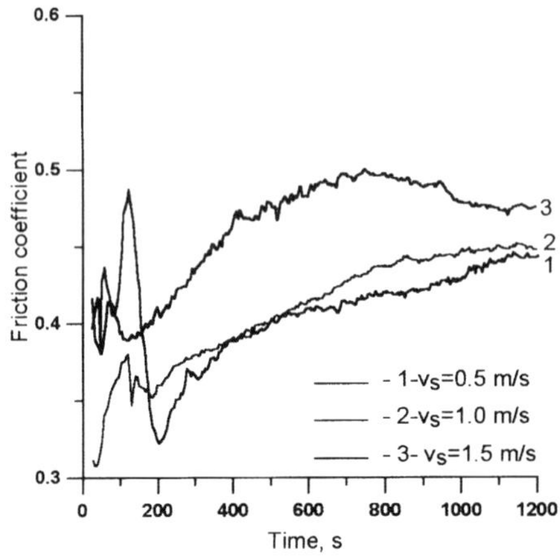

. **Fig.2 Recordings of friction coefficient vs. time for P30 uncoated carbide-AISI 1045 steel pair**

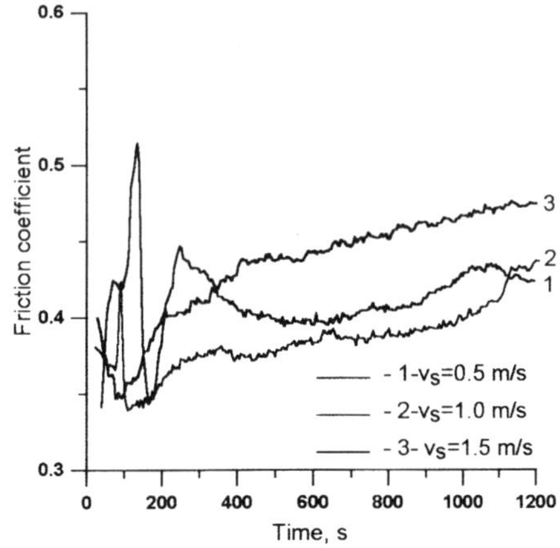

. **Fig.3 Recordings of friction coefficient vs. time for TiC/TiN coating-AISI 1045 steel pair**

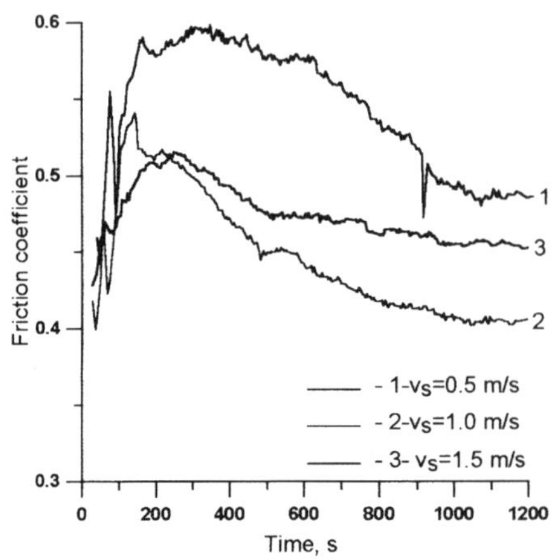

Fig.4 Recordings of friction coefficient vs. time for TiC/Al_2O_3/TiN coating-AISI 1045 steel pair

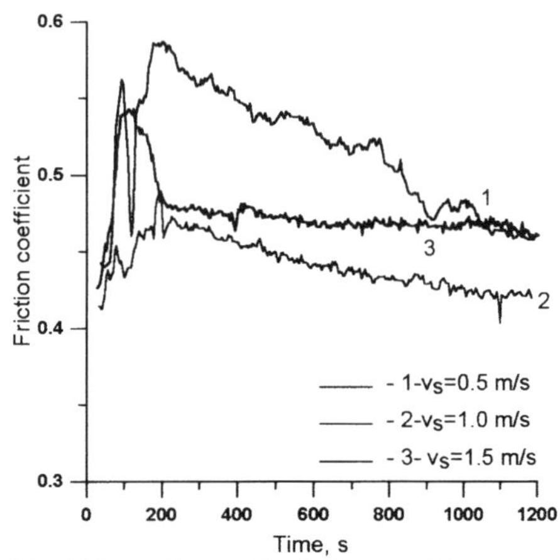

Fig.5 Recordings of friction coefficient vs. time for TiC/Ti(C,N)/Al_2O_3/TiN coating-AISI 1045 steel pair

As shown in Figs. 2 and 3, the friction coefficient curves consist of a running-in period interrupted after 100-200s, a period in which μ increases (Figs.2 and 3) or decreases (Figs.4 and 5) distinctly, and a steady-state stage with stabilised friction at the end of each test. Initially, μ increases rapidly with time and then its value is largely dependent on the sliding speed. There is a different behaviour of coatings containing the intermediate Al_2O_3 ceramic layer, especially in the initial stage of testing (Figs. 4 and 5). In these cases, a further increase in the friction coefficient is observed rather unexpectedly after the running-in period. The friction coefficient values are considerably lower for the highest sliding speed of 1.5 m/s used. This is probably an effect of the changes in the friction forces and due to the thermal softening of the countermaterial. Therefore, for stabilised friction observed in the final stage of the sliding tribo-test, as seen in Figs.4 and 5, the friction coefficient fluctuates between 0.4 and 0.5.

The effect of coating structure, at lower and higher sliding speeds, is presented in Figs. 6 and 7. At the lowest sliding speed of 0.5 m/s this influence is basically observed after 100s, but earlier the friction coefficient fluctuated widely. Conversely, for the highest speed of 1.5 m/s the stabilization of μ is achieved after 400s or more. In addition, it should be noted that the friction coefficient practically remains constant at 0.47 for all cutting materials tested. This can also suggest that coatings were fully removed because of abrasive action and the metal-uncoated carbide pair is only involved in sliding contact. The experimental evidences show that the influence of the preparation of the disk surface on the frictional process seems to be of secondary importance.

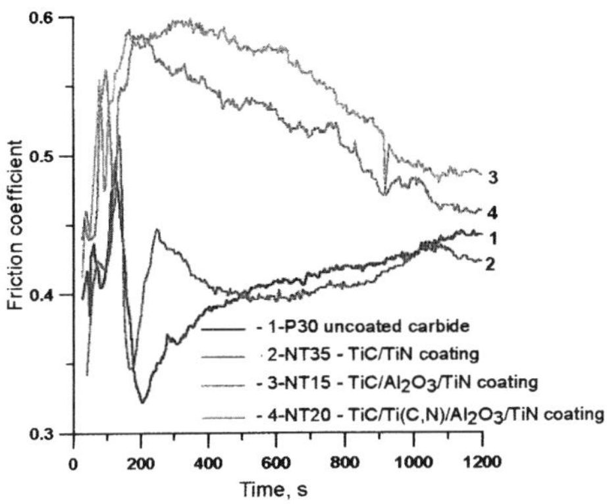

Fig. 6 Recordings of friction coefficient vs. time for tested cutting tool materials. Sliding speed = 0.5 m/s

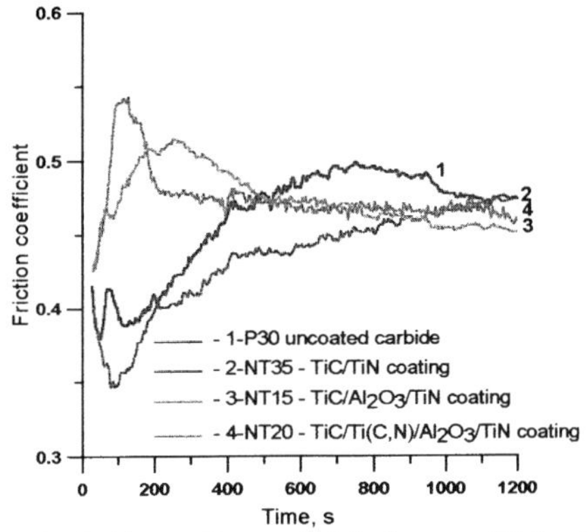

Fig 7 Recordings of friction coefficient vs. time for tested cutting tool materials. Sliding speed = 1.5 m/s

4 MEASUREMENT AND QUANTIFICATION OF WEAR

After the sliding tests, optical microscopic examinations of the wear scars produced on the cylindrical surfaces of the pins were carried out in order to select the prevailing wear mechanisms of coatings and the substrate. Roughly, three types of wear mechanisms can be distinguished. They are as follows:

1. Abrasive wear of the materials performed under the elastic contact conditions. Loose particles agglomerated in lumps close to the scar edge were observed in this case (Fig.8a).

2. Abrasive wear of the partly plastified material of the disc that is in contact with the hard coating, as illustrated in Fig. 8b.
3. Local plastic flow of the countermaterial over the slightly worn coating observed at the highest sliding speed applied (Fig.8c).

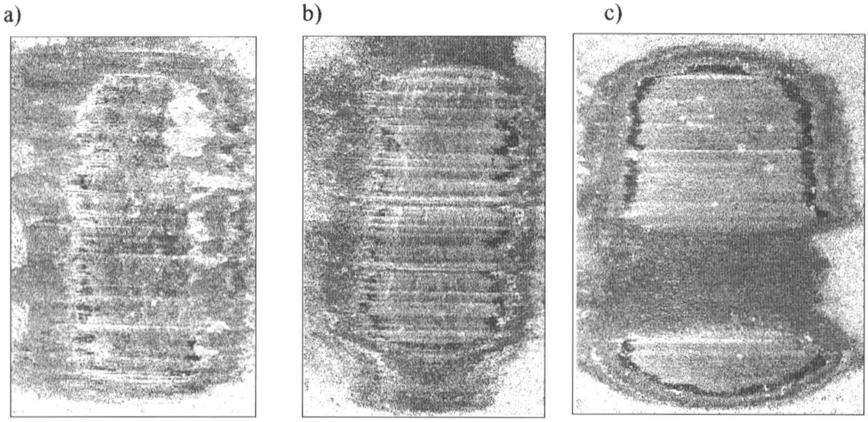

Fig. 8 Typical wear scars observed at different sliding speeds: a) v_s=0.5 m/s; b) v_s=1.0 m/s; c) v_s=1.5 m/s

The length and width of the wear scar produced on the tip were measured by an optical scope, and its depth was assessed by means of a light cross-sectional method. The results obtained were used to compute the specific volumetric wear rate given by the equation:

$$k = \frac{V}{d \times F_n} \tag{1}$$

The k, expressed in $mm^3/(mN)$, was found to be a good factor, which allows comparison of the wear resistance of coatings tested in relation to the reference. Values of the wear rates determined for the sliding time equal to 1200s by means of Eqn.1 are listed in Table 1. The influence of the sliding speed on the amount of wear when keeping the normal load at 50 N, is shown in Fig.9.

Table 1 Specification of the wear rates for tool materials tested (t = 1200s)

Pin material	v_s=1.5 m/s; d=1800m		v_s =1.0 m/s; d=1200m		v_s =0.5m/s; d=600m	
	V $mm^3 \times 10^{-3}$	k $\times 10^{-7}$ $mm^3/(mN)$	V $mm^3 \times 10^{-3}$	k $\times 10^{-7}$ $mm^3/(mN)$	V $mm^3 \times 10^{-3}$	k $\times 10^{-7}$ $mm^3/(mN)$
P30	21.0	2.33	16.6	2.77	19.2	6.37
TiC/TiN	3.5	0.38	12.7	2.11	12.5	4.16
$TiC/Al_2O_3/TiN$	31.5	3.51	26.8	4.46	27.0	9.08
$TiC/Ti(C,N)/Al_2O_3/TiN$	16.12	1.79	22.8	3.81	24	8.05

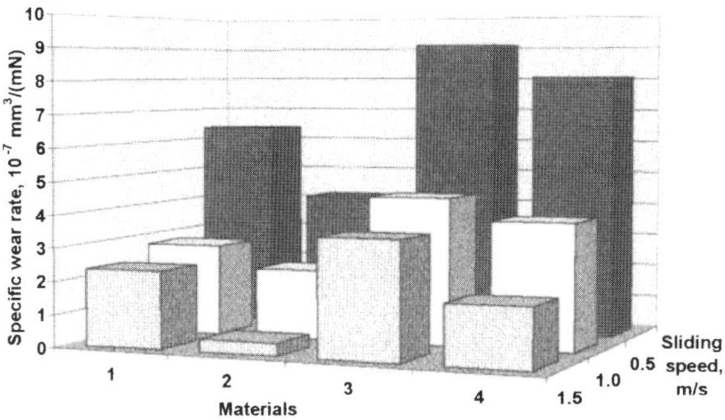

PIN MATERIALS:
1 – P30 uncoated carbide
2 – TiC/TiN
3 – $TiC/Al_2O_3/TiN$
4 – $TiC/Ti(C,N)/Al_2O_3/TiN$

Fig. 9 Comparison of the wear rates for tool materials tested at different sliding speeds

The following conclusions can be formulated based on the wear quantification results:
- the specific wear rate decreases with the elevation of the sliding speed for all tribo-pairs,
- coatings containing the intermediate ceramic layer indicate less wear resistance, especially in comparison to TiC/TiN layered. This evidence was observed at the three sliding speeds applied.
-it should be noted that the volume of material to be removed from the pin is practically independent of the sliding speed. However, it seems to be correlated with the sliding time because of distinct variations of the friction coefficient curves.

5 CONCLUSIONS

It was documented that the friction coefficient curves consist of a running-in period interrupted after 100-200s, a period in which μ increases or decreases distinctly and a steady-state stage with stabilised friction at the end of each test.

The experimental results show that due to intensive sliding friction and wear the use of the ceramic layer does not increase the wear resistance of coated inserts. High values of the friction coefficient, $\mu \approx 0.6$, can suggest an intensive abrasive action of loose ceramic particles at the contact surface. These hard particles are torn out from the trailing edge of the crater formed and then they remove intensively the exposed substrate material. Probably the sliding speeds applied in this investigation, 0.5 m/s$<v_s<1.5$ m/s, are too low to increase the contact temperature to a level high enough to more intensive plastic flow of the subsurface layer of the countermaterial (the chip material) over the top hard layer deposited on the pin.

The minimum wear rate was not yet achieved in this investigation. Its practical realization needs to be performed at a higher sliding speed than the maximum one applied, i.e. higher than 1.5 m/s.

REFERENCES

[1] Holmberg, K., Matthews, A., Ronkainen, H., "Coatings tribology-contact mechanisms and surface design", *Tribology International*, vol.32, no.1-3, 1998, pp. 107-120.
[2] Olsson, M.O., Söderberg, S., Jacobson, S., Hogmark, S., "Simulation of cutting tool wear by a modified pin-on-disc test", *Int. J. Mach. Tools Manufact.*, vol.29, no.3, pp. 377-390.

[3] Meiller, M., Lebrun, J.L., Touratier, M., Ryckelnck, D., "Friction law for tool/workpiece contact area in dry machining", in Proceedings of the International Workshop on *Friction and Flow Stress in Cutting and Forming*, ENSAM-Paris, Paris, 2000, pp. 101-109.
[4] Hedenquist, P., Olsson, M., "Sliding wear testing of coated cutting tool materials", *Tribology International*, vol.23, no.3, 1991, pp. 143-150.
[5] Wiklund, U., Wänstrand, O., Larsson, M., Hogmark.,S., "Evaluation of new multilayered physical vapour deposition coatings in sliding contact", *Wear*, vol.236, 1999, pp. 88-95.
[6] Grzesik, W., "The role of coatings in controlling the cutting process when turning with coated indexable inserts", *J. Mater. Proc. Technol.*, vol. 79, 1998, pp. 133-143.
[7] Grzesik, W., "The influence of thin hard coatings on frictional responses in the orthogonal cutting process", in Proceedings of the International Workshop on *Friction and Flow Stress in Cutting and Forming*, ENSAM-Paris, Paris, 2000, pp. 83-90.
[8] Dimarogonas A., "*Computer aided machine design*", Prentice Hall, New York, 1989.
[9] Zalisz, Z., "Antifrictional properties of selected exploitative additives for lubricants", in Scientific Bulletins of Lodz Technical University, no.765, 1996, pp.177-184.

Three-dimensional functional parameters for characterization of the bio-engineering surface at different stages of wear

S J XIAO, X Q JIANG, and **L BLUNT**
Centre for Precision Technologies, University of Huddersfield, UK

ABSTRACT

For bio-engineering components, their qualities especially wear properties cause serious concern. So the correct characterisation of these surfaces at different stages of wear is critically important. The bearing ratio curve has been referred to as a fundamental resource for functional wear of the measured surface. In the case of bio-engineering surfaces, their bearing ratio curves show very little difference concerning their shapes and features due to their high precision. An improved parameter set derived from the bearing ratio curve has been proposed recently. These parameters contain a great deal of information about the material and void volume and enhance the ability to characterise functional properties of a surface in light of the experiments on a batch of ceramic and metallic femoral hip joint heads at different stages of wear.

1. INTRODUCTION

With the increase in use of bio-engineering components such as the replacement hip joint systems to improve the health of human beings, their qualities especially their wear properties cause serious concern. This is due to the fact that early failure requires further surgery and makes the conditions for successful replacement worse. To reduce the risk of short lifetime or failure of the implants, surfaces should be strictly tested in conditions close to those encountered in the human body. At this point, the correct characterisation of those surfaces at different stages of wear is critically important. The research outlined here seeks to verify the ability of the 3D surface roughness parameters to characterise functional properties of these bio-engineering surfaces.

The bearing ratio or Abbott-Firestone curve is the area/length ratio of a total contacting area/length at a given truncation level over sampling area, which can be reflected naturally in the statistics of a surface topography. The height distribution function and amplitude probability density function of a surface can be derived from this curve. It is widely used in the measurement of engineering surfaces, for assessment of the tribological property of a surface and evaluation of the wear effect of a contacting surface moving relative to it, and also used as a criterion for assessing a surfaces ability to carry a load or resist wear. But in the case of bio-engineering surfaces, their bearing ratio curves show very little difference concerning their shapes and features due to their high precision. An improved parameter set derived from the bearing ratio curve has been proposed recently. These parameters contain a great deal of information about the material and void volumes and enhance the ability to characterise functional properties of a surface.

In this paper a series of experiments on a batch of ceramic and metallic femoral hip joint heads at different stages of wear have been introduced to fully verify the new parameter set. The result shows that the new parameters are able to clearly show the development of wear features that coincide with the functional use of the surfaces.

2. FUNCTIONAL PARAMETERS

2.1 The Primary Functional Parameter Set

The bearing ratio curve has been referred to as a fundamental resource for wear properties of the measured surface. With regard to 3D surface functional characterisation, a functional index family, based on bearing area ratio curve, has been early proposed in the primary parameter set as follows:-

2.1.1 Surface Bearing Index *Sbi*

This is the ratio of the RMS deviation over the surface height at 5% bearing area.

$$S_{bi} = \frac{S_q}{\eta_{0.05}} = \frac{1}{h_{0.05}} \qquad (1)$$

where $h_{0.05}$ is the normalised surface height at 5% bearing area.

2.1.2 Core Fluid Retention Index *Sci*

This is the ratio of the void volume of the unit sampling area at the core zone (5% ~ 80% bearing area) over the RMS deviation.

$$S_{ci} = \frac{V_v(h_{0.05}) - V_v(h_{0.8})}{(M-1)(N-1)\cdot\Delta x\cdot\Delta y}\cdot\frac{1}{S_q}, \qquad 0 \prec S_{ci} \prec 0.95\cdot(h_{0.05} - h_{0.8}) \qquad (2)$$

2.1.3 Valley Fluid Retention Index *Svi* This is the ratio of the void volume of the unit sampling area at the valley zone (80% ~ 100%) over the RMS deviation.

$$S_{vi} = \frac{V_v(h_{0.8})}{(M-1)(N-1)\cdot\Delta x\cdot\Delta y}\cdot\frac{1}{S_q}\ , \quad 0 \prec S_{vi} \prec 0.2\cdot(h_{0.8}-h_{\min}) \tag{3}$$

In the analysis, the bearing area ratio curve is divided into three parts as shown on the left in Fig.1. The peak roughness was defined as a range of heights embracing 5% of the bearing area, whilst the core and valley roughness were defined to cover 75% and 20% of the bearing area respectively.

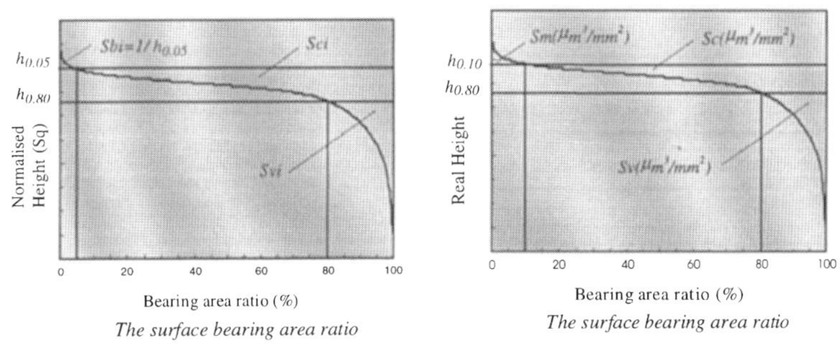

Fig.1 The basic concept of index and volume family

2.2 The Improved Functional Parameter Set

The index family is designed for a twofold purpose. One of these is to identify the shape features of different types of surface waveforms by observing the location of three functional indexes in two pattern planes, S_{vi} vs S_{bi} and S_{vi} vs S_{ci}. The other is an attempt to deal with the topographical features of the surfaces, for instance, bearing, running in and fluid retention properties. However, the range of values of these parameters for the same pattern surface was designed to be very small and not enough to identify and separate the worn stage of the surface, especially for high precision surfaces. So three improved functional parameters derived from the volume information of a bearing ratio curve have been proposed recently which were designed to increase the information for the wear and bear features of the surface while inheriting the function of pattern recognition of the surface. This parameter set is the so-called volume family, which is based on an assumption that the peak material embraces 0~10% of the bearing area whilst the core and valley ranges cover 10~80% and 80~100% of the bearing area on the right in Fig.1.

2.2.1 Materiel Volume of the Surface S_m

The material volume is defined as the material portion enclosed in the 10% bearing area, normalised to unit sampling area.

$$S_m = \frac{V_m(h_{0.10})}{(M-1)(N-1)\cdot\Delta x\cdot\Delta y} \tag{4}$$

where

$$V_m(h_{0.10}) = \frac{\Delta x \cdot \Delta y}{9}\left\{16\sum_{j=1}^{\frac{N}{2}-1}\sum_{i=1}^{\frac{M}{2}-1}\eta(x_{2i},y_{2j}) + 8\left[\sum_{j=1}^{\frac{N}{2}-1}\sum_{i=2}^{\frac{M}{2}-1}\eta(x_{2i-1},y_{2j}) + \sum_{j=2}^{\frac{N}{2}-1}\sum_{i=1}^{\frac{M}{2}-1}\eta(x_{2i},y_{2j-1})\right]\right.$$

$$+4\left[\sum_{j=2}^{\frac{N}{2}-1}\sum_{i=2}^{\frac{M}{2}-1}\eta(x_{2i-1},y_{2j-1}) + \sum_{i=1}^{\frac{M}{2}-1}[\eta(x_{2i},y_1)+\eta(x_{2i},y_{n-1})] + \sum_{i=1}^{\frac{N}{2}-1}[\eta(x_1,y_{2j})+\eta(x_{m-1},y_{2j})]\right] \quad (5)$$

$$+2\left[\sum_{i=2}^{\frac{M}{2}-1}[\eta(x_{2i-1},y_1)+\eta(x_{2i-1},y_{n-1})] + \sum_{j=2}^{\frac{N}{2}-1}[\eta(x_1,y_{2j-1})+\eta(x_{m-1},y_{2j-1})]\right]$$

$$\left.+[\eta(x_1,y_1)+\eta(x_1,y_{n-1})+\eta(x_{m-1},y_1)+\eta(x_{m-1},y_{n-1})]\right\}, \qquad \eta(x,y) \geq h_{0.10}$$

The material volume is not only a geometrical descriptor of the surface, but also has significant functional implications. The material volume may reflect wear and the running-in properties. A plateaued surface, such as an automotive bearing surface which needs a good load bearing capability and good lubrication retention, will have a high material volume; whereas a spiked surface, such as a bored surface, will have a low material volume.

2.2.2 Core Void Volume of the Surface Sc

A core void volume is enclosed from10% to 80% of surface bearing area and normalised to unit sampling area.

$$S_c = \frac{V_v(h_{0.10}) - V_v(h_{0.8})}{(M-1)(N-1)\cdot \Delta x \Delta y} \quad (6)$$

where

$$Vv(h) = V_v(h_{max}) - (M-1)(N-1)\Delta x \cdot \Delta y(h_{max} - h) + V_m(h) \quad (7)$$

2.2.3 Valley Void Volume of the Surface Sv

The valley void volume of the unit sampling area is defined as the void volume at the valley zone from 80% to 100% of the surface bearing area.

$$S_v = \frac{V_v(h_{0.80})}{(M-1)(N-1)\cdot \Delta x \Delta y} \quad (8)$$

The void volumes are proposed here to provide a direct inspection of lubrication and fluid retention of surfaces. For the same reason a plateaued surface will give a high the valley volume whereas a spiked surface will give a large core volume.

3. THE VERIFICATION OF THE IMPROVED FUNCTIONAL PARAMETER SET

A comparative study has been executed. In order to fully verify the properties of the improved functional parameters, the volume family is compared with index family. Fig.2 and Fig.3 show results gained from a batch of ceramic and metallic femoral heads in different wear

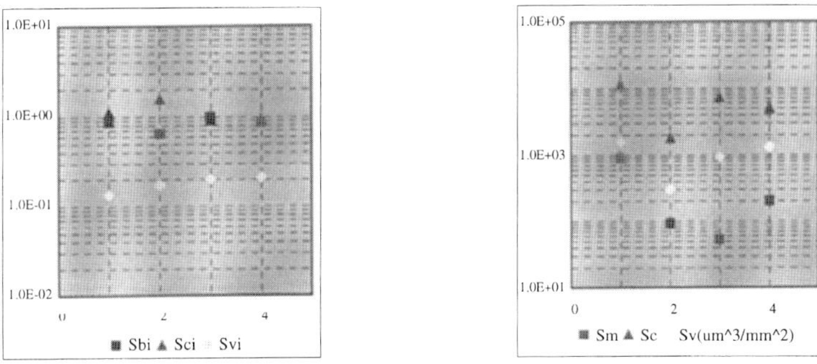

(a) Index family

(b) Volume family

Fig.2 Functional parameter sets for ceramic heads in different wear stages

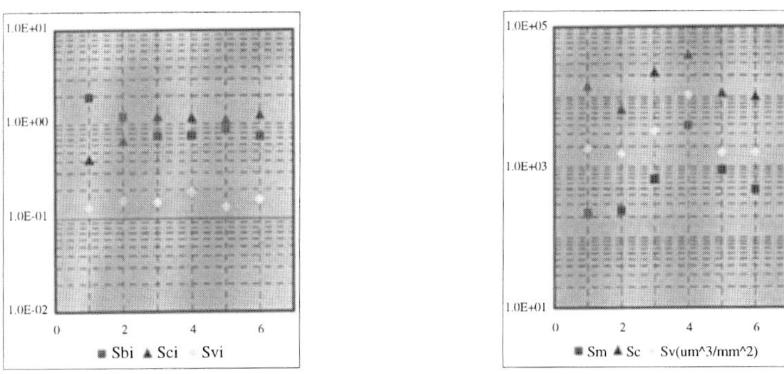

(a) Index family

(b) Volume family

Fig.3 Functional parameter sets for metallic heads in different wear stages

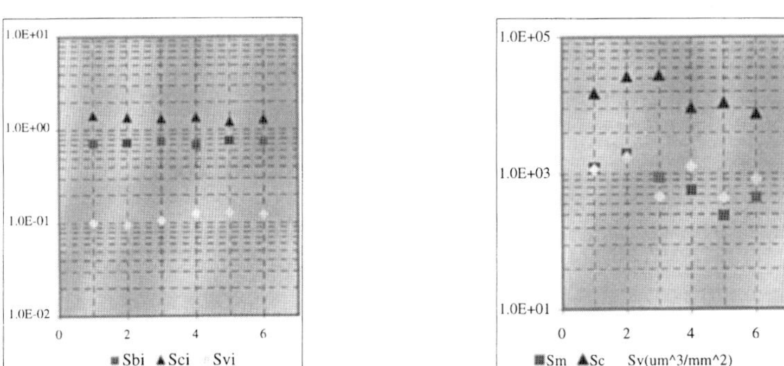

(a) Index family

(b) Volume family

Fig.4 Functional parameter sets for a pair of lapped and DLC coated prototypes in different positions
(Positions 1, 2 and 3 are for lapped surface and positions 4, 5 and 6 are for coated surface)

running-in, unworn, slightly worn and worn. In terms of ceramic heads, the *No*.1 is unworn; *No*.2 has been used and *No*.3, and *No*.4 are worn, whereas for metallic heads, the *No*.1 and *No*.2 have been used; *No*.3 and *No*.4 are slightly worn, and *No*.5 and *No*.6 are worn. A point that needs to be noted is that both ceramic and metallic heads were derived from the same manufacturing procedures, respectively. Fig4 gives a comparison of the functional parameter values of a pair of metallic plate prototypes in different positions. The positions 1, 2 and 3 are for lapped surface with S_q, $6.64E-03\mu m$, $7.03E-03\mu m$ and $8.35E-03\mu m$ whereas the positions 4, 5, and 6 for coated surface with S_q, $1.4E-02\mu m$, $2.3E-02\mu m$ and $1.7E-02\mu m$. Fig.2*a* and Fig.3*a* do not provide information concerning wear and tribological phenomena, and their values of the index parameters are still similar to each other whether unworn, slightly worn or even highly worn. Fig.4*a* gives confusing results for the different surface textures or different accuracy levels. Practical work indicates that by using the index parameters it is impossible to clearly understand the nature of the development of surface topography. Fig.2*b* and Fig.3*b* indicate the variation of the volume parameters in a running-in procedure in orthopaedic joint prostheses. It can be found that in most cases during wear, the variation of volume parameters has the same trend in spite of the scales of the values being different. The wear and tribological properties of femoral heads are described clearly. Again checking Fig.4*b*, the result demonstrates that the volume parameters have enormous practical significance, and they can be used to numerically evaluate the bearing area ratio.

4. CONCLUSIONS

Approved by these experiments, the values of the index parameters are similar to each other whether different types of surfaces or different stages of wear are considered. This makes it difficult to interpret the load bearing capability and lubrication retention properties of these bio-engineering surfaces. On the other hand, from the view of the point of tribology, the results do not provide information concerning wear and tribological phenomena, and their values are still similar to each other whether unworn, slightly worn or even highly worn. Practical work indicates that by using the index parameters it is impossible to clearly understand the nature of the development of surface topography.

The volume family addresses the disadvantages of the index family. The practical work has shown that it can characterise not only the common functional properties of surfaces, for instance, area and volume geometrical properties, but also interpret wear and tribological properties in a running-in procedure. Therefore, the volume family has enormous practical significance, and can be used to numerically evaluate the bearing area ratio of surfaces.

5. ACKNOWLEDGEMENT

The authors would like to thank the European Community for support funds on this work under the program SMT CT98-2209. The authors would also like to express very special thanks to DePuy Johnson & Johnson in Leeds for supplying the specimens for measurement and analysis.

REFERENCES:

1. X.Q.Jiang, L.Blunt and K.J.Stout, Wavelet framework Representation for surface analysis, *Proceedings of The Fourth International Symposium of Measurement Technology and Intelligent Instruments* 2sd-4th Sept. 1998, University of Miskolc, H-3515, ISBN 963 8455 578, 165-172

2. X.Q.Jiang, L.Blunt and K.J.Stout, Three-dimensional surface characterization for orthopedic joint prostheses, *Journal of Insti.Mech. Engrs*, Part H, 213(1999) 49-68

3. X.Q.Jiang, L.Blunt and K.J.Stout, Development of a lifting wavelet representation for characterization of three-dimensional surface topography , *Proc. of Royal Society* (Ref.No.99PA078)

4. ISO/TS 16610-4, 1999, Data extraction techniques by sampling and filtration Part 4 – Spline Wavelets (Under preparation).

5. K.J. Stout, P.J. Sullivan, W.P. Dong, E. Mainsah, N. Luo, T. Mathia, and H. Zahyouani, *The development of methods for the characterisation of roughness in three dimensions,* 1st edn, Commission of the European Communities (ISBN 0 70441 313 2), 1993

6. DIN Measurement of surface Roughness; Parameters R_k, R_{pk}, R_{vk}, M_{r1}, M_{r2} for the description of the Material Portion in the roughness, German standard, DIN 4776,1990

Laser remelting of 75/25 Cr_3C_2/NiCr coatings

E FERNÁNDEZ, J MATEOS, R VIJANDE, and **J M SIERRA**
Área de Ingeniería Mecánica, Universidad de Oviedo, Gijón, Spain

ABSTRACT

A CO_2 laser has been used to fuse based Ni-Cr Cr_3C_2 coatings with the purpose of homogenizing their microstructure and eliminating their porosity. Later on, in order to clarify the effect of laser irradiation on tribological properties, tests of layer control and wear tests have been carried out on the samples treated with the laser. The microstructure obtained after laser treatment was more homogeneus, the microhardness was increased and the porosity was eliminated. The remelted coating are also more wear resistant under dry contact conditions.

1. INTRODUCTION

One of the methods of extending the service lives of engineering components subject to wear is through the application of a layer of wear resistant coating on the sliding surface. Ceramics are used a lot for this purpose in tribological applications, due to their high hardness and chemical stability.

Plasma spraying is one of the most appropriate techniques for the deposition of ceramic coatings, since the temperature reached in the plasma is very high (above 18000^0 C [1]) and little risk of distortion of the base material exists (substrate temperature does not go above 300^0C [2]). However, the sprayed layers usually contain defects, such as high porosity, lack of homogeneity and weak substrate-coating adherence [3]. In recent decades laser surface treatments have been used with the purpose of eliminating such defects [4-7]; however, studies are not plentiful on the tribological behaviour of laser-treated coatings [8-10].

In this paper we report the results of an experimental study of some tribological properties of chromium carbide coatings. Coatings were plasma sprayed onto a metallic base and later laser remelted. Measurements of the porosity, coating-substrate adherence, microhardness and sliding wear resistance have been carried out, with the purpose of analyzing the results of such treatment.

2. EXPERIMENTAL PROCEDURE

Plasma spraying was carried out with a 756 METCO system. The sprayed material was a Ni-Cr chromium carbide cermet (75% Cr_3C_2, 20% Ni and 5% Cr). The coatings were deposited onto AISI 1043 steel blocks, of 15.75 x 10.16 x 6.35 mm. The mean thickness of the coatings was 0.4 mm.

Laser treatment was carried out with a CO_2 laser, ROFIN SINAR model RS 1700 SM; this equipment has a nominal power of 1700 W and it can operate in continuous or pulsed mode. The traverse distribution of energy in the laser beam corresponds to the TEM_{10*} mode (*doughnut* mode), although it can be modified by means of an integrator beam.

The porosity was measured by the standard test method for determining volume fraction by systematic manual point count, according to standard ASTM 562-89. This method consists of counting the number of pores that coincide with the knots of a grid placed in the ocular of the microscope. A significant number of measurements was used to produce an unbiased statistical estimation of the volume fraction of the pores.

The coating-substrate bonding strength was measured according to ASTM-C633 standard: five cylindrical specimen tests were covered on one of their bases with the ceramic material and later they bonded by means of an adhesive to other counter-specimens; later on, each pair of samples underwent a tensile test in normal direction to the joining plane. The used adhesive was EC-2214, manufactured by 3M.

The microhardness was measured in the traverse section of the coatings, according to standard ASTM E384; a Vickers indenter was used with a load of 300 g. Five specimen tests were analyzed, taking measures to five depths and carrying out five indentations for each depth.

The wear tests were of the block-on-ring type with linear contact, under dry conditions. These tests were carried out at room temperature, following the standard ASTM G77; the coated blocks were matched with rings of 50 mm of external diameter, manufactured with tempered and hardened AISI 1043 steel. Before the tests were carried out, the blocks were ground until they got an average surface roughness equal to 0.2 μm; the average surface roughness of the rings was equal to 1.1 μm. Two series of tests were conducted, with the purpose of analyzing the influence of the load and sliding speed on the wear resistance. The values of these parameters are shown in Table 1. The sliding distance was 6300 m (40000 cycles).

Table 1. Load and sliding speed used in wear tests

1st series		2nd series	
v (m/s)	P (N)	v (m/s)	P (N)
0.4	465	1.3	200
	333	0.8	
	200	0.4	
	57	0.2	

3. RESULTS AND DISCUSSION

3.1. Laser treatment

The first tests were carried out with laser beam in mode TEM_{10*}.We used a range of energy densities from 275 to 3600 J/cm^2. Using high densities there appeared holes in the coating and sometimes it comes off the substrate; reducing the energy density alterations in the microstructure could not be contrasted and the porosity was conserved. This behaviour is similar to the one observed in other carbides coatings tried with laser [7].

The elimination of the porosity was obtained using an integrator beam to modify the distribution of the energy of the laser beam; melted areas, with homogeneus microstructure and without porosity were obtained with these parameters: power: 1100 W; scanning speed: 800 mm/min; spot diameter: 3 mm; protective gas: Ar (0.5 bar). Figures 1 and 2 correspond to coatings without and with laser treatment, respectively.

3.2. Layer control

3.2.1. Porosity

The average porosity in the plasma sprayed coatings was equal to 9.2% in volume (standard deviation s = 5.1, 95% confidence limit similar to 1.33). The laser remelted coatings have a practically null porosity, as one may observe in figure 2.

3.2.2. Coating-substrate adherence

The laser treatment increased the adherence between the ceramic layers and the metallic base; the average coating-substrate bonding strength in the untreated coatings was 56.7 MPa (standard deviation s = 3.2); the treated coatings offered an average resistance superior to 59.3 MPa. We could not determine the exact value, because the breaking did not take place in the coating-substrate union, but in the union with the adhesive.

The coating-substrate bonding strength increment has probably been caused by the incidence of the radiation on the metallic base; this incidence may have been the cause of the diffusion of Fe in the coating material and a metallurgic union may have taken place; the analysis of metallic elements has confirmed the presence of iron in the remelted coatings.

3.2.3. Microhardness

The measure of the microhardness yielded a mean value $HV_{0.3}$ = 561 in the plasma sprayed coatings and $HV_{0.3}$ = 655 in the laser treated coatings; this represents an increase of 17% and it indicates that laser remelting can improve the performance of sprayed coatings in wear resistant applications. The high residual stress, wich is caused by the quick heating and quenching wich take place during laser irradiation, could be related with the increment of the coatings microhardness.

The hardness variation with the distance to the coating-substrate interface is shown in figure 3; one can appreciate that the hardness stays practically constant in the thickness of all the treated coatings; this may be a consequence of the highest homogeneity having taken place because of the treatment. On the other hand, the hardness of the metallic base has been increased: the mean values are $HV_{0.3}$= 279 and $HV_{0.3}$= 460 in plasma sprayed and laser remelted samples, respectively; this would confirm that the metallic substrate has been affected by the radiation.

3.3. Wear tests

Both materials, treated and non treated, showed a very similar behaviour during the dry slidind wear tests: un initial period (wear-in) exists during which the wear rate varies in an irregular way. Once the initial period has taken place, the lost of material takes place in an almost uniform way and the wear rate is practically constant. Figure 4 shows the wear rate evolution of a laser remelting coating during a test, which was carried out with 200 N load and 0.4 m/s sliding speed. The variation of wear rate was determined by measuring the material losses every 500 cycles.

Laser remelted coatings are more wear resistants than plasma sprayed coatings; figure 5 provides an example, where the load is 57 N and the sliding speed 0.4 m/s. The difference is probably due to the influence exerted by the increase in hardness; the biggest homogeneity in the treated coatings could also have influence, since the formation of abrasive particles is less probable than in the plasma sprayed coatings.

Figure 6 allows to verify that the material loss of the laser remelted coatings varies almost proportionally to the applied load. The influence of the sliding speed on the wear is shown in figure 7; in the range of tested values, the data obtained in the tests carried out with a constant load seem to be adjusted satisfactorily to a curve of second degree; this fact would corroborate the studies of several investigators on the wear of the ceramics [11,12].

In all of the tests the ceramic coatings showed bigger wear resistance than steel. The relationship among the losses of volume of both materials varied between a maximum of 62.9 (57 N, 0.4 m/s) and a minimum of 2.9 (200 N, 0.8 m/s). In the figure 8 can be observed that in the wear-in period the steel has a much bigger wear rate that the ceramic and that during the steady state the wear rates (the same slope) of both materials are similar. This behavior could be justified by the formation of high hardness wear debris during the wear-in, which would have an abrasive effect on the steel; later on, the metallic and ceramic particles will accumulate in the wear tracks. Thus, areas will exist in which the metal-metal contact prevails and the wear rates of both ceramic coating and steel will be similar.

The average value of the friction coefficient has been 0.49 (standard deviation similar to 0.06); the measured maximum value was 0.63 (200 N, 0.2 m/s) and the minimum 0.41 (465 N, 0.4 m/s). The evolution of the friction coefficient during the tests is similar to that of the wear rate: initially it presents marked oscillations and their value is stabilized later on; these processes, however, do not often reach the stationary state at the same time and they are not in phase with each other: a high coefficient of friction does not imply necessarily high wear rate (figure 9); these results coincide with the investigations carried out by Blau [13].

4. CONCLUSIONS

1. Laser remelting is an appropriate technique to improve the quality of plasma sprayed chromium carbide coatings. To obtain satisfactory results it is necessary to use an integrator beam.

2. Laser treated coatings are more homogeneous than plasma sprayed coatings and their porosity is practically null; the treatment also increases the hardness, the coating-substrate adherence and the wear resistance.

4. In the range of tested values, the wear of the treated coatings increases in a linear way with the applied load; the influence of the sliding speed is not so clear, although the obtained results seem to be adjusted satisfactorily to a curve of second degree.

5. The ceramic coatings have been more wear resistant than steel in all the tests; the wear rate of the steel has been greater that of the coatings during the running in; in the steady state the wear rates (the same slope) of both materials are similar.

REFERENCES

1. Nicoll A. R. "Protective coatings and their processing thermal spray". Proc. Hight temperature materials and coatings. 1984. Finland. pp. 24-29.

2. Hocking M.G., Vasantasree V. and Sidky P.S. "Metallic & Ceramic Coatings". Longeman Scientific & Technical. Essex. 1989.

3. Bushan B. and Gupta B. K. "Handbook of Tribology". McGraw Hill. New York. 1991.

4. Mordike B. L. and Sivakumar R. "Laser melting of plasma sprayed ceramics coatings". Surface Engineering. 1988. 4(2). pp. 127-140.

5. Hatefeuille L., Beylat L., Carron D. and Puig T. "Amelioration des propiétés de surface par refusion laser". Proc. Conference Laser de puissance et traitements des matériaux. Ecole de Printemps. France. 1991. pp. 27-31.

6. Wang K., Lin J. and Han D. "Microstructure and properties of laser remelted cobalt based and nickel based plasma srayed coatings". Surface Engineering. 1996. 12(3). pp. 235-238.

7. Mateos J., Cuetos J.M., Cadenas M. and Montes H. "Tratamiento con laser de CO_2 de recubrimientos de carburo de tungsteno proyectados por plasma". Anales de Ingeniería Mecánica. 1997. 4. pp. 113-119.

8. Cuetos J.M., Fernández E., Vijande R. and Rincón A. "Plasma sprayed coatings treated with laser: tribological behaviour of Cr_2O_3". Wear. 1993. 169. pp. 173-179.

9. Fernández E., Cuetos J.M., Vijande R. and Rincón A. "Comparison of wear behaviour of plasma sprayed Al_2O_3 coatings with and without laser treatment". Tribology International. 1996. 29 (6). pp. 477-485.

10. Mateos J., Cuetos J. M., Fernández E. and Cadenas M. "Plasma sprayed tungsten carbide coatings after laser surface treatment". Proc. 1998 Powder Metallurgy World Congress & Exhibition. 1998. Spain. 1. pp. 531-537.

11. Fernández, J.E., Rodríguez, R., Wang Y., Vijande R. and Rincón A. "Sliding wear of a plasma-sprayed Al_2O_3 coating". Wear. 1995. 181-183. pp. 417-425.

12. Denape J. and Lamon J. "Le comportement en frottement sec des céramiques á hautes performances". Proc. 4éme Congrés Européen de Tribologie, Soc. Francaise de Tribologie. 1985. Elsevier. Amsterdam.

13. Blau P.J. "Friction science and technology". Marcel Dekker Inc. New York. 1996.

FIGURES

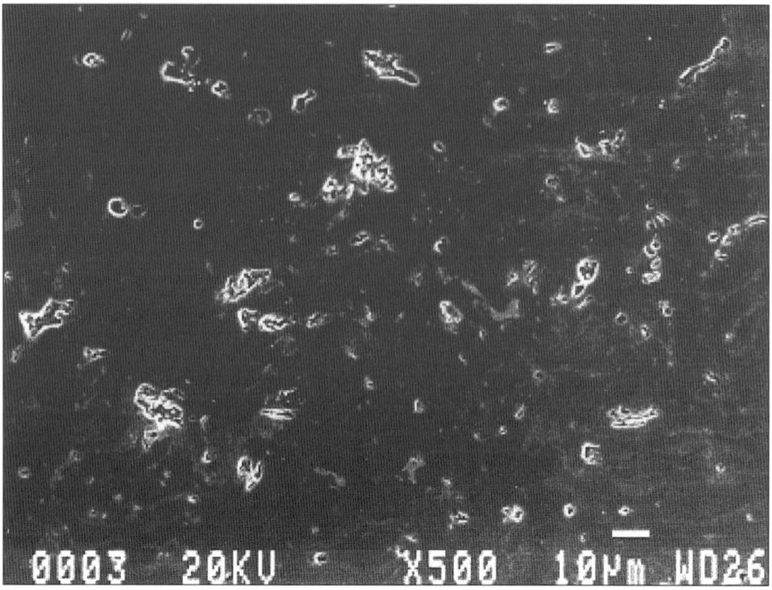

Figure 1. SEM microgrph of a plasma sprayed coating.

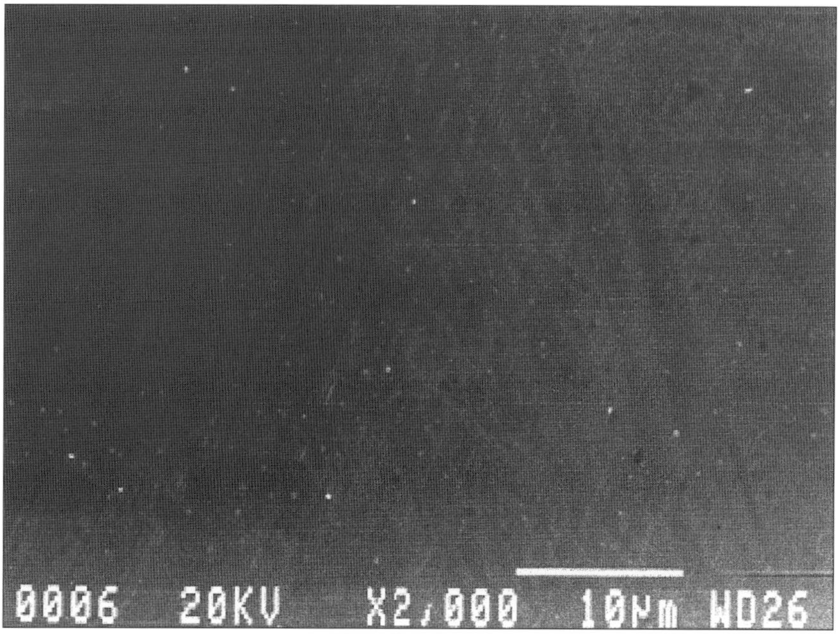

Figure 2. SEM micrograph of a laser remelted coating.

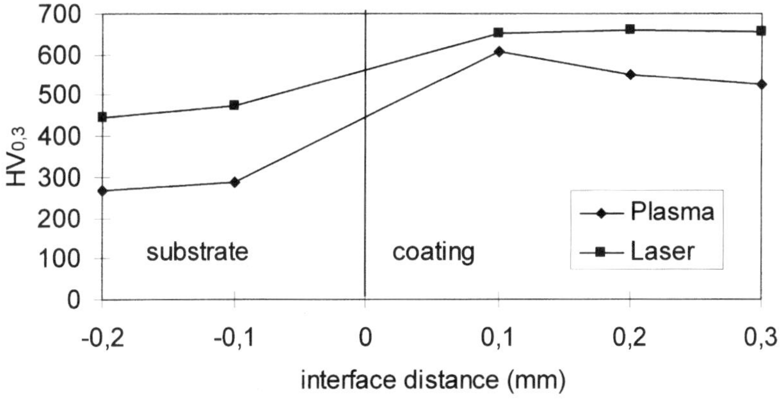

Figure 3. Microhardness vs. depth

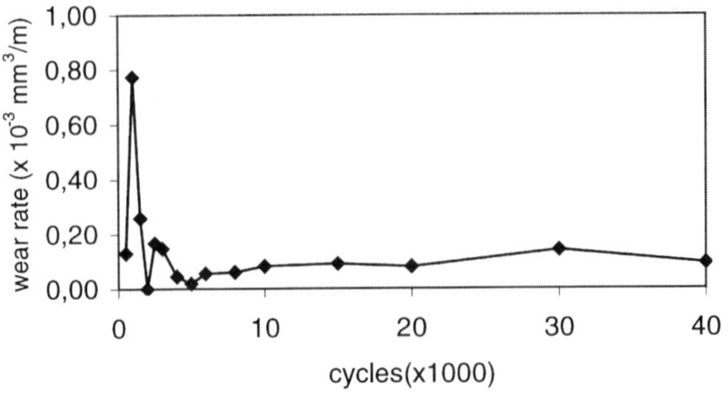

Figure 4. Wear rate vs. slided distance. 200 N, 0.4 m/s. Laser remelted coating

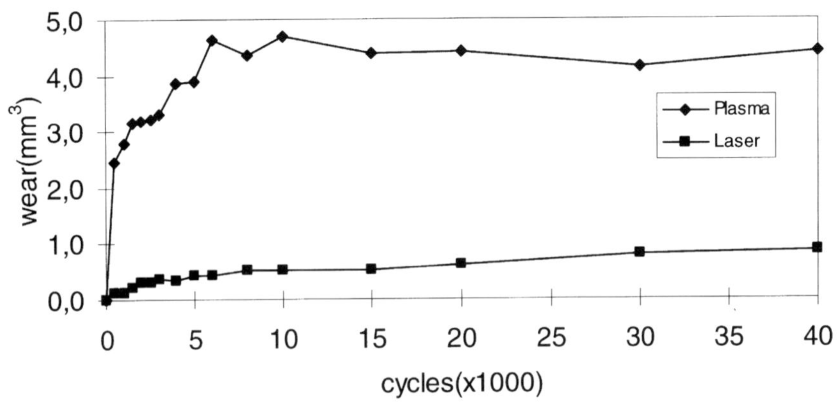

Figure 5. Wear comparison between plasma sprayed and laser remelted coatings

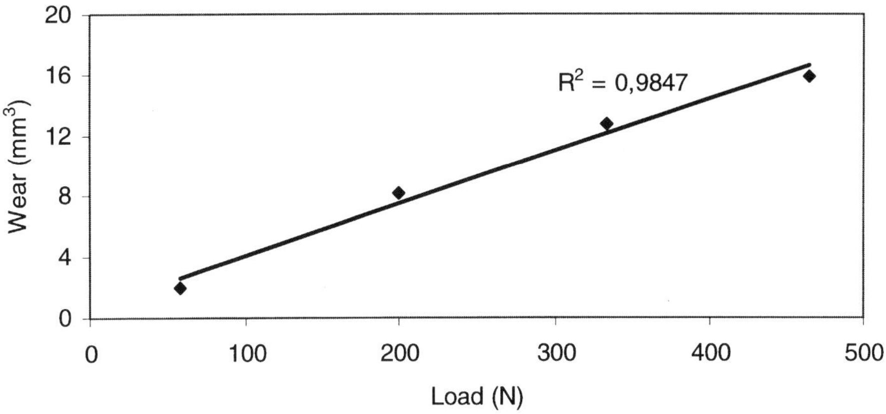

Figure 6. Wear volume vs. Load. v = 0.4 m/s

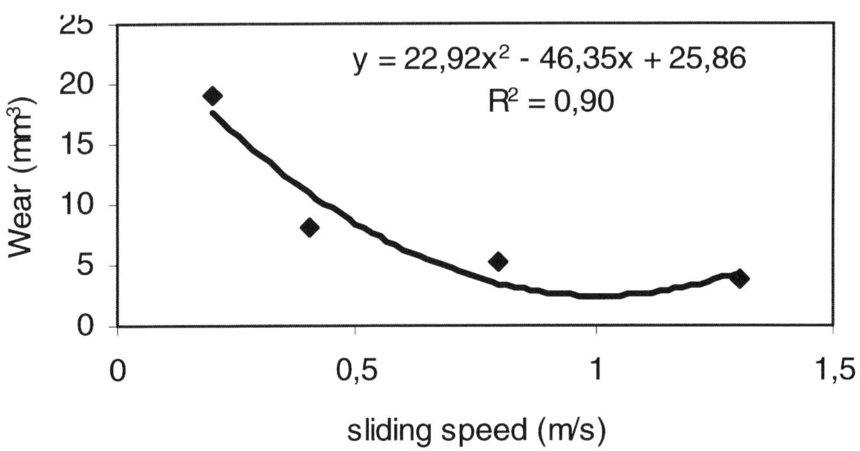

Figure 7. Wear volume vs. sliding speed. Load = 200 N

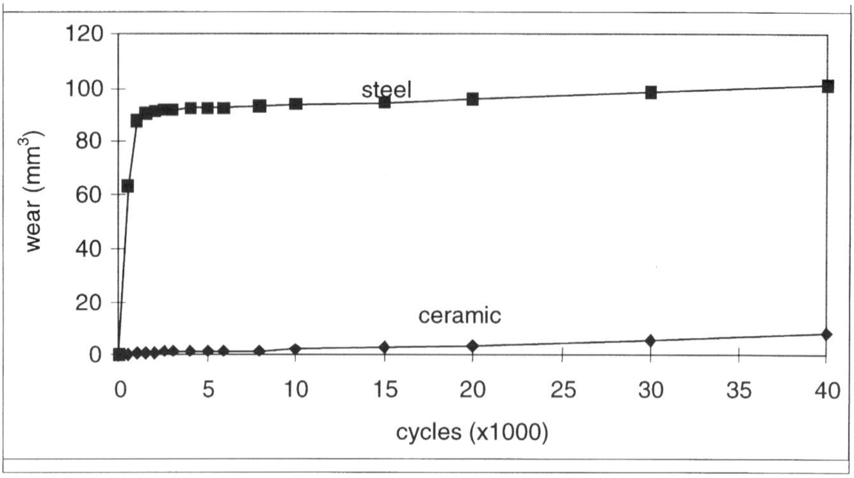

Figure 8. Wear comparison between ceramic coating and steel. 200 N, 0.4 m/s

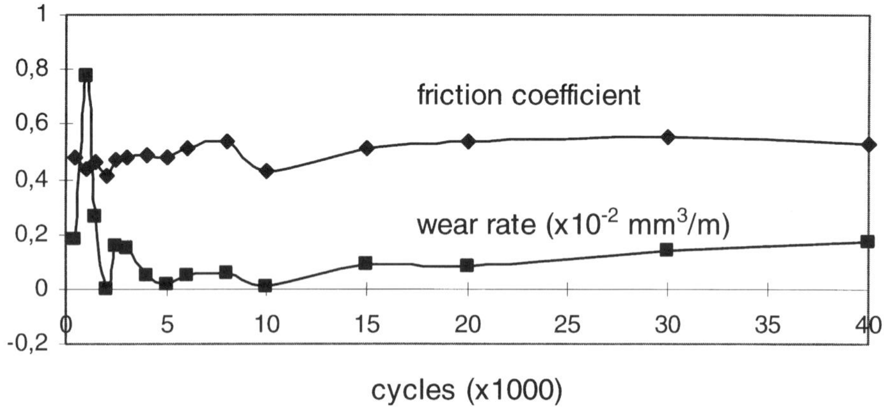

Figure 9. Comparative evolution of the friction coefficient and the wear rate

Lubricants

The influence of the environmental process conditions on the machining of aluminium alloys

J A WILLIAMS and **I M HUTCHINGS**
Department of Engineering, Cambridge University, UK
W Y H LIEW
School of Mechanical and Production Engineering, Nanyang Technological University, Singapore

ABSTRACT

Metal cutting is conventionally carried out under conditions of flood lubrication both to provide adequate cooling of the tool insert, and so give acceptable tool lives, but also to lubricate the potentially high-friction interface between the rake face of a cutting tool and the adjacent freshly formed chip surface. Notwithstanding this, there is increasing interest in machining either dry or under conditions of mist lubrication stimulated not least by concern over the problems associated the handling and disposal of cutting fluids which often contain environmentally harmful components such as the elements chlorine or sulphur. In an effort to identify the essential lubricating aspects of fluid activity, as opposed to its role as a coolant, we have carried out experiments involving the orthogonal machining of precipitation hardened aluminium alloys, principally 2014 in controlled environments in which the feed (that is the depth of cut) speed and temperature have been varied while using a variety of tool materials and lubricating species. The results indicate that there can be unexpectedly subtle, but significant, interactions between the metallurgy of the work-piece, the nature of the surface of the tool insert and the surrounding environment which are not wholly consistent with conventional theories of lubricant activity. The implications of these observations for the complex tribological system constituted by the combination of work-piece, tool surface and local environment are discussed.

1 INTRODUCTION

In orthogonal machining chip formation takes place by a process of intense plastic shearing in a region known as the primary shear zone extending from the tip of the cutting tool to the free surface as indicated in Fig. 1. The chip has a freshly created and so chemically uncontamintaed surface and, as it flows up the rake face of the tool, it is subject to a very high normal stress. Under these conditions, strong adhesion may occur between chip and tool giving rise to additional shear in the region of the chip adjacent to the tool surface: this region is often known as the secondary shear zone. It is generally accepted that the most significant attribute of successful cutting lubricants – in contrast to their action as coolants - is their ability to penetrate the chip-tool interface and reduce the extent of this rake face adhesion through the formation of low shear strength boundary films. The effectiveness of a number of organic compounds in the vapour state has been demonstrated by several authors [1-5]. In general, as the cutting speed is increased, so the effectiveness of any lubricant becomes attenuated and this has conventionally been attributed to the reduction in interface penetration and thus the growth of intimate contact and adhesion between the chip and the tool. Tetrachloromethane (carbon tetrachloride CCl_4) in both liquid and vapour form has long been known for its success as a cutting lubricant and its action is generally associated with the formation of a low shear strength layer of chloride at the interface; its particular effectiveness is thought to be due to its combination of high saturated vapour pressure and small molecular size. More recently, Wakabayashi et al [3] have demonstrated that vapours of other organic species such as ethanol (C_2H_5OH) can also be effective in reducing the interaction between the chip and the tool surface of aluminium (specifically Al 2014-T4) and lead to significant improvements in surface finish, similar to those seen in CCl_4 vapour. While health and environmental safety considerations now rule out the practical or industrial use of CCl_4 its comparatively simple chemical nature and well established physical properties make it a useful model lubricant which can be used to test or validate the various models of lubricant action that have been postulated in the literature.

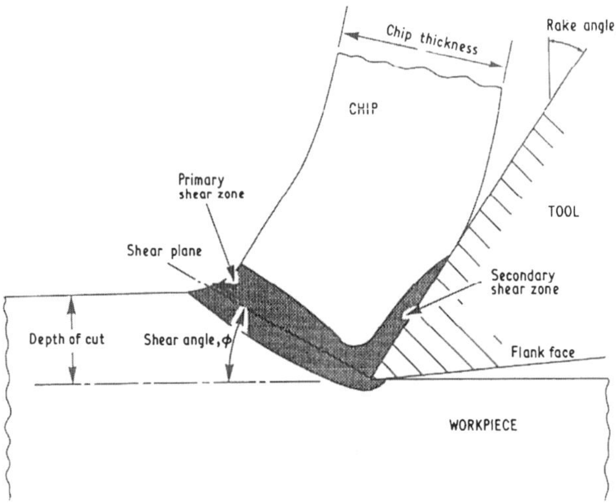

Fig. 1 Nomenclature of orthogonal cutting.

2 EXPERIMENTAL

The controlled environment cutting apparatus used in the investigation has been described more fully elsewhere [3,6]. The cutting tool is attached to a strain gauge dynamometer which continuously measures the cutting force components parallel and perpendicular to the direction of relative motion between the tool and the workpiece. In all the experiments described the depth of cut was set to 100 μm and the workpiece was 3 mm in thickness. The cutting chamber could be evacuated to a pressure of less than 10^{-4} Pa by a diffusion pump backed by a single stage rotary pump and vapours and gases admitted into the evacuated chamber from either a flask of liquid or a gas cylinder through an adjustable leak valve. Prior to testing the cutting chamber was flushed thoroughly with dry nitrogen and baked to improve the cleanliness of the vacuum and tests were carried out only once a steady gas pressure within the chamber had been established. The tool tips were manufactured from a single bar of M2 high speed steel (BS 970 grade BM2) and were held in a standard tool holder at a rake angle of 6° (and clearance of 5°) having been ground and polished to be sharp i.e. with negligible nose radius.

Figure 2 illustrates the forces involved in orthogonal machining. The total cutting force R can be resolved into components T (cutting force) and N (thrust force) which are in directions respectively parallel and normal to the cutting direction. R may also be resolved into the components F (friction force) and W (normal force) parallel and normal to the tool rake face, thus

$$F = T\sin\alpha + N\cos\alpha \quad \text{and} \quad W = T\cos\alpha - N\sin\alpha \qquad (1)$$

where α is the tool rake angle, as defined in Fig. 1. In this study, the coefficient of friction μ defined by the relation $\mu = F/W$ has been used to describe the state of rake face friction, and changes in its value used to indicate the influence of gaseous and other lubricants on the cutting process.

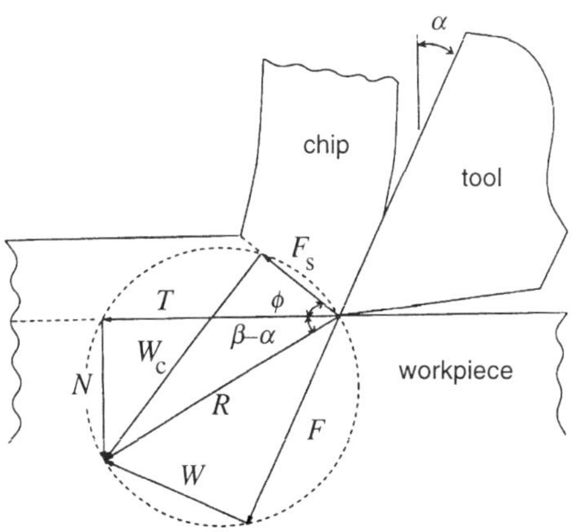

Fig. 2 Forces involved in orthogonal machining for a workpiece moving from left to right relative to the tool. The diagram shows the total force exerted by the tool which can be resolved into either the *F* and *W* or *T* and *N* respectively parallel and perpendicular to the rake face of the tool or the direction of motion.

Most experiments were performed with workpieces of aluminium grade 2014 (composition (wt%) 4.4% Cu, 0.8% Si, 0.8% Mn, 0.5% Mg, balance Al). This age-hardening alloy is widely used in aerospace applications because of its high strength and good machinability. Samples were received in standard T4 (solution treated but not aged) and T6 (solution treated and aged) conditions. Further samples were also solution treated at 500°C before quenching and ageing at 160°C as indicated in Table I, which also shows the Vickers hardness of all the samples.

The progress of the age-hardening process is clearly illustrated by the samples TA (under-aged), TB (aged to peak hardness) TC and TD (increasingly over-aged). Progression through this series is associated with the formation and coarsening of (predominantly) copper-aluminium precipitates. After quenching from solution, the alloy contains regions of solid segregation, or clustering, which are known as Guiner-Preston type 1 or GP(1) zones. Local strains generate an increase in overall hardness and this can be further increased by additional ageing, which leads to the ordering of larger clumps of copper atoms and the formation of GP(2) or θ'' structure (see [7]). These precipitates are perhaps a thousand times the volume of

GP(1) zones and the material in this form is said to be in commercial condition T6. Additional ageing times lead to the precipitation of definite platelets of $CuAl_2$ or θ' now only partially coherent with the matrix and with still further ageing the equilibrium phase ($CuAl_2$) known as θ is formed; because these precipitates are no longer coherent with the matrix, hardness and mechanical strength fall. Continued treatment for even longer times leads to over-ageing, causing coarsening of the precipitates and still further decreases in hardness. Figure 3 illustrates these changes in hardness, and thus yield stress, with ageing time. T4 and T6 are commercial heat treatments, those marked TA, TB, TC and TD represent additional laboratory treatments used in this investigation though here we restrict the presentation to results obtained with materials T4 and TB (equivalent to T6).

Table I: Designations, heat treatment and hardness of Al 2014 samples

Designation	Ageing	Hardness HV	Comments
T4	Not aged	138	Received in T4 condition
T6	Artificially aged	153	Received in T6 condition
TA	10 h at 160°C	145	Under-aged
TB	18 h at 160°C	152	Peak hardness
TC	168 h at 160°C	145	Over-aged
TD	672 h at 160°C	132	Over-aged

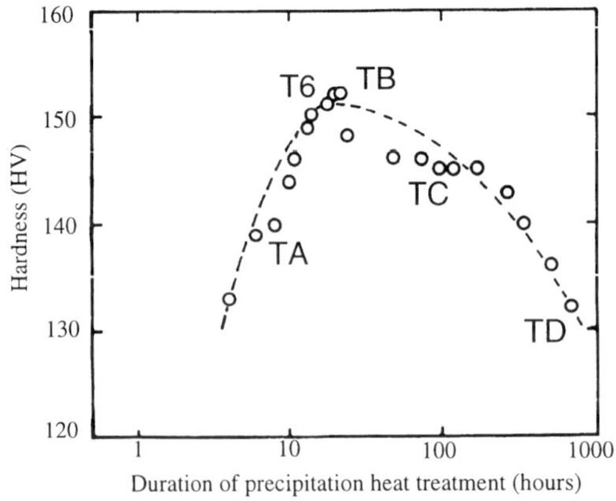

Fig. 3 Variation of hardness with ageing time for aluminium alloy Al 2014.

3 RESULTS

Figure 4 shows the rake face coefficient of friction as a function of the cutting speed for workpieces of the two heat treatments, T4 and TB, cut under vacuum with hss tooling inserts. Over the part of the speed range from 1 to 100 mm/s both traces are essentially constant showing no not significant differences.

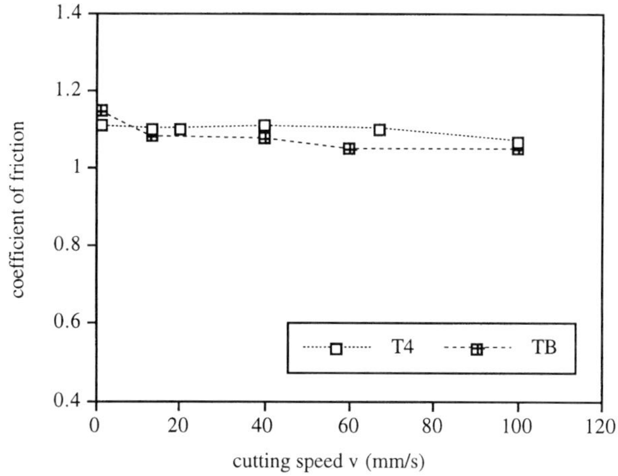

Fig. 4 Variation of μ rake face coefficient of friction with cutting speed V for Al 2014 cut in vacuum with hss tooling.

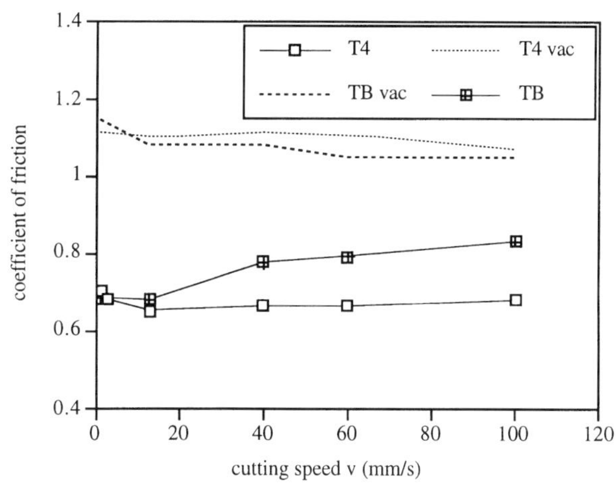

Fig. 5 Variation of μ rake face coefficient of friction with cutting speed V for Al 2014 cut in an environment of 2 kPa CCl_4 with high speed steel tooling: T4 and TB represent different heat treatments.

Figure 5 shows the corresponding data measured under a pressure of 2 kPa of CCl_4 vapour. The curves from Fig. 4 are shown dotted in each of these plots for comparison. At very low cutting speeds both material are well lubricated by CCl_4 vapour and give much reduced values of rake face friction. However, for the alloy in condition TB the lubricating effect of CCl_4 reduces as the cutting speed is increased. This behaviour is in contrast to the results from the same material under condition T4, for which CCl_4 vapour at this pressure retains its lubricating effect throughout the speed range tested (with a friction coefficient of 0.65 compared to a vacuum level of 1 to 1.1). Under 2 kPa of ethanol vapour both materials behaved in a similar fashion showing reduced rake face friction throughout the range of speed investigated, although the values obtained were not as low as those achieved under CCl_4.

Further experiments were carried out using Al 2041-T4 specimens and the same two vapours in which the cutting speed and the environmental pressure were varied independently. Typical results are shown in Fig. 6 for CCl_4. The rake face coefficient of friction was calculated and plotted against lubricant vapour pressure. At each cutting speed, an increase in the pressure of the lubricant vapour led to a decreases in μ which followed a roughly linear relationship on a semi-logarithmic plot. Above a characteristic pressure, the value of μ remained constant (in the cases shown, at ca. 0.67) and this pressure, designated P^*, was determined from the point of intersection of the two linear regimes. The value of P^* provides a significant measure of the efficacy of vapour lubrication since it provides the minimum concentration of vapour to achieve 'full' lubrication. Values of P^* determined in this way for material Al 2041-T4 in environments of CCl_4 and C_2H_5OH are plotted in Fig. 7, from which it can be observed that the way in which this critical pressure level increases with cutting speed varies with the chemical nature of the vapour.

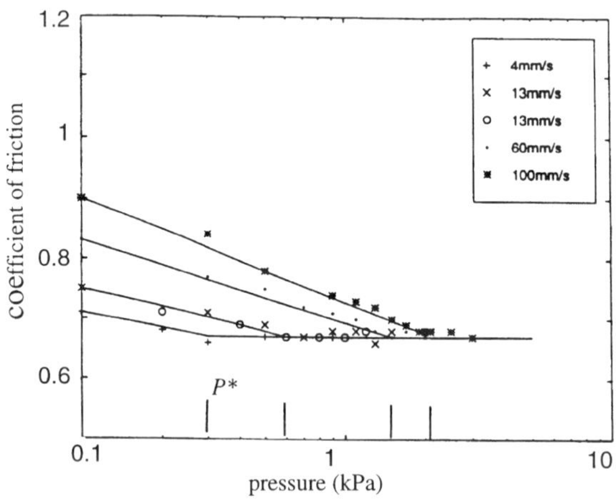

Fig. 6 Variation of μ rake face coefficient of friction environmental pressure of CCl_4 vapour for Al 2014 -T4 cut with hss tooling.

4 DISCUSSION

Reconciliation of the results shown in Figs. 4 to 7 (and of other experiments carried out under similar conditions) with the theoretical models of lubricant activity present serious difficulties. The 'classical' picture of lubricant action on the rake face of cutting tools postulates that the rate-controlling step is the transport or movement of lubricant molecules at the interface between the chip and the tool surface against the direction of chip flow [8-10]. It is assumed that the driving force for this transport is the pressure difference between the surrounding atmosphere and the conditions close to the tip of the tool. 'Full' lubrication is assumed to occur when the rate of transport of lubricant matches the rate of its consumption within the cutting zone by reaction with the freshly produced chip surface. An estimate of this rate can be made by assuming that a monolayer of lubricant molecules will suffice to saturate the chip surface and inhibit interfacial adhesion. The rate of transport will be proportional to the pressure difference between the ends of the channels through which the lubricant moves and to their conductance; this, in turn, can be estimated assuming the flow to lie either in the viscous or the Knudsen regimes, depending on the relative magnitudes of the mean free path of the gas molecules and the dimensions of the channels down which they move. In the case of tetrachloromethane under the conditions examined, Knudsen flow is more likely and, on this basis, it has been shown [9,12] that the expected relationship between P^* and cutting velocity V is of the form

$$P^* = k_2 t V \Theta^{1/2} \tag{2}$$

where t the depth of cut, Θ the temperature within the cutting zone and k_2 a constant for the particular system which will depend on the size and mass of the lubricant molecules and the depth of the cut to which the molecules penetrate under the driving pressure P^*.

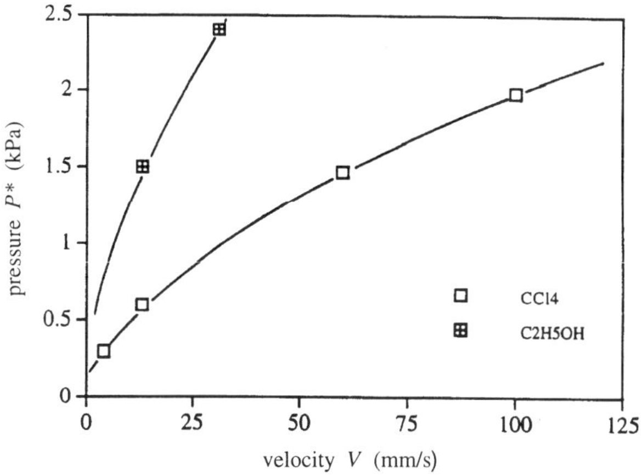

Fig. 7 Variation of critical pressure P^* with cutting speed V for Al 2014-T4 cut in environments of CCl_4 and C_2H_5OH vapour with hss tooling.

In making use of this equation it is important to note that the temperature Θ within the cutting zone will be higher (in some cases significantly higher even at modest cutting speeds) than the bulk temperatures of the workpiece and the tool. Shaw has shown [13] that the temperature rise generated in the cutting process itself can be estimated from the empirical relationship

$$\Delta\Theta \propto V^{0.5} t^{0.3} \tag{3}$$

and we have used this expression to evaluate this effect.

If the experimental values of P^* are plotted against $tV\Theta^{1/2}$, where Θ is the absolute temperature within the cutting zone obtained by adding the appropriate temperature increase $\Delta\Theta$ from eqn. (4) to 293 K, then the points do not lie on a single curve suggesting no simple linear proportionality between the observed values of P^* and the theoretical predictions based on the limiting factor being one of lubricant supply. In this regard these observations confirm those of Wakabayashi et al. [12] who suggested an alternative view of lubricant activity in which it is assumed that the rate-controlling step is reaction with the freshly formed chip surface rather than that of transport at the interface. In such a model, it is supposed that lubrication occurs by the formation of a reaction product (in the case of lubrication by tetrachloromethane, probably a chloride) resulting from chemisorption of the lubricant molecules on the metal surface, and that this adsorption and reaction process is thermally activated with an activation energy Q. Further, if we now suppose that the molecular transport of the lubricant to the reaction zone is sufficiently efficient for the effective local pressure to be equal to the external ambient, then it is possible to generate an alternative expression for P^* to that of eqn (3) viz. [12]:

$$P^* = \gamma \frac{V\Theta^{1/2}}{t} \exp(-Q/R\Theta) \ . \qquad (4)$$

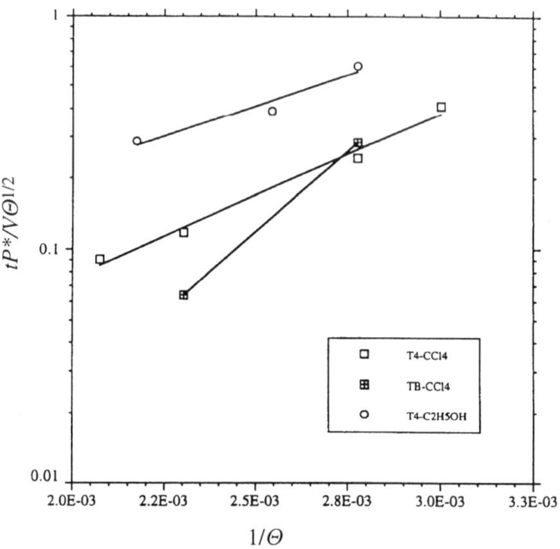

Fig. 8 Dependence of the quantity $P^*t/V\Theta^{1/2}$ on $1/\Theta$ for Al 2014-T4 and TB cut in environments of CCl_4 and C_2H_5OH vapour with hss tooling.

Equation (4) suggests that a plot of $\ln(P^*t/V\Theta^{1/2})$ against $1/\Theta$ should be linear and the experimental data examined above has been plotted in this way in Fig. 8. Reasonable linear fits are obtained for each set: from the slope of the plots, albeit based on a limited set of data points, it is possible to estimate the numerical values of Q, as shown in Table II.

The differences in values of activation energies for the aluminium alloy solution treated for different times would seem to reflect the observed difference in frictional response to the presence of CCl_4 vapour on high speed steel tools illustrated in Fig. 4. It appears that the matrix-coherent $CuAl_2$ inclusions in TB have some form of passivating effect so inhibiting the friction reducing capacity of the lubricant species. Both the lower activation energy associated with ethanol and the similarity of its behaviour with both specimen materials suggest that its effect is due principally to chemisorption rather than reaction.

Table II: Estimated values of activation energies from cutting tests

workpiece material	lubricant	Q kJ/mol
2014-T4	CCl_4	13.4
2014-TB	CCl_4	26.2
2014-T4(& TB)	C_2H_5OH	10.0

The experiments have confirmed the special capability of tetrachloromethane CCl_4 to lubricate extreme contacts between aluminium alloys and steel surfaces which has been long established [14]. Recent work which has involved monitoring the growth of the chloride film on iron surfaces with time and temperature in very carefully controlled conditions has shown that the behaviour of CCl_4 is different in kind, not just in degree, from other chlorinated short-chain hydrocarbons [15,16]. Under more benign conditions, i.e. lower loads and temperatures, the lubricating film produced by CCl_4 on an iron surface is predominantly $FeCl_2$ as with compounds such as $CHCl_3$, CH_2Cl_2 etc. However, with more extreme conditions of load and sliding it appears that a further additional regime of protection is available with tetrachloromethane involving the deposition of an interfacial layer containing both iron chloride and other carbonaceous species [17]. Activation energies under these circumstances for CCl_4 are quoted as being of the order of 38 kJ/mol. While this activity can make a contribution to the effectiveness of CCl_4 on hss tools (even at the modest machining speeds used in work presented here, where bulk temperature rises are not large, the intense interfacial conditions may generate local flash temperatures of the magnitude required) it is clear from the observations made on the effect of the heat treatment given to the specimens of Al-2014 that there are additional effects when the lubricant is available at an interface between two dissimilar metals: the observations suggesting some additional and supplementary mechanism involving the form of the copper alloying constituent.

5 CONCLUSIONS

Experiments involving the low-speed machining of an aluminium alloy (Al 2014) under controlled pressures of tetrachloromethane and ethanol vapours have demonstrated a reduction in cutting forces with increasing vapour pressure until some critical pressure P^* above which no further reduction was observed. The value of this pressure depends on cutting velocity and workpiece temperature.

Previous theories of lubricant activity have assumed the rate controlling step to be that of the transport of lubricant molecules at the chip-tool interface. The observed dependence of P^* on cutting variables in this work cannot be explained on this basis and is more reconcilable with the limiting process being one of chemical reaction or chemisorption of the lubricant species at the interface. The rate of these processes appears to be affected by the heat treatment to which the alloy has been subjected suggesting that the rate of reaction is influenced by the dispersion of the copper alloying element in the aluminium matrix.

Acknowledgements

We are grateful to the Cambridge Commonwealth Trust for the provision of financial support to WYHL and to Marcel Dekker Inc for permission to reproduce data in Figs. 3 to 8.

REFERENCES

1 Rowe, G. W. and Smart, E. F., "Vapour lubrication in friction and low speed metal cutting", Proc. Inst. Mech. Engrs., **181** (Part 3O), 1967, 248-257.

2 Williams, J. A. "The role of lubricants in metal cutting", PhD Thesis, University of Cambridge, 1975.

3 Wakabayashi, T., Williams, J. A. and Hutchings, I. M., "The action of gaseous lubricants in the machining of an aluminium alloy by titanium nitride coated tools", Surface and Coatings Tech., **57**, 1993, 183-189.

4 Rowe, G. W. and Smart, E. F., "Experiments in lubricant breakdown in friction tests in cutting metal on a lathe",. Proc. Instn Mech. Engrs., **179** (Part 3J), 1965-6, 229-241.

5 Shaw, M. C., "The metal cutting process as a means for studying the properties of extreme pressure lubricants", Ann. N. Y. Acad. Sci., **53**(4), 1951, 962-978.

6 Wakabayashi, T, "Studies in the lubricating action and performance of metal cutting fluids", PhD Thesis, Keio University, 1994.

7 Polmier, I. J., "Light Alloys", Edward Arnold, London, 1989.

8 Williams, J. A. and Tabor, D., " The role of lubricants in machining", Wear, **43**, 1977, 275-292.

9 Smith, T., Naerheim, Y. and Lan, M. S., "Theoretical analysis of cutting fluid interaction in machining", Trib. Int., **21**, 1988, 239-245.

10 Cassin, C. and Boothroyd, G., "Lubricating action of cutting fluids", J. Mech. Engng. Sci., **7**(1) 1963, 67-78

11 Mori, S., "The relationship between absorption activity and boundary lubrication", J. Jap. Soc. Lub. Engrs., **33**(8), 1988, 585-590.

12 Wakabayashi, T., Williams, J. A. and Hutchings, I. M., "The kinetics of gas phase lubrication in the orthogonal machining of an aluminium alloy", Proc. Instn Mech Engrs., **209** (3J), 1993, 131-136.

13 Shaw, M. C., "Metal Cutting Principles", Clarendon press, Oxford 1984.

14 Wolfe, K. J. B., Kinman, M. D. and Lennard, G., "A preliminary investigation of the effectiveness of various chlorinated hydrocarbon compounds as cutting oil additives", J. Inst. Petrol., **40**, 1954, 253-256.

15 Kotvis, P. V., Huezo, L., Millman, W. S. and Tysoe, W. T., "The surface decomposition and extreme-pressure tribological properties of highly chlorinated methanes and ethanes on ferrous surfaces", Wear, **147**, 1991, 401-419.

16 Kotvis, P. V., Huezo, L. and Tysoe, W. T., "Surface chemistry of methylene chloride on iron; a model for chlorinated hydrocarbon lubricant additives", Langmuir, **9**, 1993, 467-474.

17 Kotvis, P. V., Lara, J., Surerus, K. and Tysoe, W. T., "The nature of the lubricating films formed by carbon tetrachloride under conditions of extreme pressure", Wear, **201**, 1996, 10-14.

The suitability of food grade cooking oil to act as a biodegradable chainsaw lubricant in forestry and arboriculture

P PRESKETT
Forestry Consultant and Researcher, Bangor, UK

ABSTRACT

Use of mineral oil as chain lubricant is declining in forestry. The alternative offered by suppliers in the UK and Europe is rapeseed oil-based together with a tailored additive package, the presence of which may preclude their use in environmentally sensitive environments. A comparative series of field and laboratory tests of three commercial chain lubricants and a cooking oil is reported here. This led to the conclusion that rapeseed oil of food grade is a suitable replacement for both mineral oil-based and biodegradable lubricants available for chainsaw use as the chain lubricant. Previously reported reductions in consumption were not substantiated.

1. INTRODUCTION.

The current use of mineral oil-based chainsaw lubricants is set to decline in the face of trends in environmental awareness in forestry. The move towards environmentally acceptable practice has been limited by a higher cost compared to mineral oil-based lubricants. The high cost is, in the view of the author, based on studies claiming up to 70% reduction in oil consumed.

Biodegradable oils used as chainsaw lubricants are generally vegetable oil based although some synthetic oils have been studied. It should be noted that chain lubricants differ from two-stroke oils used in chainsaws. Chain lubricant is held in a reservoir and is pumped to the moving chain whereas the two-stroke lubricant is mixed with fuel and held in a separate reservoir. Four-stroke engines are not used in chainsaws. The reasons for the poor market penetration of biodegradable chain lubricants to date has been their significantly higher cost compared to mineral-oil based lubricants and most operators continue to rely on the latter as a lubricant of choice. A further problem is the toxicity of the additives which preclude their use in environmentally sensitive areas such as water abstraction zones, fisheries and wetlands. Additives may be, for example, organo-metallic compounds (e.g. zinc dialkyldithiophosphate) used as anti-wear agents or polyacrylates used as a pour point depressant. The presence or absence of additives is difficult to establish due to commercial confidentiality. Only two companies were prepared to release details of their additives and then only in generic terms. Yet without these additives, it is claimed [1], [2], [3] vegetable oils cannot be expected to perform as well as other lubricants, such as synthetic and mineral oil derived lubricants, which contain similar additive packages.

Additives in cooking oil are food-grade additives. Only the addition of α-tocopherol (vitamin E) at 200 ppm as an antioxidant is made for example to Safeway vegetable oil [4] and the addition of polymethylsiloxane which may be found as an anti-foaming agent in commercial frying oil in conjunction with α-tocopherol [5]. Certainly, the addition of an anti-oxidant would increase the effective operating range of a cooking oil. Temperatures are in excess of 100° C in cooking applications and oil is often retained for future use, passing through many heating and cooling cycles before disposal. Little oil, of whatever base, is retained by a chain moving at 60 mph, hence the term total loss lubrication.

Part of the work discussed will show that additives are unnecessary in vegetable oil-based biodegradable lubricants and mask the ability of the vegetable oils to act as an effective lubricant in their own right.

2. BACKGROUND

As a forestry contractor working in Germany for 10 years and five years UK forestry experience, the author has used professionally both mineral oil- and vegetable oil-based lubricants. Concern over cost of biodegradable chainsaw oil led to the question that if a vegetable oil is used as a base, why should cooking oil not provide the same performance? It seemed to have the same properties and after using cooking oil as a chain lubricant successfully for six years the opportunity arose, through the study reported here [6], to quantify and report its effectiveness. Four oils were selected for a comparative field test: Husqvarna Bio Oil, B&Q 20/50 engine oil, Powerlube mineral oil-based chain lubricant and Safeway vegetable oil (food grade rapeseed oil).

Concern is growing regarding the continued use of mineral based oils in forest harvesting and arboriculture due to their adverse environmental effects. This has led to the development of biodegradable oils for use in environmentally sensitive areas and for operator safety. Biodegradable lubricants have been available for a number of years for both chainsaws and other total loss lubrication systems. In addition, these oils are claimed to have added health and safety benefits for operators. Their introduction has been well taken up in the more environmentally conscious countries in Europe, such as Austria, Germany and Scandinavia.

The UK has recently seen Forest Enterprise (which manages forty per cent of the UK's forests) specifying biodegradable chain lubricant for its in-house harvesting equipment and this requirement is soon to be extended to contractors. Although eminently laudable, there remains a majority of chainsaw users continuing to use mineral oil-based chain lubricant.

Some foolhardy operators will even use waste engine oil. The mineral oil-based lubricants used may be in the form of a cheap, readily available engine oil (favoured by small contractors) or tailored lubricants, with additives to improve performance such as tackifiers to decrease fling-off from the saw. To put the above in context, a chainsaw operator in forestry will use, on average, around 400 litres of chain oil each year in production harvesting.

It is claimed that biodegradable vegetable oil based lubricants reduce oil consumption by forty per cent and it is on this basis, in the view of the author, that the major suppliers of biodegradable lubricants justify the significantly higher cost compared to those of mineral oil-based lubricants. To illustrate this, biodegradable chainsaw lubricant retails at prices of £77.08, £60.00 and £44.41 respectively for Husqvarna, Stihl, and Plantoil per twenty litre containers. Compare this to the same quantity of mineral oil-based chain lubricant and the price differential becomes clear. From Husqvarna and Stihl, respectively, twenty litres retails at £33.84 and £48.00, based on these supplier's 1998 catalogue prices for 25 litres. Engine oil at a major DIY retail outlet may be as low as £15.96 per twenty litres. It is clear that chainsaw operators, seeking to reduce costs will go for the cheapest option. While they may recognise that engine oil is not as good as a tailored oil, sawchains and bars are throwaway items on a chainsaw, and their perception may well be that the service life of these components is unlikely to be greatly affected by the lubricity of the oil. A key feature supporting this application of vegetable oil lubrication to chainsaws is the consumable nature of the cutting attachments and their short service life. In contrast, the cost of twenty litres of the Safeway oil was, in 1998 prices, £11.40.

3. PREVIOUS STUDIES

3.1. Consumption

Studies have reported, <u>in all cases</u>, reductions in oil consumption when using a particular manufacturer's biodegradable oil. Sources have reported a reduction of between 30% and 50% [1], [7], [8] [9] [10]. However, in all these cases, reductions have been achieved by the simple task of reducing the flow rate of the oil delivery pump so that the biodegradable oil was pumped at the lowest setting and the mineral oil-based lubricant was pumped at the highest setting. To add to this, it must be noted that the oil pump is normally adjusted to take account of changing oil requirements for a longer or shorter bar and sawchain combination. A 15" bar and chain need considerably less oil than a 24" bar and chain on the same saw regardless of the oil used. One report even claims a 70% reduction in oil used [11] in studies carried out in Finland. Cutting at the consumption quoted (i.e. between 0.01 and 0.029 litres/m^3 of timber felled, debranched and cross cut and a daily consumption of 1.5 litres) would require an average tree size of 15-20 m^3 - with no branch removal or cross-cutting. From experience, such levels of daily production are not achievable in European forests where the study was based. An average chainsaw operator should be able to achieve an output of 1 m^3 felled, debranched and cut timber for one tank of fuel.

Perhaps more revealingly, the studies noted that the biodegradable oils were twice the price of mineral oil-based chain lubricants and emphasised that this could be offset by the lower consumption. Subsequent reviews and life cycle analyses have consistently reported these "reductions" almost as an article of faith and the claims have become an integral part of the marketing and pricing strategy of biodegradable chain lubricant suppliers. Closer examination show the claims to be based more on semantic interpretation rather than acceptable scientific study.

3.2. Health effects and user perceptions.

Generally comments from the studies were favourable highlighting cleaner clothes and hands [8], another [11] reporting no significant changes in field studies. Both sets of workers reported that the hotter parts of the saw (exhaust etc.) became encrusted when using biodegradable oils. This observation is in agreement with that of [1] which includes an additional recommendation of the regular cleaning of temperature critical components such as engine cylinder cooling fins.

Skin and eye irritation were also apparently reduced [8] although this conflicts with recent evidence [12] suggesting the inclusion of enzymes in the oil to increase the rate of biodegradability on exposure to the sun may also cause severe skin reaction in operators. The lower vapour pressure of biodegradable oil was suggested to reduce inhalation of fumes or mist caused by throw-off from the chain [1]. It was further stated that biodegradable oil is of a similar composition to that of human skin and also claimed a reduction in skin irritation and excema. The use of the biodegradable oil as a hand lotion is also reported [11].

3.3. Feasibility of lubrication with vegetable oils.

Vegetable oils, commonly termed triglycerides, are composed of three long chain molecules (fatty acids) which are bonded to a glycerol group. The ability of a fatty acid to act as a lubricant was demonstrated in 1920 [13]. This work showed that stearic acid was able to reduce the coefficient of friction for a sliding metallic surface with a layer only a few molecules thick. The action of a lubricant in such thin layers was termed boundary lubrication in 1936 [14]. Following on from this work, a mechanism was postulated for boundary lubrication with a fatty acid [15] where the positive polarity on a tail of the fatty acid molecule held an attraction for the negative polarity of a metal surface and that this thin layer kept the surfaces apart. However, a more likely mode of action is thought to be a chemical attack of the surfaces, leading to the formation of metal soaps which acted as lubricants up to their respective melting points [16]. The metal soaps, under conditions of heat and pressure, melt and become the lubricant film between the asperities of the nanometric peaks and valleys in the machined metal faces of sliding surfaces.

4. MATERIALS AND METHODS

4.1. Field tests

The saw used was a Husqvarna 272XP, commonly used in forestry production cutting, fitted with a new 15" Sandvik guide bar and chain for each oil examined. The positive displacement pump was adjusted to the lowest setting and the manufacturer's stated oil tank volume of 400 ml. was verified. New drive sprockets were also used for each bar/chain combination but were not assessed for wear or dimensional change. The oil pump was set to its' lowest flow position for all four oils tested and a new spark plug fitted. To establish weights of bars at the start of the study, the bars were stripped of paint and manufacturer's oils. The dimensions measured at three identical positions across the face of the guide bars and across the guide bars rails in six identical positions where it was considered burring and wear was likely to occur. Measurement was carried out using electronic calipers with a reported accuracy of 0.02 mm. Chains received similar pre-treatment to establish weight and chain dimensions were measured horizontally across the rivet heads on the leading and trailing edges of alternate chain strap links on the right hand side of the chain. This gave 56 data points per chain per tank of fuel.

Eight tanks of fuel and chain lubricant from each oil was used to cut branch wood to firewood dimensions of 30 cms length. The wood was cut under field conditions from a number of 250 year old sessile oaks (*Quercus petrea* L.), felled by storm damage in North Wales in December 1997. Samples were randomly taken from all trees and mean values for wood density and moisture content were calculated together with volumes of wood cut calculated from forestry management tables. Following dismounting of the chain and bar from the powerhead after each tank of fuel and cooling to ambient temperature, dimensional change of the chain and bar were measured. After the seventh tank, chains were not measured as wear was difficult to establish between the sixth and seventh tank of fuel. Final dimensions of the chains were therefore a combination of tanks seven and eight.

Two thermocouples were used; one fitted through the oil reservoir cap and a second inserted between a pair of strap links. These established the operational temperatures of the oil reservoir and chain immediately after the saw ran out of fuel and the chain stopped moving for all eight tanks.

Following this subroutine, the quantity of oil remaining in the reservoir was transferred by draining the oil reservoir over a funnel to a measuring cylinder for approximately ten seconds. This was carried out as the operational design of chainsaws allows for a residual quantity of oil to be left in the tank. A margin of error to allow for cling of the oil on the walls of the reservoir was established prior to the tests by measuring oil dripping from the reservoir after this ten second drain, allowing an accurate calculation of oil consumption per tank of fuel. The volume of fuel was not measured and the fuel tank was simply refilled. Residual oil was flushed from the oil reservoir after each oil tested with petrol and allowed to evaporate from the hot saw prior to the following oil being tested with a new bar and chain. This procedure took advantage of the time taken for cooling to ambient temperature and subsequent measurement of the chain and bar.

After all tests had been completed, the four bars and chains were washed in acetone, and boiled in neat bleach followed by a household detergent solution to remove pitch and extractives from the wood adhering to the chains and bars. Following rinsing and drying in an oven, they were re-weighed to calculate percentage loss of material.

4.1. Laboratory examination

4.1.1. Viscosity

Viscosity of all four oils was established using a Brookfield viscometer, model DV-II+ Calculating Digital Viscometer in a range of temperatures from -20^0 C to +90^0 C. Using a spindle LV1 at 50 rpm, 200 ml of oil was measured into a 200 ml glass beaker having an internal diameter of 65 mm and height of 950 mm. Cooling was provided by a Wobser UK20DL Kryostat below -5^0 C. Above this temperature, an ice bath was used followed by a hot water bath as the temperature range was increased. The spindle and legs of the viscometer were immersed in the cooling or heating baths to minimise heat transfer to the samples. The two mineral oils required the spindle to be changed to LV2 and the speed reduced to 10 rpm at temperatures between -10° C and -20° C, the samples having too high a high viscosity to measure causing an error function being displayed with the LV1 spindle.

Time-dependant, low temperature behaviour of the oils was established by more pragmatic means of storage for 24 hours in a domestic freezer at -25°C with regular observations.

4.1.2. Density

Density was calculated by using, for each oil, a 50 ml volumetric flask, placed on a four point decimal balance and the balance tared. 50 ml of oil was filled into the flask and the weight recorded. Density was calculated by weight over volume and converted to kg/m^3.

4.1.3. Gas liquid chromatography and thin layer chromatography

Gas liquid chromatography (GLC) was used to confirm the fatty acid profile of the Safeway rapeseed oil and the Husqvarna. GLC, although a lengthy process, allowed the identification of the fatty acids from the values of known standards and quantitative calculation of their relative proportions in the samples. When contrasted with the density found for these two oils, it would then be possible to make an assumption as to the presence or absence of additives in the Husqvarna oil. Thin layer chromatography (TLC), a rapid, qualitative technique, was used to confirm the results of the GLC between the two samples.

4.1.3.1 Sample preparation

2 ml samples of Safeway cooking oil and Husqvarna biodegradable oil were prepared by saponification of the oil using 3 ml of 0.5 molar methanolic NaOH. The mixture was heated to obtain an homogenous solution then cooled to yield saponified fatty acids. Added to this was 2 ml of BF_3 methanol, the mixture being then boiled for three minutes, cooled and poured in a separating funnel. 20 ml of saturated NaCl solution and 25 ml of hexane was added and the mixture shaken. After settling, a cloudy precipitate was drawn off and discarded leaving the product of esterified fatty acids held in the hexane. This was drawn off and 1g of anhydrous MgSO4 was added as a drying agent. The mixture was subsequently boiled to drive off the hexane, cooled, filtered and the containers stoppered.

4.1.3.2 GLC

Using a Pye Unicam 204 chromatograph, a 10 μl sample was introduced via an injection port. As the sample passed through a heated column, low molecular weight compounds were rapidly detected, namely the shorter chain fatty acid esters. Retention time in the column was longer for the heavier long chain fatty acid esters. The apparatus was allowed to run overnight to clear the column and the second sample was introduced. The output on a chart recorder was compared with known standards prepared by qualified staff of the University of Wales, Bangor. The conditions of the chromatograph remained unchanged between the samples.

4.1.3.3. TLC

Both samples were spotted on a silica plate to a line drawn 7 mm from the bottom of the plate. The plates were placed in beakers containing a small quantity of hexane as a carrier solvent to a depth of about 4 mm. The bottom of the plate stood in hexane and the beaker was covered. The solvent was allowed to climb the plate until the solvent front was 5 mm from the top of the plate. The plate was removed and placed in a jar containing elemental iodine and left to develop.

5. RESULTS AND DISCUSSION

5.1. Field tests

5.1.1. Guide bar wear

Unexpectedly, the performance of the Husqvarna oil in preventing wear was, on the guide bar, unacceptable for operational use due to severe burring of the guide bar rails. This is suggested to be due to the presence of additives, derived essentially from applications for mineral oil based lubricants, hindering the ability of fatty acids to perform as boundary lubricants. Bar wear was demonstrated by a percentage weight loss to the guide bar of 0.956% with Husqvarna oil. The bar used with cooking oil shows weight loss of 0.239%, a level closer to those found with the two mineral oil lubricants of around 0.18 %.

5.1.2. Chain wear

All four chains tested with each oil were found to have similar dimensional change as the test progressed from 7.1 mm to 6.9 mm. Little variation was found in the rates of wear. Concomitant losses in weight at the end of the tests were also found. Such wear and weight loss is entirely normal for chains. This is surprising however given the weight loss for the bar used with Husqvarna oil. A probable cause suggested is the relative metal hardness of the surface of components. After the field data was examined, hardness values for the components was sought from the manufacturers. The guide bar rails, where the majority of wear was evident, was reportedly induction hardened with a Rockwell (RC) value of 78-81. The strap links were reported to have a RC value 49-52. Unfortunately, further examination of the effect of hardness fell outside the limited time and budget dedicated to this study.

5.1.3. Operational temperatures

Temperatures of the chains and oil reservoirs were also consistent through the test of all oils. A range of mean chain temperatures of 65.9^{O} C for B&Q oil to 71.38^{O} C for the Safeway oil was established. Oil reservoir temperatures did not achieve more than 70° C. The daytime air temperature was recorded during the tests and varied between 13° C and 14° C. Although rapeseed oil begins to polymerise at 60^{0} C, it is considered to be inconsequential given that the time on the chain is measured in fractions of a second.

5.1.4. Consumption

Based on the hypothesis that a cooking oil would lubricate as well as currently available analogues and suggest lower consumption, this study showed that consumption reduction was not of the order of magnitude as suggested by the literature. When considering the results for consumption, it was found that the vegetable oils had mean consumption values of 53% (Safeway) and 56% (Husqvarna) which were observed to positively correlate with values found from the calculated densities of the oils. Likewise, such a relationship was found for the mineral oils with values of 60% (B&Q) and 66% (Powerlube). This suggests the rheological properties of the oils determine the flow rate of the oil with the pump set at a given setting. Furthermore, when the calculated values of wood density (947-1095 kg/m^3) and moisture content (43-50%) were examined in relation to consumption, it was shown that these two factors had not influenced consumption. This was further confirmed by the total volume of wood cut (37m^3) over 32 tanks of fuel. Under the operational conditions of the study, this was around 1m^3 per tank of fuel, typical of a forestry production cutter's output.

5.2. Laboratory tests

5.2.1. GLC, TLC and density

Evidence for the additive content of the commercial biodegradable oil was indicated by the qualitative GLC method which demonstrated that the cooking oil had a near identical fatty acid profile to that of the biodegradable oil examined and confirmed the fatty acid profile of the cooking oil which was given by Safeway stores (Tutill, 1998). This was contrasted with the calculated density of the cooking oil, 917 kg/m^3 which was significantly different to that of the biodegradable oil found to be 938 kg/m^3. This suggests the inclusion of compounds of a molecular weight at variance with that of the constituent fatty acid profile found for cooking oil and Husqvarna oil. To put the above in context, the density for B&Q and Powerlube oils was 879 kg/m^3 and 894 kg/m^3 respectively.

5.2.2. Brookfield viscometer and temperature performance

The viscosity of cooking oil is less than that of a Husqvarna biodegradable oil and the mineral oils at 40^0 C. But at 70^0 C, found to be the average operating temperature of the chain and oil reservoir, the Brookfield viscosity for all four oils examined was around 40cP (centipoise). At 90° C, all four oils had values in a range of 10-20 cP.

With low temperature observations, the viscosity of the cooking oil was found to be 486 cP at -15° C and poured freely but with a cloudy appearance. The other oils (B&Q, Powerlube and Husqvarna) had Brookfield viscosities of 3528 cP, 3930 cP and 1692 cP respectively at -15° C. Visually, the mineral oils displayed a treacly consistency suggesting the change of spindle and speed from -10° C had no significant effect. The Husqvarna oil poured well at these low temperatures but was clouded, although to a lesser degree than the cooking oil.

Time dependant behaviour was altogether different for the cooking oil which over a twenty four hour period, turned to a hard waxy block. This was the effect of progressive waxy crystallisation of the constituent fatty acids which increased after the first eight hours. The remaining oils showed no time dependant behaviour and from observation appeared to have the same characteristics after 24 hours as during the Brookfield viscosity assessment.

6. CONCLUSION AND RECOMMENDATIONS

Shelf life of cooking oils make its use ideal for a small contractor looking to reduce his or her costs and to move away from DIY store cheap engine oil or expensive chain oils. Rapeseed oil, commonly labelled in the UK as pure vegetable oil, has a fatty acid profile suitable for use as a lubricant in chainsaws. The oil pump can stay at the same low rate regardless of whatever oil is used, provided that the bar and chain are suited to the rated power of the saw powerhead. A longer bar/chain combination is most sensibly accompanied by increasing the oil flow.

The operating temperature of the oil reservoir and chain together with the viscosity of cooking oil indicate that it is suitable for use at temperatures above freezing. Unagitated, it forms a hard waxy block through a process of waxy crystallisation in the container after just a few hours below freezing. But for an arboricultural or forestry contractor in UK conditions, this presents less of a problem than for users in operating environments where temperatures fall below freezing for extended periods of time. For a small contractor, a supply of oil is readily available with purchases made at the start of the working day obviating the need for bulk storage. However, oil does harden with age on exposure to air and if the saw is not used on a regular or daily basis it can become sticky on the bar and chain. In extreme cases after long term storage (>1 year), it may seize the oil pump. However this can be avoided by running one tank of mineral oil through before long term storage or flushing with petrol. Note also that the use of used cooking oil cannot be recommended due to higher levels of peroxide compounds, products of free-radical chain reactions in oil exposed to high temperatures in air.

Tests of biodegredation focus on the primary breakdown products and look for either methane of carbon dioxide given off as micro-organisms "feed" off the oil. What remains is not considered. Complex synthetic polymers and organo-metallic compounds make up these additive packages and may be up to 10% of the oil. If consideration has to made of the environmental impact of the oils adequate information on oil additives needs to be made available and the use of local environmental impact assessments may prove to be of use to the arboriculturalist as much as they are to forestry managers. More importantly, the issue of user health has to be faced and the use of risk assessments require much fuller information to be effective than is currently afforded to managers of chainsaw operations. If food grade cooking oils are to be deployed in environmentally sensitive operations in total loss applications, standard tests of biodegredation as well as other factors such as toxicity, allergenic properties and seal compatibility need to be fully examined and openly reported on. This will ensure acceptability from purchasers and potentially extend the application to fully mechanised harvesting where tree felling and conversion to products are undertaken by dedicated, purpose built machinery using considerably greater quantities of chain lubricant daily.

The use of vegetable oils was commonplace as a lubricant prior to the use of mineral oils and it seems that our search for improvements on natural products has, in this instance, been based on additive technology derived from mineral oil-based lubricants. Here it is the rôle of tribologists, using their multidisciplinary science to develop compatibility between the two opposing factors of safety and efficacy to develop an acceptable product for both the operator and the environment. The final conclusion is that cooking oil is a suitable and effective replacement for all currently available chainsaw lubricants.

ACKNOWLEDGEMENTS

Gratitude is expressed for support and encouragement of the Biocomposites Centre, University of Wales, Bangor. Thanks are also due to the School of Agriculture and Forest Sciences, University of Wales, Bangor, Sandvik Saws who supplied the chains and guide bars and to Husqvarna and Plantoil for the supplying some of the lubricants used.

REFERENCES

[1] Makkonen, I., "Environmentally compatible oils", FERIC Field Note General 39, Forest Engineering Research Institute of Canada, Eastern Division, Quebec,. 1994.

[2] Harald, S., "Vegetable oils as lubricants" in Proceedings, Gateway to Renewable Industrial Feedstocks-GRIF '97 held at the Central Science Laboratory, Sand Hutton, York. ACTIN, Leatherhead, 1997, Paper 3.

[3] Harald, S; Lal, K; Lawate, S., "Varieties of rapeseed oil and derived products for use in fuel and lubricants" in; "Rapeseed Today and Tomorrow". Proceedings, 9th. International Rapeseed Congress, vol. IV. Organising Committee for the 9th. International Rapeseed Congress. 1995.

[4] Tutill, C., Personal communication. Product Quality Assurance, Safeway Stores, Hayes, 1998.

[5] Kristott, J., Personal communication, Food Research Association, Leatherhead, 1998.

[6] Preskett, D., "On the suitability of food grade cooking oil as a saw chain lubricant-comparative field tests and laboratory analysis", Unpublished BSc. thesis, School of Agriculture and Forest Sciences, University of Wales, Bangor, 1998.

[7] Jokai, M., "The use of vegetable oil for chain and bar lubrication", FERIC Field Note General 35, Forest Engineering Research Institute of Canada, Western Division, Vancouver, 1993.

[8] Gardh, R.; Halfvarsson, R., "Vegetable chain oil is better than mineral oil", translation of Skogforsk Resultat (undated), supplied by Dr. Pat Wightman, SAC, Aberdeen, 1998.

[9] Harding, J., "Biodegradable chain oil" Forest Authority, Technical Development Branch, Ae Village, Dumfries, 1994.

[10] Rajamaki, J.; Vuollet, E., "On the use of vegetable oils for lubricating chain saw chain" (abstract), Metsa teho-Katsaus, 13/94, 1994, p. 4.

[11] Kytö, M., "Biodegradable vegetable oil-based lubricants in timber harvesting. A field test", Technical Research Centre of Finland, Laboratory of Fuel and Process Technologies, 1993.

[12] Potter, P., Personal communication, Dryad Furnaces, Machynlleth, Wales, 1999.

[13] Langmuir, --1920 (*no title given*), cited in Bowden and Tabor, (*ibid*).

[14] Hardy, W.B., "Collected Works", Cambridge University Press, Cambridge, 1936.

[15] Bowden, F.P. and Tabor, D., "Friction and lubrication",. Methuen, London, 1956.

[16] Bowden, F.P.; Gregory, J.N. and Tabor, D., "The effect of temperature on boundary lubricants". Nature, 156:97, 1945.

Wear and friction in boundary lubricated contacts

E KASSFELDT
Division of Machine Elements, Luleå University of Technology, Sweden
T NORRBY
Statoil Lubricants Research and Development, Nynäshamn, Sweden
J RIEGLERT
SCA Packaging Munksund AB, Piteå, Sweden

SYNOPSIS

In this study, the anti-wear performance and friction characteristics of a synthetic ester (SE) and a polyalphaolefin (PAO)-based lubricant in boundary lubricated, highly loaded sliding steel-steel contacts was examined. In addition to tests on the pure base fluids, tests with two types of anti-wear additives were also carried out, for a total of six samples.

The friction force was measured continuously during the experiment, and the wear was evaluated after the test. The wear scars were examined with Scanning Electron Microscopy (SEM) recordings. Comparable boundary lubrication conditions are to be found in hydraulic pumps and motors. Surface analysis showed that all ester mixtures resulted in very smooth wear scars with no sign of smearing at the edges, nor scuffing. All polyalphaolefin mixtures except pure polyalphaolefin also resulted in very smooth wear scars. Some signs of smearing at the edges of the wear scars were detected for the polyalphaolefin. The wear measurements with fluids containing additives showed that the use of synthetic ester resulted in higher wear rates of the steel surfaces compared to polyalphaolefin, irrespective of the type of anti-wear additive used. Friction measurements during the wear tests showed that the ester mixtures gave lower friction compared to polyalphaolefin, although the wear rates were substantially higher.

The increased (adhesive) wear could be a result of increased competition for surface sites, leading to less surface coverage by additive molecules; or increased (chemical) wear by corrosive action of the lubricant molecules; or to lowered chemical reactivity due to increased solvatisation of the additive molecules, inhibiting the formation of surface reacted layers.

1 NOMENCLATURE

Nomenclature, symbols and abbreviations used in this paper.

CI	Confidence interval
H	Indentation hardness, [N/m^2]
K	Wear coefficient
Q	Wear volume/sliding distance, [m^2]
R_q	Surface roughness, [μm]
W	Normal load, [N]
a, b	Index for surface a and b
h	Film thickness, [μm]
k	Number of means
min	Minimum
n	Number of tests
q	Statistical range
Λ	Film parameter
α	Significance level
ν	Degrees of freedom associated with s
IIDN(0, s^2)	Identically, independent distributed in a normal distribution with zero mean and variance s^2.

2 INTRODUCTION

Growing environmental awareness has increased the use of environmentally adapted lubricants (EALs) in many applications. To reduce the possible environmental impact of leakage, the forest industry in Sweden now strives to run all their hydraulic machinery with such fluids [1]. In mobile hydraulic systems much effort has been put into reducing the weight of component and to reduce the flow rates of hydraulic oil. This is often achieved by minimising the piston area. However, in order to maintain the actuators load capacity, the system pressure has to be increased which results in higher system temperatures and greater demands on pump design.

The objective of the work presented in this paper was to compare the anti-wear performance of a synthetic ester (SE) and a polyalphaolefin (PAO) based fluid in boundary lubricated contacts. Each base fluid was tested in its pure form, as well as blended with two similar types of anti-wear additives with differing chemical composition. Earlier work has shown that there is a difference in the anti-wear characteristics of synthetic esters and mineral oils [2]. Van der Waal in 1985 [3] introduced the Non-Polarity Index (NPI) in an effort to better explain differences between important base fluid related parameters and properties. One conclusion from van der Waal's work is that a difference in wear rate may be explained by interference of the polar ester groups with anti-wear additives, for esters with very non-polar characteristics (NPI 40-70). If the ester groups are too polar, they, and not the anti-wear additives, will preferentially cover the metal surfaces. This can lead to an increase in wear rate under severe conditions, because of the low anti-wear efficiency of the polar ester groups blocking the surface. However, in a hydraulic fluid of ISO VG 46, the ester base fluid has a less polar, more oil-like NPI (ca. 185), and so direct competition for surface sites is not likely to be the main explanation for increased wear in contact lubricated by "oil-like" esters.

Rather, solubility (and *solvatisation*) of additive molecules, and the resulting *activity and chemical reactivity*, paired with other parameters like better heat transfer will alter the conditions in the contact to a significant extent. One possible effect of this could be that the additives *remain passive*, due to lowered reactivity and lower contact temperature, in a wear regime where the additives should snap into action. When metal surfaces come into contact, welding occurs at the asperity contacts with a resulting increase in friction and wear. This is in contradiction to study [2], where the coefficient of friction was still lower for the rapeseed oil, synthetic ester, and mixture of rapeseed and synthetic ester, compared to the mineral oil, even though the wear was substantially higher. Also in this study, wear and friction are not co-variant for all fluid/additive combinations, and *substantial wear* is found also where the friction is *low*.

3 BASE FLUIDS AND ADDITIVES

The fluids tested were combinations of two different base oils and two different anti-wear additives. The anti-wear additives were based on nitrogen-, sulphur- and phosphorus-containing organic molecules. The pure base fluids were also tested without any anti-wear additives. The different mixtures are presented in Table 1. The SE base fluid was an ISO VG 46, and the PAO-based fluid was blended to ISO VG 46 as well. The same amount of additives was added to each of the base oils. Additional base oil data is presented in Table 2.

Table 1. Tested fluids.

Mixture code	Base oil	Anti-wear additive
PAO-A	Polyalphaolefin	A (S, high P)
SE-A	Synthetic ester	A (S, high P)
PAO-B	Polyalphaolefin	B (high S, P)
SE-B	Synthetic ester	B (high S, P)
PAO	Polyalphaolefin	None
SE	Synthetic ester	None

Table 2. Physical data for tested fluids.

	Polyalphaolefin	Synthetic ester
Density @15°C [kg/m^3]	829	920
Viscosity @40°C [mm^2/s]	46	46
Viscosity @100°C [mm^2/s]	11.6	9.5
Non-polarity index	215*	185

*estimated for a 75:25 mixture of PAO and ester base fluid.

4 EXPERIMENTAL METHODS

Wear and friction measurements were carried out in a Plint and Partner High Frequency Friction Machine (TE 77B). Surface characterisation was carried out with the help of a SEM and using a Stylus profilometer.

4.1 Wear measurements

The wear characteristics of a steel-steel contact were established from tests carried out with the contact area flooded with oil and the oil bath held at constant temperature. In all tests the contact geometry was a line contact of the cylinder-plate type, created by using the flat end of a ø10 mm roller as the plate and a ø6 mm roller as the cylinder. The steel cylinder and plate were standard bearing rollers with a hardness of 850 Vickers and surface roughness (Ra) of about 0.03 μm. The machine moves the steel cylinder in a reciprocating sinusoidal motion relative to the plate, with a amplitude of 2.5 mm and frequency of 5Hz, for a total of 22 hours. Figure 1 shows the main parts of the test apparatus and the contact geometry. The wear scar on the cylinder was more pronounced compared to the wear scar on the plate, and therefore chosen for analysis. The wear scar on the cylinder from each oil test was measured and the volume of material removed calculated. In all tests the temperature was held at 80 °C, the maximum relative sliding speed was 0.08 m/s and the normal load was 200 N, corresponding to a maximum pressure of 780 MPa according to Hertz. However, this pressure is only present in the beginning of the wear test due to the growing wear scar, which gradually increases the contact area and thus lowers the contact pressure.

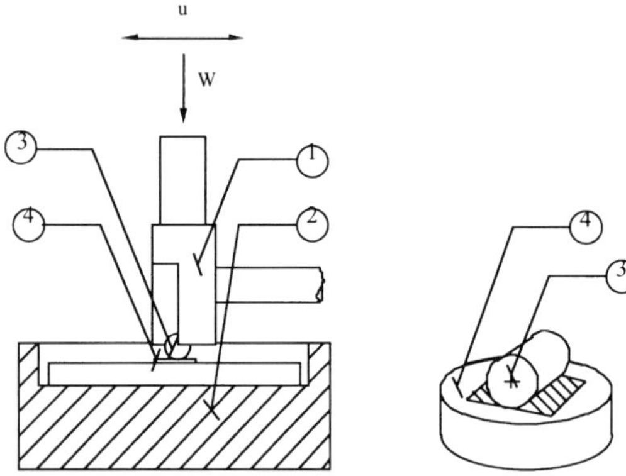

Figure 1. Moving specimen carrier (1) with upper specimen (3) and oil bath (2) with fixed specimen (4).

The film parameter was calculated in order to determine the lubrication regime at the beginning of each wear test, equation 1 [4].

$$\Lambda = \frac{h_{min}}{\sqrt{R_{q,a}^2 + R_{q,b}^2}} \qquad (1)$$

The film parameter was 0.1, which indicates that boundary lubrication was the dominant lubrication mechanism.

4.1.1 Wear

To determine the severity of the wear, the Archard wear equation was applied and the wear coefficient calculated, equation 2 [5].

$$K = \frac{QH}{W} \qquad (2)$$

In order to compare the wear coefficient resulting from the different oil mixtures, Tukey's paired comparison procedure was used, equation 3 [6].

$$CI = K_i - K_j \pm \frac{q_{k,v,\alpha}}{\sqrt{2}} s \sqrt{\frac{1}{n_i} + \frac{1}{n_j}} \qquad (3)$$

The method compares the mean wear coefficient for all the oil combinations, and indicates whether any real differences in the mean wear coefficient for the different oils are present.

4.2 Wear scar characterisation

The wear scars were examined in a SEM and using a Stylus profilometer. The SEM gives important visual information about the kind of wear mechanism that has occurred whilst the profilometer gives standard numerical values of the surface roughness. The edge of the wear scar was examined in the SEM in order to detect any tendencies of smearing and/or changes in surface roughness between the wear scar and the undamaged surface. The surface roughness was measured perpendicular to the sliding direction for both for the wear scar and the undamaged surface.

4.3 Friction measurements

The friction force was measured continuously during each test. The test equipment measures the friction force using a Kistler 9203, piezoelectric force transducer. The force amplifier gives a DC voltage corresponding to the rms value of the friction force during one stroke cycle, with a resolution of 0.001 N. The friction force was sampled using a datalogger once every minute. The mean coefficient of friction and the difference between the maximum and minimum coefficient of friction for each test was calculated and compared for the different oils using the statistical method described in section 4.1.1.

5 RESULTS

The results presented consist of wear and friction measurements followed by surface analysis of the wear scars.

The contact pressure at the beginning of each test was about 780 MPa. Since the wear scar expands during wear, the contact pressure reduces. The calculated mean contact pressures at the end of each test are presented in Table 3.

5.1 Wear

Figure 2 shows the mean wear coefficients, K, for the tested oils. Table 3 shows the mean wear coefficients, K, for the tested oils, the number of measurements per oil, and the individual standard deviations.

Tukey's paired comparison procedure, and many other statistical methods, are based on the assumption that the errors are independent and identically distributed in a normal distribution with mean zero and variance s^2, or "IIDN(0, s^2)" written in statistical terms. Error analysis showed dependence on the mean value of wear coefficient. Furthermore, the errors were not normally distributed.

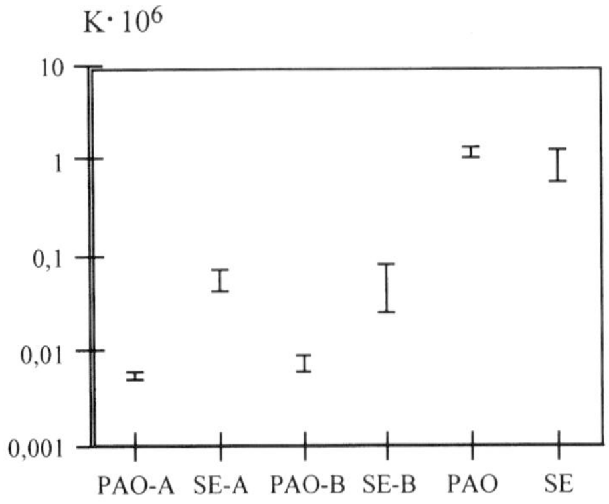

Figure 2. Mean wear coefficient, K, with 95% individual confidence intervals.

Table 3. Analysis of wear.

Mixture code	$K \cdot 10^6$ *	Standard deviation$\cdot 10^6$	Mean contact pressure at end of test (MPa)	Ra-value
PAO-A	0.0055	0.00081	328	0.023
SE-A	0.056	0.026	145	0.033
PAO-B	0.0075	0.0020	274	0.023
SE-B	0.053	0.035	108	0.036
PAO	1.2	0.17	42	0.141
SE	0.91	0.43	47	0.015

*Average of 4-5 measurements.

The dependence on mean wear coefficient can be explained by the fact that oils that gave high wear also showed a large spread (standard deviation) in mean wear compared to oils that gave little wear.

In order to use Tukey's paired comparison procedure, the measurements must be transformed in some way so that the errors are IIDN(0, s^2). From analysis of the standard deviations, it was decided to use the $\log_{10}$ transform as a variance stabilisation transformation. A new analysis of variance with Tukey's paired comparison procedure was conducted with the transformed wear coefficient responses.

From the analysis it can be seen that there were no differences in wear coefficients between PAO-A and PAO-B, between SE-A and SE-B, and finally between PAO and SE. PAO-A and PAO-B resulted in lower wear coefficients compared to SE-A and SE-B, and much lower wear coefficients compared to the pure fluids, PAO and SE. SE-A and SE-B had lower wear coefficients compared to PAO and SE.

5.2 Friction

Figure 3 shows the coefficient of friction (μ) during each wear test for the six samples.

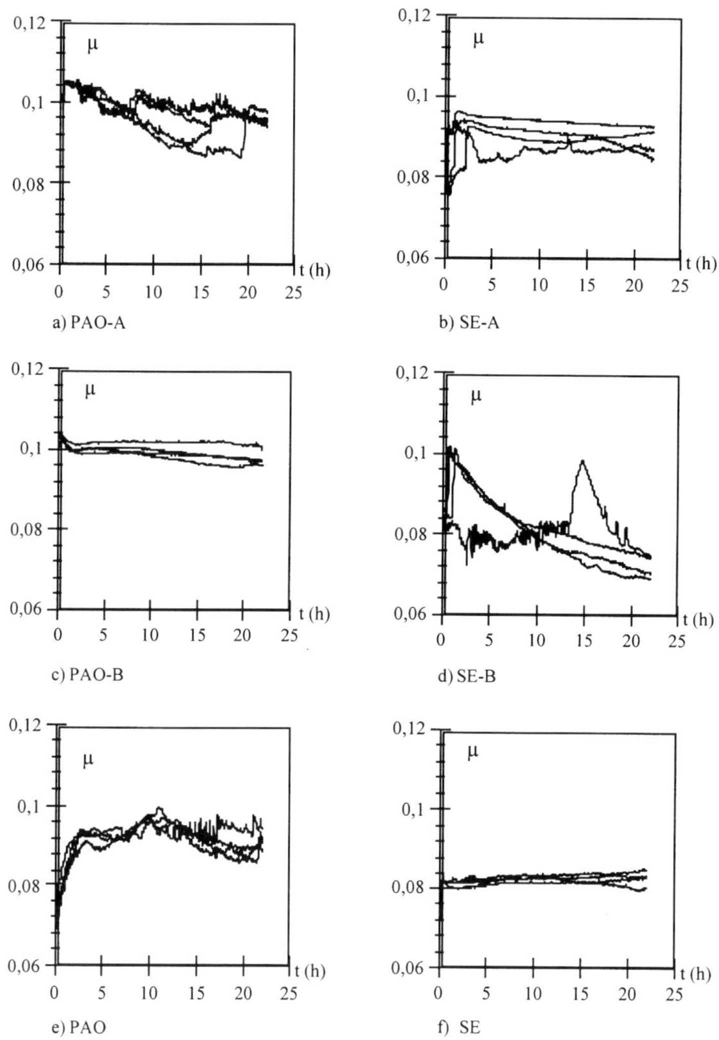

Figure 3. Coefficient of friction during the wear test for a) PAO-A, b) SE-A, c) PAO-B, d) SE-B, e) PAO, f) SE.

All fluid mixtures containing anti-wear additives showed a very rapid increase in friction at the beginning of the tests. This is followed by a slow decrease in friction and, for some oils, further fluctuations in friction occur. The pure ester gave a very constant coefficient of friction in spite of the growing wear scar. This indicates that no transition in lubrication regime occurs, i.e. boundary lubrication was still dominant at the end of the test.

5.3 Surface analysis

The SEM pictures in Figure 4 show a small part of the worn cylinder specimens. The left hand side of each picture is the undamaged surface and the right hand side a part of the wear scar.

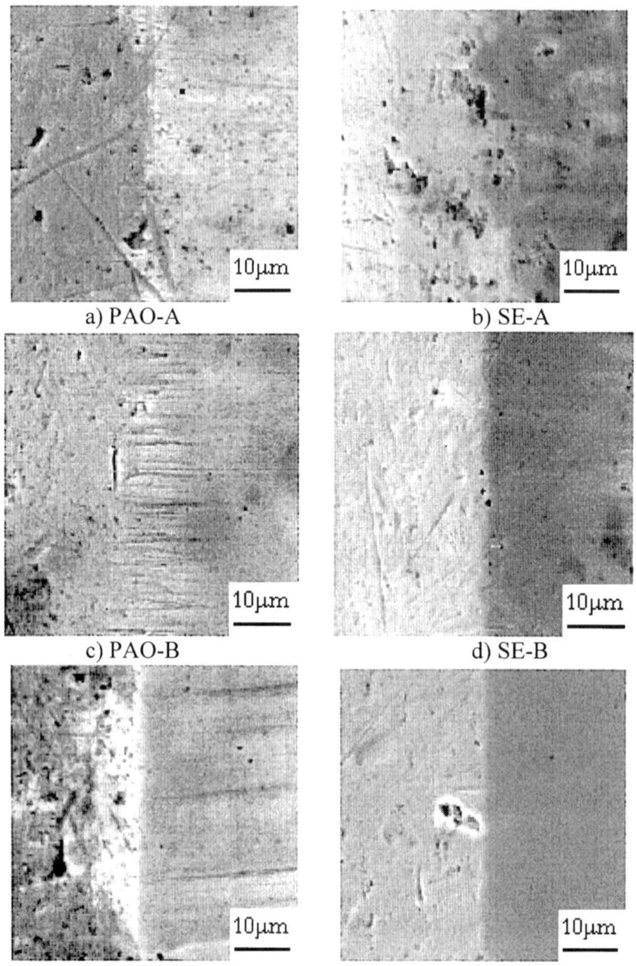

Figure 4. SEM pictures of wear scar edges for each test fluid.

The surface roughness measurements in Table 3, show that PAO-A and PAO-B result in smoother wear scars compared to SE-A and SE-B, and are equal in roughness to the undamaged surface. Pure ester clearly results in the smoothest wear scar whilst PAO gave the roughest wear scar.

A concurrent study of the SEM pictures and the surface roughness measurements show that:

1. The use of **PAO-A** results in a very smooth wear scar with no sign of scuffing or smearing at the edge of the scar. The edge is somewhat more irregular compared to the edges of the larger wear scars. Sharper edges are, however, created with large wear scars due to the larger difference in slope between the wear scar and the undamaged surface.
2. **SE-A** gave a smooth wear scar. The edge showed a very irregular dark and light area. This is, however, thought to be an anomaly and is therefore not analysed further.
3. **PAO-B** results in a very smooth wear scar with no signs of scuffing. The scar edges showed signs of smearing, probably resulting from sheared anti-wear additive boundary layers.
4. **SE-B** gave a smooth wear scar. The edge is very sharp, with no signs of smearing or scuffing.
5. **PAO** results in a locally smooth wear scar. Local seizure and signs of abrasive wear are also present.
6. **SE** results in an extremely smooth wear scar with no sign of scuffing or smearing at the edge of the scar.

6 DISCUSSION AND CONCLUSIONS

The different combinations of base fluids and additives all give different results, see Table 4. Generally, the PAO-A/B are better performing than any of the other combinations, with regards to wear results, with both additive combinations. PAO-A gives the best protection against wear, and PAO-B gives somewhat more wear, but a more constant friction. This indicates the beneficial formation of a readily sheared sacrificial layer, presumably formed from the high-sulphur additive and the metal iron surface. In SE-A, it appears that the additives work better under the test condition than in SE-B, where larger variations in wear scar width is found. Also, the friction shows larger fluctuations than in SE-A. For the pure base fluid samples SE and PAO the amount and type of wear damage is much more extensive.

Table 4. Summary of results and observations

	Wear	**Contact pressure at end of test**	**Friction**	**S***	**P***	**Comments**
PAO-A	Low	High	High, declining	Lo	Hi	Good protection, lowered friction after running-in
SE-A	Medium	Relatively high	Medium, constant	Lo	Hi	Friction varies between tests
PAO-B	Low	High	High, constant	Hi	Lo	Formation of sacrificial layer
SE-B	Medium	Relatively high	Strongly declining	Hi	Lo	Formation of sacrificial layer
PAO	High	Low	Medium, constant	-	-	Marked running-in behaviour
SE	high	Low	Low, constant	-	-	Chemical corrosion?

*Sulphur and phosphorous levels, based on additive treat rate and typical concentration

One additional mechanism for wear in case of SE might be caused by free fatty acids, liberated from the ester base fluid in a decomposition reaction, during harsh conditions in the contact. Free fatty acids would act as friction modifiers, giving low friction. The friction modifier action involved the formation of metal soaps, which constitutes a form of chemical corrosive wear. This effect only appears in the additive-free SE, and is not seen in PAO although that sample is a 25:75 mixture of ester and PAO. Thus, some differences in the contact conditions must be sufficient to allow this mechanism to operate only in pure ester.

Obviously, *any* of the additive combinations are better than no additives at all. The origin of the "graded response", found for the other four combinations, suggest that there is room for improvement in the composition of additive packages. Also, the chemical interaction between the ester base fluid and the two different additive combinations overall result in a less successful anti-wear performance. Input from studies like this should be taken into account when designing future generations of additives, specifically development for use in environmentally adapted lubricants and other fluids based on unconventional base oils (UCBOs). We believe that the performance of esters may be improved by the right additives, as the formation of readily sheared sacrificial layers is one important mechanism for wear protection in this type of sliding contact.

7 REFERENCES

[1] Norrby, T. and Kopp. M. "Environmentally Adapted Lubricants in Swedish Forest Industry- A Critical Review and Case Study", in Proceedings of the 12th International Colloquium-Tribology 2000-Plus , Vol 1, Ed. Bartz, W. J., Technische Akademie Esslingen, Ostfildern, Germany, 2000, pp. 161-168.

[2] Rieglert, J. and Kassfeldt, E. "Performance of Environmentally Adapted Hydraulic Fluids at Boundary Lubrication", in Proceedings of the 23rd Leeds-Lyon Symposium on Tribology, Vol 32, Ed. Dowson D., Elsevier, Amsterdam, Netherlands, 1997, pp. 467-474.

[3] Van der Waal, G. "The Relationship Between the Chemical Structure of Ester Base Fluids and their Influence on Elastomer Seals and Wear Characteristics", J. Syn. Lubr. 1(4), 1985, pp. 280-301.

[4] Hamrock, B. J., "Fundamentals of Fluid Film Lubrication", McGraw-Hill, New York, 1994 (ISBN 0-07-025956-9).

[5] Hutchings, I. M., "Tribology: Friction and Wear of Engineering Materials", Arnold, London, 1992 (ISBN 0-340-56184-X).

[6] Box, G. E. P., Hunter, W. G. and Hunter, J. S., "Statistics for Experimenters : an Introduction to Design, Data Analysis, and Model Building", Wiley, New York, 1978 (ISBN 0-471-09315-7)

Test Methods

Abrasion test to improve lifetime of excavator teeth

E FERNÁNDEZ, Mª R FERNÁNDEZ, V MERAYO, and **J M CUETOS**
Faculty of Engineering, Oviedo University, Gijón, Spain

ABSTRACT

The cost of wear of small parts and components by abrasion represents a considerable economic factor in mining industry. The supply of surface treated materials is also very high and generally, the only way to select them is by testing on real working conditions which constitutes a long and costly process.

In this paper some coating materials has been analysed in the laboratory and the results contrasted with excavator teeth under real conditions of work. As a result, it has been possible to establish of a method to select materials in a faster and more economic way.

1. INTRODUCTION "EXPLAIN TRIBOLOGY" THE NATURE OF THE PROBLEM.

The abrasive waste produces the premature failure of many components in the extraction machinery in the mining industry in general, with an important economic cost. In this failure different wasting processes are involved, according to the nature of the abrasive material, the type of loads, and the situation in which the work is carried out.

The direct impact of metallic components with the floor requires the employment of alloys that have, at the same time, good characteristics of ductility and resistance to the abrasion. High values of superficial hardness are needed in those surfaces over which the extracted material moves and even higher value of hardness in the materials of the equipment to mill minerals. Due to these reasons, the necessity to make a careful when selecting a resistant material to the waste (Roda, 1992).

The abrasive nature of most of the minerals produces a significant waste in machines that transport or process this material. This project seeks to give answers to the practical problems of the elevated waste rates that are presented in the excavator teeth of the mining excavators and in the chutes of the mills. The materials that are normally used in the excavators teeth are hard alloys that don't always possess enough anti-wear properties (Olson et al, 1995); therefore, in this work it is sought to evaluate the characteristics of these alloys.

In order to find a solution to this problem, several cast iron alloys such as chromium-niobium, chromium-niobium-vanadium, chromium-niobium-molybdenum-tungsten, and boron alloys have been tested through Standard ASTM 105-89.

2. ABRASION TESTS TO IMPROVE LIFETIME OF AN EXCAVATOR TEETH.

Finding a method of lengthening the lifetime of the teeth of an excavator resulted in a research project between the University of Oviedo and Endesa Company which tried to find materials that could be used in abrasive conditions. The research plan had two main stages:

a) Carrying out an abrasion test in the laboratory through a standard testing machine. In this case the standard testing method for abrasion ASTM G105-85 was selected, which use wet sand and a rubber wheel.

b) Later when the laboratory tests had been finished a further, test was carried out in an opencast mine.The materials were tested in real working conditions in order to validate the experimental results.

3. MATERIALS OF THE TREATED SURFACE AND METHOD OF APPLICATION.

The toothed excavators work both with, and without load; so, for this equipment it is necessary to choose a tough steel to withstand cutting stress, and enough hardness to withstand considerable wear. The steel must be of a type that can be welded into special parts to protect against abrasion wear (Roda, 1992). Two types of materials were selected:

a) Steel alloys of chromium, vanadium, niobium.

b) Coated surfaces formed from a welding ferrite matrix which contain tungsten carbide of varying shapes and sizes.

The teeth used in our case are similar to the shape of a "root" as we can see in figure 1. The coating was carried out with semiautomatic electric welding machine according to the manufacturer specifications. The welding process was carried out twice. Firstly it was done directly over the base metal so it had a high percentage of dilution and was not representative of the real composition of the coating material. In the second layer dilution problems were not presented, especially withcarbon. It is this second layer that is really tested by abrasive wear.

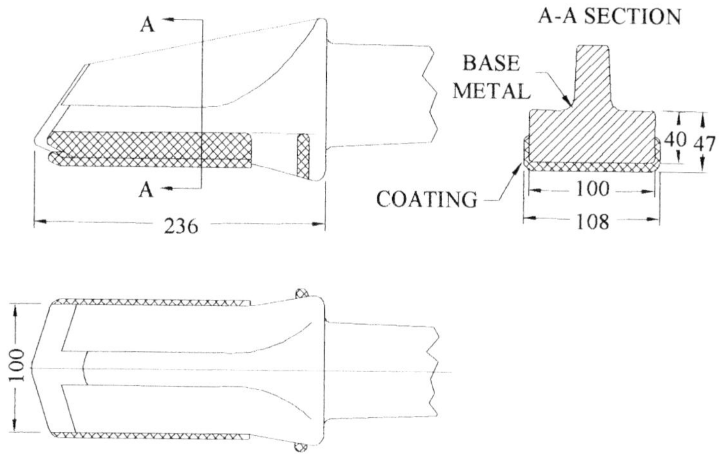

Figure 1: Scheme of one tooth of excavator.

3.1. Materials Chosen.

The materials chosen to make the surface extra-hard and to give it high resistance to wear can be divided into five groups:

- Cast iron alloyed with chromium and niobium.
- Cast iron alloyed with chromium, niobium and vanadium.
- Cast iron alloyed with chromium, niobium, molybdenum and tungsten.
- Cast iron with boron.
- Treated surface in carbide of tungsten in a Nickel or ferrite matrix.

Preliminary test (hardness, micro-hardness and metalographic analysis through optic microscope) were carried out. The material MR9 was analysed with wave electronic microscope.

Hardness tests were carried out with a Durometrer MACROMET in scale Rockwell C (load 150 kg), and micro-hardness with a microdurometrer MICROMET in scale Vickers (load 50-200 grams).

The average of three measures carried out at the same level for each layer was registered as the macro-hardness value. In order to measure the micro- hardness 5 measurements were carried out in the eutectic matrix and also in the carbide. We do not have information about micro-hardness in the case of material MR5. Tables 2 and 3 shows the obtained values.

Table 1: Selected materials for rehearsals.

Code	Alloy (%)	Characteristic	Hardness (HRC)
MR1	20% Cr – 6 % Nb	High abrasion resistance	60.8
MR3	11% Cr- 6.5% Nb – 6% V	Good performance at high temperatures due to the vanadium presence.	64.7
MR4	Boron alloy	It doesn't present differentiated borides although this element provides an significant increase of the hardness.	65
MR5	22 % Cr – 7% Nb - 3% Others	It should be pointed out the increase of hardness that produces the presence of dispersed carbides of niobium in front of that of the chromium carbides which are present in practically all these alloys	62.1
MR8	20.5% Cr -6.5% Nb <Si Mn		62.7
MR9	20.2 Cr 6.1 Nb 6.8 Mo 1.8W	The presence of molybdenum and tungsten carbides doesn't seem to contribute to a significant increment of hardness, at least at room temperature, in front of the type of MR5 alloys.	62.7
MR14		Small content in carbon, alloyed with Cr and addition of small percentages of Mo, B and V.	61.5

Table 2. Macro Hardness (HRC)

Material	Layer 1	Layer 2	Material	Layer 1	Layer 2
MR1	57.1	62.2	**MR8**	61.6	62.7
MR3	66.4	66.6	**MR9**	61,2	62,7
MR4	64,1	65,6	**MR14**	64.3	66.1
MR5	58,8	62,1			

Table 3. Micro Hardness (Vickers 0.3 N).

Material	Eutectic	Carbides	Material	Eutectic	Carbides
MR1	622				
MR3	593/806	1065	**MR9**	624,4	845,5
MR4	777,0	1002	**MR14**	863	1045

4. TEST ABRASION.

To carry out this abrasion test, a method using vulcanised rubber and wet sand was chosen, Standard ASTM G105-89, in which steel wheels with different degrees of elasticity and different hardness (SHORE) turn against the test sample under a constant load.

The test time is predetermined and it is constant for the different wheels. At the end of each test the weight loss was measured. The abrasive chosen for the abrasion test was silicon sand. The test samples were prismatic blocks 54 mm long, 20 mm wide and between 20 and 24 mm thick. They were cut from steel plates (ST-52) 120x100x15 mm which were previously treated twice. These treated blocks were cut with a metal-cutting saw and planed in the treated surface to make the surface flat and smooth. The original plate and the state of one tested specimen can be show in figure 2.

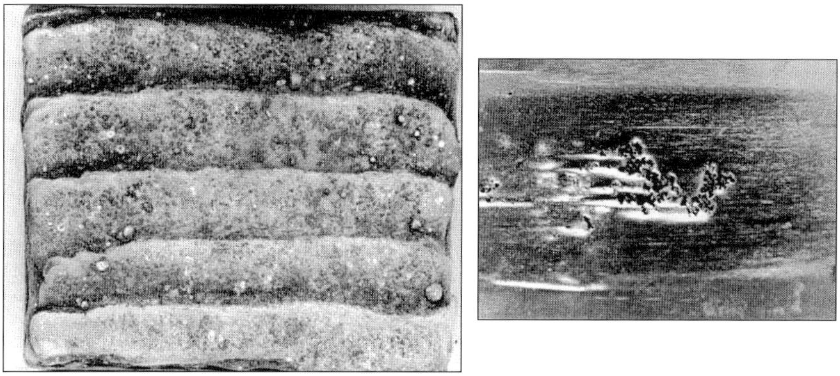

Figure 2: Plate of which 6 test specimens are obtained and superficial state of one of them after the test.

4.1. Result Test

In the earliest research stage it was observed that the presence of superficial damage had influenced greatly the weight losses. If the sample had broken, the loss of weight was 100% more than if had not. For each material six specimens were tested. Graph 1 shows the total weight lost in grams (as medium value of six tests) and the standard deviation for each tested material. It can be compared with graph 2, which shows the total losses of three superficial faultless specimens.

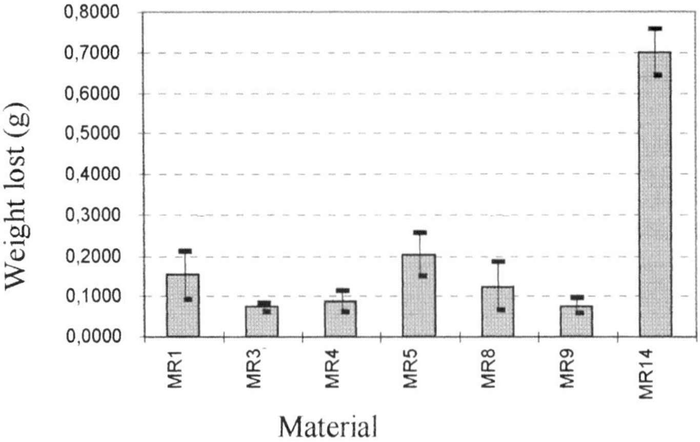

Graph 1: Loss of weight for each material (6 specimens for each value).

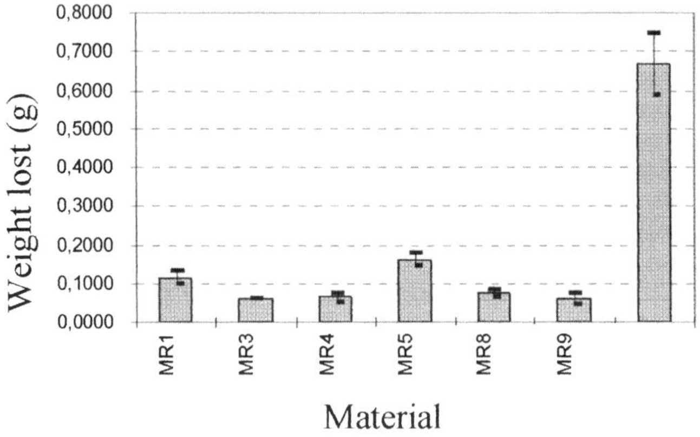

Graph 2 Loss of weight for 3 samples, which had the superior treatment on the surface.

5. DISCUSSION.

5.1 Effects of adding boron.

The addition of boron to the treatment improves the hardness (HRC) by about 3 or 4 points compared with cast iron. In the graphs 1 and 2 can be seen the good performance of the material MR4 compared with MR1, this is probably due to the superior hardness of the borides in front of the chromium and niobium carbides of MR1.

5.2 Effects of adding different components to the Cr-Nb alloys.
The addition of other elements does not increase the hardness, although the presence of tungsten carbides should have a positive effect in the hardness increment, as much in the eutectic phase as in the carbides. It must be pointed out the good resistance to abrasion of MR9 and MR3 at room temperature, slightly better than MR4.

5.3 Effect of superficial state.
In comparing graphs 1 and 2 it can be observed that the reduction in the deviation of the weight lost for the same material due to the elimination of superficial defects, discontinuities and pores. Therefore, especial care must be taken in applying, in order to increase its useful working life.

5.4 Mechanism of wear.
The wearing process followed in this kind of test is not clear because the shown waste surfaces are little differed. This is possibly due to the presence of varying sizes of abrasive particles (from 0 to 0.7 mm), resulting in the smallest ones rolling over the test surface, polishing and smoothing off the edges of the marks that have been left by the wearing process.

In figures 3 and 4, some indications of the aforementioned process can be seen, around the defects on the surface. Figure 3 shows grooves and lines similar to mechanisms of ploughing and cutting. In softer parts of the matrix or above subsuperficial discontinuities, cavities or pores can be seen possibly formed by the pressure of the biggest particles as they are exposed to abrasion, and increasing the material loss therefore.

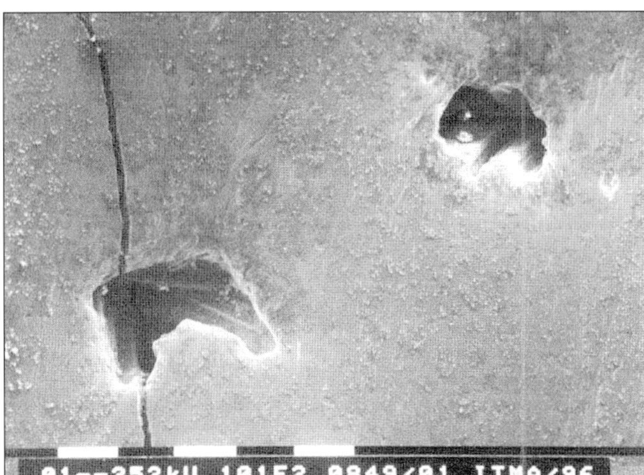

Figure 3: Small superficial defects that can be the beginning of the waste holes located in the test tube.

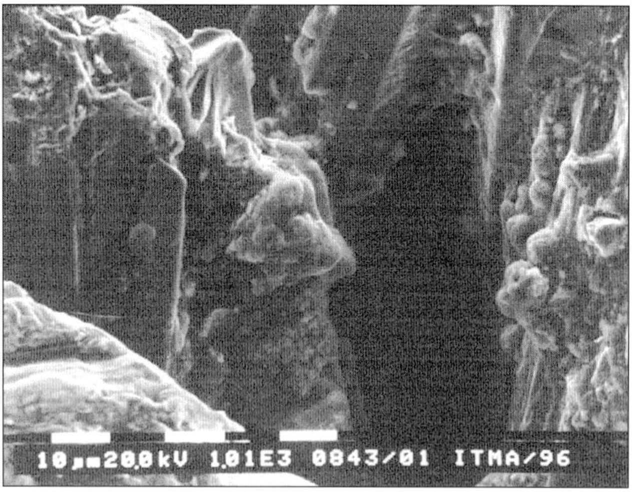

Figure 4: Removed carbides: hole left by a carbide when coming off.

In the lower part of the figure 4, a large hole left by a loosened carbide crystal, which was transversal to the surface, can be seen. This suggest that the softer material around the tungsten carbide crystal was worn first; the crystal lost its support, and was ripped away.

6. FIELD TESTS.

Analysis was carried out on 216 teeth (18 of each material) that were mounted onto the first 12 buckets of a bucket wheel of 20 buckets. These teeth were chosen from a batch of 3000 new teeth, rejecting those that displayed significant surface imperfections and replacing them with others. In figure 6 the excavator tooth scheme is shown.

The coating of the teeth was always carried out by the same operator to minimise the possible imperfections due to the coating technique. The intensity, fumes, sparks, etc. of the welding process were all noted.

Three tests were carried out, each for a different period of time. In each test, 6 teeth of each material were mounted in one of the 6 possible positions on each mounting and on different buckets in order to minimise the effect of the position, and the bucket deformation. On the 8 remaining mountings, new teeth were mounted each time to evenly distribute cutting stresses over the entire bucket wheel. Graph 3 shows the agreement between the laboratory tests and those carried out in real working conditions.

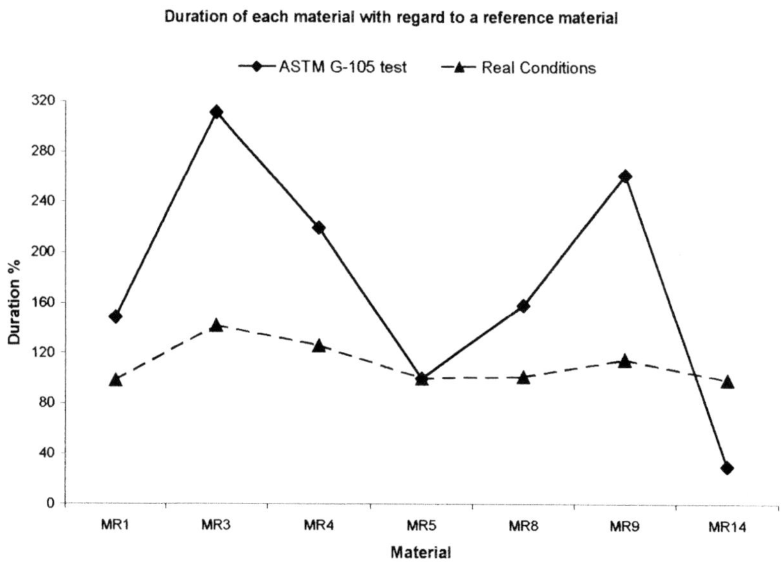

Graph 3: the agreement between laboratory and mine tests.

This allowed us to establish an economic evaluation taking the material MR5 as reference. It is logical to consider a higher cost for those materials, which provide a longer life. Taking into consideration the life and cost of the reference material, and estimating its value as 1000 ptas (about 4 pounds), the estimated price of other materials can be read from the table 4:

Table 4: Costs valued in function of the duration.

Material	Appraisal	Price	Material	Appraisal	Price
MR1	4,03	0.85· X	**MR9**	1,09	2.1· X
MR2	2,85	2.4· X	**MR13**	1,12	3.2· X
MR3	2,08	4.2· X	**MR14**	4,86	0.9· X
MR4	4,86	3· X	**MR16**	4,01	3· X
MR5	**4,85**	**X**	**MR17**	4,02	2.6· X
MR8	1,09	X	**MR18**	2,99	0.6· X

7. CONCLUSIONS

From these tests, the following conclusions have been extracted:

1. Alloys with a low carbon content (MR14) are not the best choice as abrasion resistant material.

2. Coatings based on Cr-Nb do not perform well and tend to disperse too much.

3. The alloyed layers with boron or Cr-Mb-Mo-W-V performs twice as well than C-Cr-Nb, alloys.

4. The resistance to the abrasive wear is clearly determined for the quality of the coating process.

5. There is a connection between the laboratory tests and real-life behaviour. The only exception is MR14 which presents a better behaviour than we expected.

6. A clear correspondence exists among the tests carried out in the laboratory and those that were developed under real conditions. Material MR 14 was exception since it performed better than was expected

7. The duration differences regarding the standard material are much more prevalent in the laboratory tests that under real conditions.

8. BIBLIOGRAPHY.

[1] Buckley, Donald H. *Friction, wear and lubrication in vacuum*. 1971

[2] Dowson D. *History of Tribology* 2nd ed. 1998

[3] Neale J. *Component failures, maintenance and repair : a tribology handbook* 1995.

[4] Nottingham. International conference on tribology in mineral extraction, *International Conference on Tribology in Mineral Extraction*, 1984

[5] Williams, John Austin. *Engineering tribology* 1994

Application of the method of caustics to rolling contact fatigue

T TAMIYA
Graduate School of Science and Technology, Chiba University, Japan
K SATO
Department of Urban Environmental Systems, Chiba University, Japan
T A STOLARSKI
Department of Mechanical Engineering, Brunel University, Uxbridge, UK

SYNOPSIS

The experimental method of caustics has been applied to rolling contact fatigue for determining the crack growth mechanism. The caustic images which are formed from stress fields, affected by rolling contact load and crack-tip stress concentration, are very complicated. Firstly, computer simulations for the caustic image have been done by the computational scheme combining directly caustics method with boundary element method. The size and shape of simulated caustic images, which are formed from the contact region and crack tip in disc-to-disc rolling contact, were coincident with those of experimental caustic images. Secondly, the change of relative displacement of crack faces with rotation was measured directly using an optical microscope. The results indicated that the size of caustic image formed from the crack tip is in good relationship with the relative displacement and the friction between the crack faces is important for the crack growth in rolling contact fatigue. Finally, it is shown that the caustic technique can evaluate the stress intensity factor K_{II}.

1 INTRODUCTION

In order to study the mechanisms of crack initiation and propagation under rolling contact fatigue, it is necessary to measure the magnitude and amplitude of the complex stress fields produced by rolling contact. The method of caustics is a powerful technique to evaluate such stress fields. The method is an optical and non-contact technique, and it is well known to be applicable to measure static and dynamic stress intensity factors of cracks.

Theocaris[1] has discussed the theoretical aspect and the geometry of caustic images formed by a concentrated load. Sato et al.[2-4] have expanded the theory for measuring contact forces in the disc-to-plate contact. Tamiya and Sato[5] have extended the method to the disc-to-disc contact. The above applications have been based on the Hertz's contact theory. In order to apply this method to more practical, but complicated problems, Sato[6] has proposed a scheme of caustic images analysis using the boundary element method, and has shown that the images simulated by the scheme are in good agreement with those obtained by experiments under the disc-to-plate contact.
The main purpose of this paper is the investigation of the behaviour of crack displacement and crack growth by applying the method of caustics to the stress field created in a disc with a crack under rolling contact.

2 THE THEORY OF THE METHOD OF CAUSTICS

2.1 The principle of the caustic image formation

Figure 1 is a schematic illustration how caustic image forms by light rays. When laser rays illuminate the near crack region of a plate under tensile stresses, a shadow spot is formed on a screen due to the changes of the thickness and the reflective index of the plate. This phenomenon results from the stress concentration around the crack tip. The bright image along the periphery of the shadow spot is called caustic image.
Figure 2 shows an example of caustic images formed from the stress fields under rolling contact with a crack. The bright curves are the caustic images. There are two caustic images in the figure: a cusp-like image is formed from the contact region and a circular image is formed from the crack tip. The shape and size of two caustic images change with optical, material, and contact conditions.

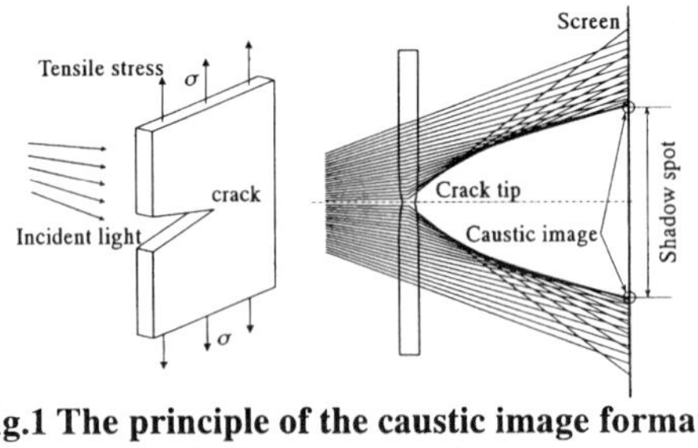

Fig.1 The principle of the caustic image formation.

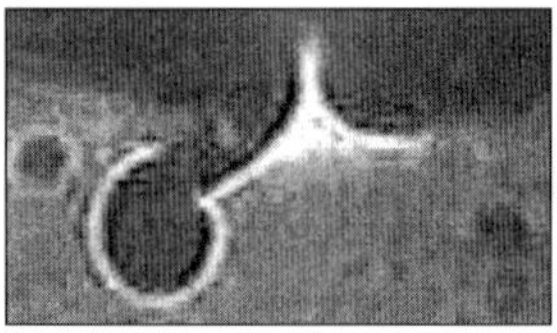

Fig.2 An example of the caustic images formed under rolling contact.

2.2 The basic equation of method of caustics

Figure 3 shows the vector relationship for representing caustic image formation. A light ray impinges the point A in a specimen. When two discs are not in contact, the light ray passes thorough the point A and arrives at the point A' on the screen. However, when the discs are in contact, the ray arrives at point B. The vector from the origin O to the point A on a specimen is the vector r, the vector from the origin O' to the point B on the screen is the vector W, and the vector from the point A' to the point B is the vector w. The relationship of the vectors on the screen are expressed by

$$W=\lambda_m r+w, \tag{1}$$

where $\lambda_m(=(z_i+z_o)/z_o)$ is an optical magnification factor, z_i and z_o are the distances from the specimen to the focus and from the specimen to the screen, respectively.
It is known that the vector w is in proportion to the gradient of stress sum, grad$(\sigma_x+\sigma_y)$, and the equation (1) is

$$W=\lambda_m r+C\text{grad}(\sigma_x+\sigma_y), \tag{2}$$

where $C(=z_o c_t t)$ is a constant depending on the arrangement of the optical set-up used; c_t is an optical constant of the material and t is the thickness of the specimen. Therefore, the caustic image can be obtained through the analyses of the vector W for many points on a specimen using equation (2).

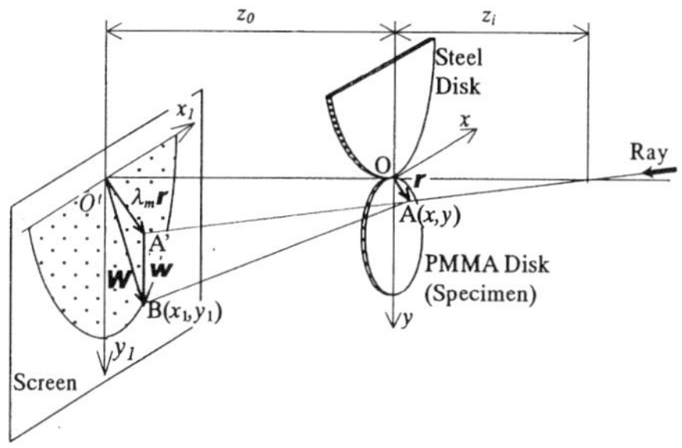

Fig.3 Vector relationship of passing and reaching points on specimen and screen.

3 EXPERIMENTS OF THE CAUSTICS METHOD

It is important to investigate the tractions (normal contact force and tangential force) and the stress intensity factors of rolling contact fatigue crack. The authors have presented the method to estimate the values of the tractions acting on contact surface under disc-to-disc contact in a recent paper [7]. Figure 4 is an illustration of the test rig used in rolling contact fatigue tests [8]. This test rig was designed and constructed by the authors to study the rolling contact fatigue with the help of the method of caustics.
The PMMA specimen disc was used in caustic experiments. The values of Young's modulus and Poisson's ratio of PMMA are 3.7GPa and 0.37, respectively. The specimen is 50mm in radius and 4mm in thickness, and has a radial crack of 2.4mm long started from the rolling surface as shown in Figure 5. This crack was taken as the radial crack initiated in rolling contact fatigue tests [8]. The specimen was in rolling contact with a steel disc of 290mm in diameter and 5mm in thickness. The crack position was defined as the angle of θ from the contact point, and the specimen was rotated from θ=+10deg to θ=–10deg. The normal contact force was 150N/mm, and the tangential force was 0 or 30N/mm.

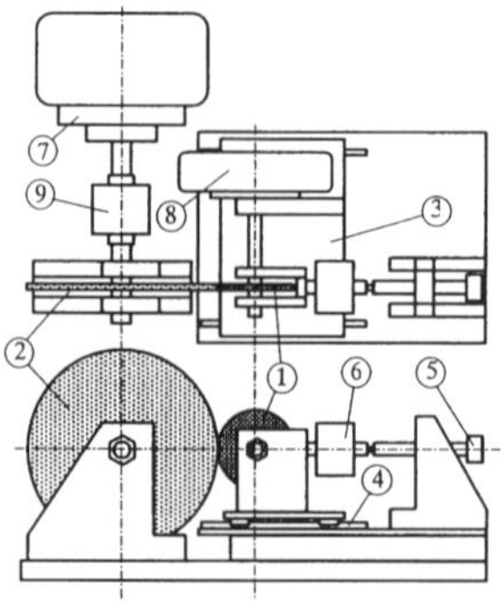

Fig.4 The test rig used in rolling contact fatigue.

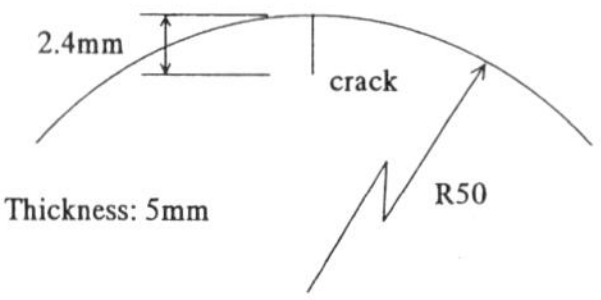

Fig.5 Details of PMMA disc specimen used in caustic experiment.

Figure 6 shows the optical set-up for the caustic experiments. This optical set-up was attached to the test rig shown in Figure 4. The light source used was a He-Ne gas laser tube with 632.8nm in wavelength. The light rays were illuminating the specimen disc as divergent light rays via a collimator and a convex lens. When the light rays were illuminating the contact region, caustic images were formed on the screen and were captured in a PC using a CCD camera. The optical magnification of this system was 7.74, because of z_i=230mm and z_o=1585mm.

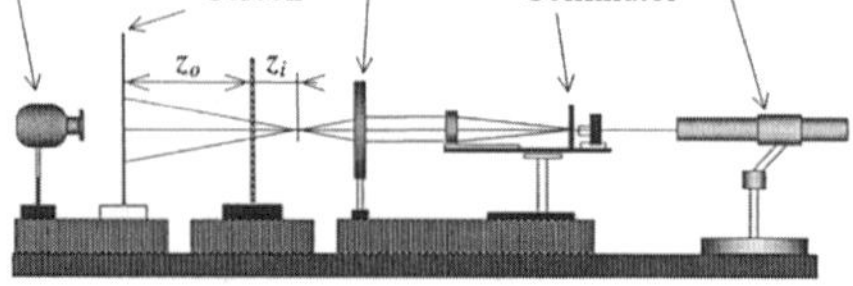

Fig.6 Optical set-up for caustic experiments.

4 CAUSTIC IMAGES ANALYSIS BY THE BOUNDARY ELEMENT METHOD

4.1 Combination of the caustics method with BEM

In the caustic image analysis, it is necessary to obtain the coordinates of B(x_1,y_1), given by W of equation (2) and to calculate the gradient of stress sum grad(σ_x+σ_y) for many points of

A(x,y) in the specimen. In this paper the value of grad$(\sigma_x+\sigma_y)$ was calculated using boundary element method. Stresses σ_x and σ_y are calculated from the following equation using displacements u_x(Q) and u_y(Q) and surface forces t_x(Q) and t_y(Q) on boundary elements [9]:

$$\sigma_{ij}(\mathrm{p})=\int_s \{D_{ijx}(\mathrm{p,Q})t_x(\mathrm{Q})+D_{ijy}(\mathrm{p,Q})t_y(\mathrm{Q})\}\, ds$$

$$-\int_s \{S_{ijx}(\mathrm{p,Q})u_x(\mathrm{Q})+S_{ijy}(\mathrm{p,Q})u_y(\mathrm{Q})\}ds \qquad (3)$$

where D_{ijk}(p,Q) and S_{ijk}(p,Q) are coefficient matrices created from surface forces t_x(Q) and t_y(Q) and surface displacements u_x(Q) and u_y(Q), and signify the effective coefficients of surface forces and surface displacements to inner point p. Also, i and j represent x and y, respectively. The sum of stresses $(\sigma_x+\sigma_y)$ at the point A(x,y) is given by,

$$\begin{aligned}\sigma_x(\mathrm{p})+\sigma_y(\mathrm{p})= &\int_s [\{D_{xxx}(\mathrm{p,Q})+D_{yyx}(\mathrm{p,Q})\}t_x(\mathrm{Q}) \\ &\quad +\{D_{xxy}(\mathrm{p,Q})+D_{yyy}(\mathrm{p,Q})\}t_y(\mathrm{Q})]ds \\ - &\int_s [\{S_{xxx}(\mathrm{p,Q})+S_{yyx}(\mathrm{p,Q})\}u_x(\mathrm{Q}) \\ &\quad +\{S_{xxy}(\mathrm{p,Q})+S_{yyy}(\mathrm{p,Q})\}u_y(\mathrm{Q})]ds \\ \equiv &\ \sigma_{\mathrm{sum}}. \end{aligned} \qquad (4)$$

Therefore, equation (2) is,

$$\begin{aligned} W &= \lambda_m r + z_o c_t t\ \mathrm{grad}(\sigma_{\mathrm{sum}}) \\ &= (\lambda_m x + z_o c_t t\ \sigma_{\mathrm{sum},x})i+(\lambda_m y+ z_o c_t t\ \sigma_{\mathrm{sum},y})j, \end{aligned} \qquad (5)$$

where subscripts $,x$ and $,y$ represent differentiations by x and y, respectively. Details of the differentiation are omitted because they are very long and complicated.

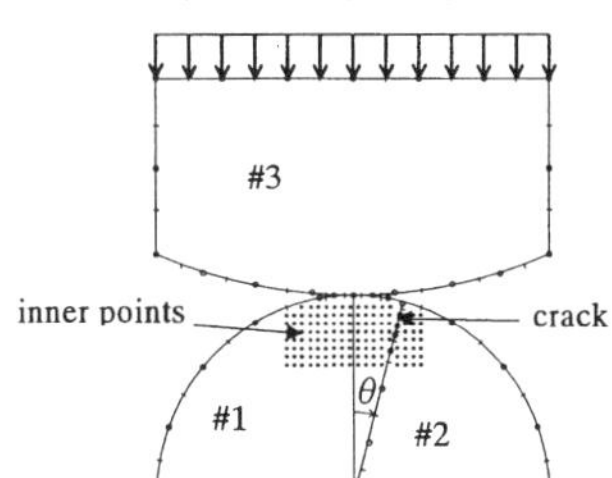

Fig.7 Model and boundary element meshes for BEM analysis.

4.2 Conditions for analysis

The boundary element model used for analyses is shown in Figure 7. The lower body is the PMMA disc and the upper body is the steel disc. The PMMA disc had a radial crack of 2.4mm long inclining at an angle θ to the contact point. The boundary element analysis was carried out for the crack positions θ ranging from θ=7deg to θ=-7deg at every 0.5 degree. The bottom surface of PMMA disc was fully constrained. On the upper boundary of the steel disc, normal force of 150N/mm was applied. Tangential force 30 N/mm or 0N/mm was applied to both sides as tangential force. The elements near the contact point and crack faces were treated as contact problem, and the friction coefficient for contact element for both discs and crack surfaces was assumed to be 0.3. To calculate the coordinate of B(x_1,y_1) by equation (5) many internal points, over 3000, were created around the contact point and the crack tip.

5 RESULTS AND DISCUSSION

5.1 Caustic images obtained by experiment and analysis

Figure 8 and 9 show caustic images obtained by experiment and analysis, respectively. Two types of caustic images can be noticed: one is the cusp-like image formed from around the contact point, and another is the circular image formed from around the crack tip. In these figures the angle θ defines the position of the crack relative to the contact point, (a) to (e) the crack is approaching the contact point, (f) is just at the contact point, and (g) to (l) is leaving the contact point.

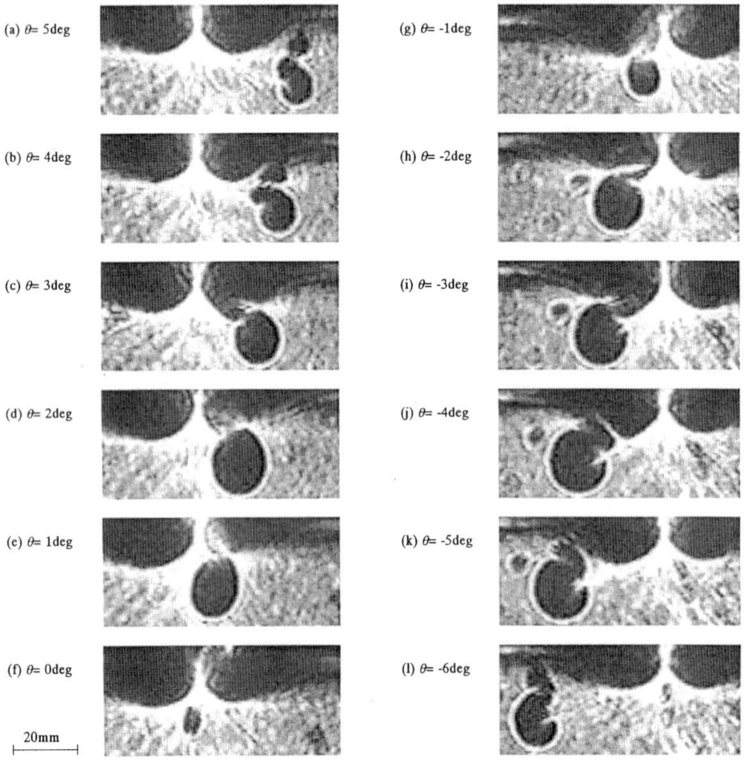

Fig.8 Caustic images obtained by experiment.

Both caustic images change with rolling contact, due to complexity of the stress state caused by the contact stress and the stress concentration at the crack tip. The caustic images are,

nevertheless, quite clear. This means that the method of caustics is applicable and useful to observe continuously the stress field of rolling contact with crack.
In order to investigate the relation between the size of circular image resulting from experiment and analysis, a diameter D_x of the circular image was measured as shown in Figure 10. Figure 11 shows variations of the diameter D_x with the crack position θ. In this figure, the crack position θ, defined in Figure 7, is changing from positive value to negative value, and the D_x is defined as negative when the crack position θ is negative.

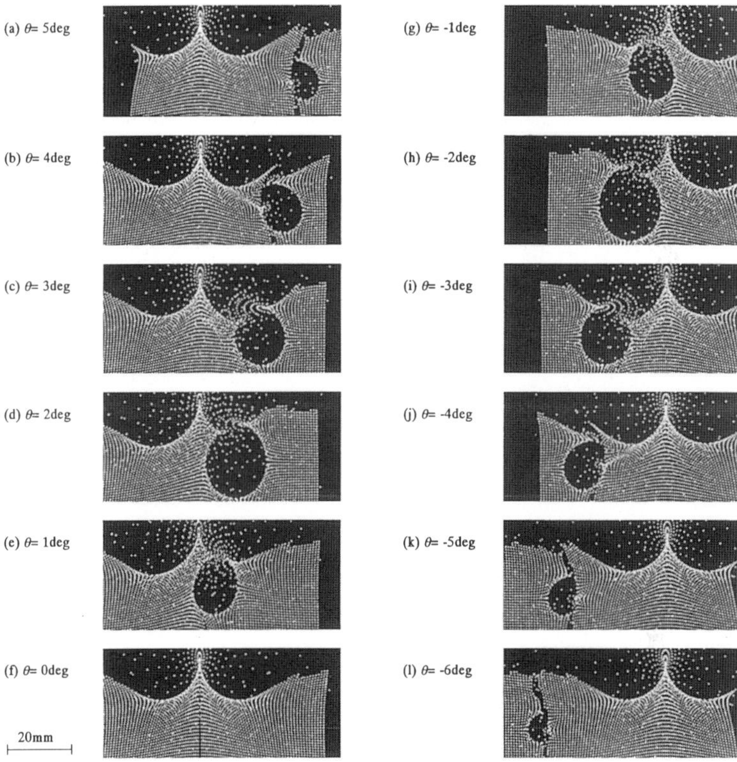

Fig.9 Caustic images obtained by analysis.

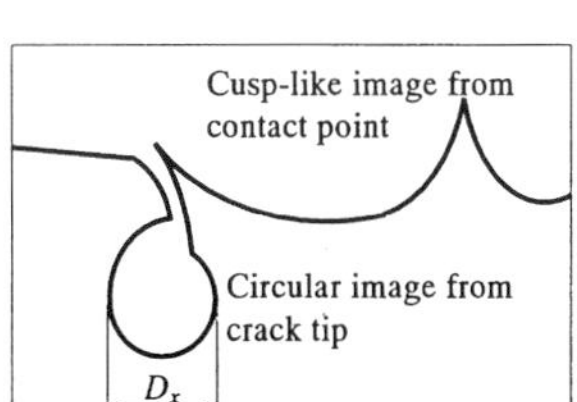

Fig.10 The position for measuring a diameter D_x of the circular image.

It can be seen from Figure 11 that in the process where the crack is approaching the contact point, the diameter of the circular image D_x increases till θ=2deg. At just under the contact point, θ=0deg, D_x=0 because the circular image disappears, and only the cusp-like image is observed (see Figures 8(f) and 9(f)). In the range of θ>0deg, the agreement between experiments and analyses is good. In the case of θ<0deg, however, a difference between experimental and simulated circular images exists as shown in Figure 11. The difference is most pronounced in the range from θ<-2deg.

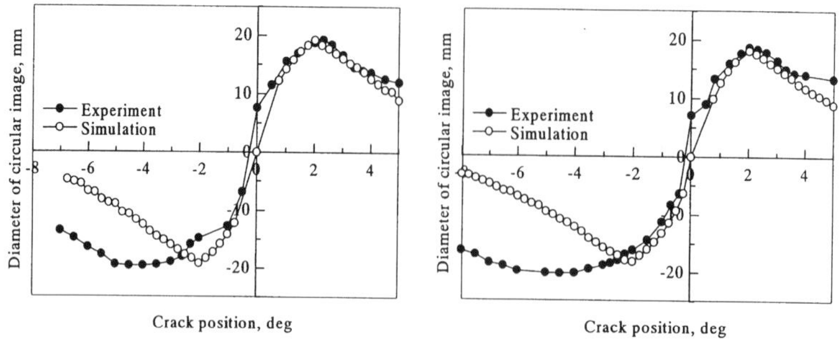

Fig.11 Variations of the diameter of the circular image D_x with the crack position.

5.2 Relationships between relative displacement on crack surface and caustic image

In order to investigate the relationship between the caustic image and crack deformation behaviour, the relative displacement of crack surfaces was measured. It is known from the analysis by BEM that the crack was never opened, and thus the investigation also had the purpose to consider the difference between experiment and theoretical analyses.

Figure 12 shows the definition of the relative displacement along crack faces. A line-mark on the specimen surface perpendicular to crack faces was made by scratching it with emery paper at the distance of 0.5mm from the crack tip. The relative displacement was defined as the value of tangential distance of the mark along the crack line. The measurement was done under *in-situ* observation using an optical microscope. During the measurement the tangential force was not applied. Figure 13 shows the variation of the relative displacement with the contact position. In this figure, analytical results obtained by BEM are also shown. The relative displacement shows a peak value at $\theta=\pm 2$deg. Experiment agrees well with theory when the crack is at θ>-2deg. On the other hand, when the crack position is θ<-2deg, it is decreasing slowly in experiment, but quite rapidly according to analysis. These changes of the relative displacements are in accord with the change of caustic images described in Section 5.1. Also, the size of caustic image depends on the relative displacement.

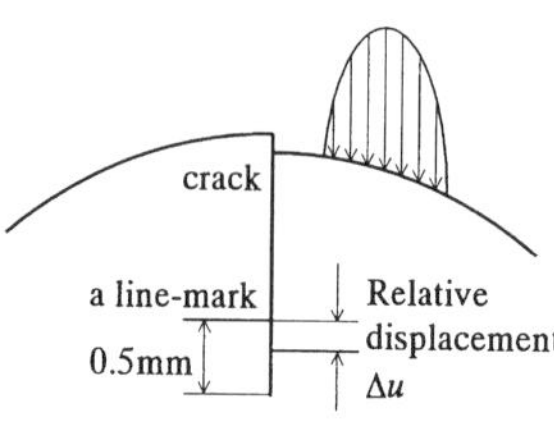

Fig.12 Relative displacement of crack faces.

This means that the effect of the friction on the crack faces is very important, and shows as the difference between experiment and analysis. This problem can be solved by introduction the loading-history analysis in BEM program.

5.3 Stress intensity factor, K_{II}, estimation

The caustic images obtained by experiments and analyses were in good agreement, although a difference existed in the case of θ<-2deg. An attempt was made to examine the relationship between the stress intensity factor, K_{II}, and the size of circular caustic image, because the mode II stress intensity factor, K_{II}, results from the relative displacement. Figure 14 shows the variation of K_{II} with the diameter of circular image formed from crack tip. In this figure, the solid lines are obtained by the least square method; and the expressions are:

$$K_{II}=1.17D_x-3.12\times10^{-19}D_x^2+1.08\times10^{-3}D_x^3 \quad \text{(for } \phi=0\text{)}, \qquad (6)$$
$$K_{II}=1.18D_x-1.24\times10^{-2}D_x^2+1.44\times10^{-3}D_x^3 \quad \text{(for } \phi=0.2\text{)}. \qquad (7)$$

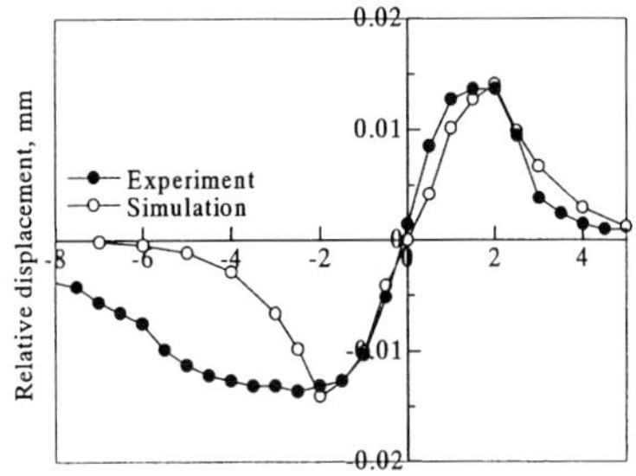

Fig.13 The variation of relative displacement with crack position.

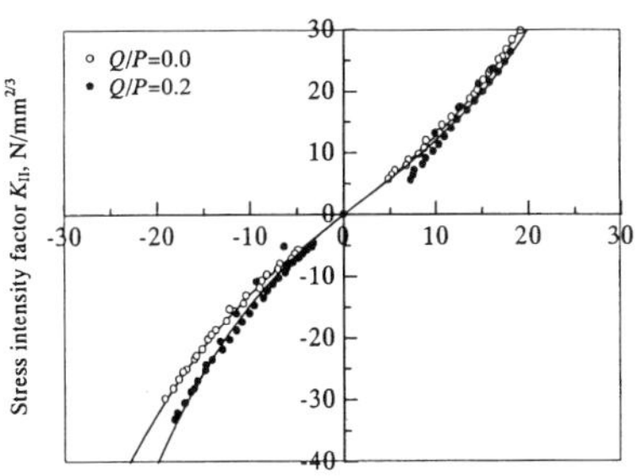

Fig.14 The variation of a stress intensity factor K_{II} with the diameter of circular image.

6 CONCLUSIONS

The analysis program of caustic images has been developed for applying the method of caustics to study rolling contact fatigue with a crack. The applicability of the program was verified experimentally. The main conclusions are as follows:

1. When the crack is approaching the contact point, circular and cusp-like caustic images, formed in rolling contact experiments, can be well simulated using the program.
2. When the crack is departing from the contact point, the difference between experimental and solved caustic image exists, because of the analysis of the loading history is not included in the program.

3. The changing of relative displacement of crack faces coincides with the changing of the size of the circular image.
4. The friction behaviour between crack faces is important for the stress field formed under rolling contact fatigue and crack growth.
5. This program is useful to simulate caustic images formed from complex stress fields and to understand the background of caustic image formation.

REFERENCES

[1] Theocaris, P.S., "Mechanics of Fracture", Vol.7, Martinus Nijhoff Pub, Hague, pp. 189-252, 1981.

[2] Sato, K., Sato, T., Itoga, H. and Namaizawa, Y., "Measurement of Contact Load by the Method of Caustics", Journal of Japan Society Non Destructive Inspection, **39**-8, pp. 648-653, 1990.

[3] Sato, K., Sato, T., Kojima, E. and Dohi, A., "Measurement of Contact Load by the Method of Caustics (Part II)", Journal of Japan Society Non Destructive Inspection, **40**-12, pp. 798-803, 1991.

[4] Sato, K. and Ito, T., "A Method for Measuring Tractions in Disc-to-Plate Contact Using Caustic Images", Journal of Japan Society of Tribologists, **40**-9, pp. 762-767, 1995.

[5] Tamiya, T. and Sato, K., "Applications of the Method of Caustics to Studies of Rolling Contact Fatigue", Journal of Japan Society of Tribologists, **43**-8, pp. 723-729, 1998.

[6] Aliabadi, M. H. and Samartin, A. (eds.), "Computational Methods in Contact Mechanics III", Computational Mechanics Publications, Southampton and Boston, pp. 331-340, 1997.

[7] Tamiya, T. and Sato, K., "A Method for Measuring Normal and Tangential Forces Under Disk-to-Disk Contact Using Reflection-Type of Caustic Images", Transactions of The Japan Society of Mechanical Engineers(A), 65, **637**, pp. 1990-1995, 1999.

[8] Tamiya, T. and Sato, K., "Surface –Initiated Crack Growth in Rolling Contact Fatigue", Transactions of The Japan Society of Mechanical Engineers(A), 65, **632**, pp. 833-839, 1999.

[9] Yuki, Y. and Kisu, H.,"Elastic Analysis by Boundary Element Method", Baifu-kan, Tokyo, pp. 73-75, 1987.

The consequences of aggressive lapping processes on the surface integrity of HIPed silicon nitride bearing balls

J KANG and **M HADFIELD**
School of Design, Engineering, and Computing, Bournemouth University, UK
R CUNDILL
SKF Engineering and Research Centre BV, Nieuwegein, The Netherlands

SYNOPSIS

Hybrid ceramic bearings give substantial performance advantages in many applications and have a positive environmental impact. Further extension of the application of ceramic balls in hybrid bearings will depend on cost effective manufacturing methods. In an attempt to accelerate the finishing process for silicon nitride ball blanks, the effect of high lapping loads and speeds (aggressive lapping) has been investigated. The aim was to determine the upper limits, if any, of the load and speed in aggressive lapping and the subsequent side effects on ball surface and sub-surface integrity. An experimental study has been carried out on a small-scale eccentric lapping machine under various aggressive lapping conditions. After lapping, balls were inspected for surface damage using a fluorescent dye penetrant technique as well as by normal optical microscopy. Surface spalls and sub-surface cracks were found on ball after lapping under the highest load of 10.87kg/ball average load. No substantial surface or sub-surface damage found on ball after lapping at load 4.37kg/ball or less. Surface spalls were found on balls after lapped at a speed of 500 rpm. No evidence of lapping induced surface damage, apart from C-cracks, was observed on balls lapped at 270 rpm. The radius of C-crack s appears to be related to the lapping speed, but further work is needed to confirm this.

1 INTRODUCTION

Silicon nitride is a ceramic material with the optimum combination of properties for use in rolling bearings. Technologies of powder manufacture and material densification have advanced to the stage that fully dense Hot Isostatically Pressed (HIPed) silicon nitride components are readily available [1]. The rolling contact fatigue life of HIPed silicon nitride is superior to that of bearing steels. The material properties - low density, high elastic modulus, corrosion resistance, and temperature resistance - give significant performance advantages in hybrid bearings and all-ceramic bearings [2]. Hybrid precision bearings with

HIPed silicon nitride balls as the rolling elements and steel inner and outer rings are used extensively in machine tools and other applications involving high speeds or extreme operating conditions.

The application of hybrid precision bearings with HIPed silicon nitride balls also has some positive environmental impacts in the sense that these bearings are less noisy, save energy and have longer service life. The only disadvantage in environmental terms is that the ceramic balls are difficult to recycle after in service, but they produce no pollution if abandoned. With more widespread use of hybrid bearings, it will be possible to re-process used balls by lapping them down to a smaller size after use. The overall influences of the application of hybrid precision bearings on the environment are therefore positive; the only restriction is the relatively high manufacturing cost of HIPed silicon nitride balls compared to steel balls. It is estimated that the lapping and finishing of HIPed silicon nitride balls and associated inspection constitutes about two thirds of the total manufacturing cost. Therefore, finishing balls to high levels of dimensional accuracy and surface quality in a cost effective manner is critical to their widespread application.

In order to investigate the loose abrasive wear mechanisms in the ball lapping process, the influences of various finishing parameters on ball surface quality and rolling contact fatigue life, a small-scale eccentric lapping machine was constructed [3]. The lapping rate of this eccentric lapping machine is much higher than in conventional concentric ball lapping machines. Taguchi Methods were used to optimise four lapping parameters—lapping load, lapping speed, diamond particle size and paste concentration [4]. The analysis of variance (ANOVA) showed that the most significant lapping parameter was lapping load, which accounted for 50% of the total, followed by lapping speed (31%). The diamond particle size and paste concentration parameters only accounted for 12% and 7% respectively. During that study, the lapping loads were varied from 0.9kg/ball to 1.85 kg/ball, which is representative of loads on balls in conventional lapping. Within that load range material removal rate increased almost linearly with load. What is not yet known are the limits for ball loads and speeds under conditions of aggressive lapping and what are the side effects of high loads and speeds on ball surface and sub-surface integrity. Since the finished balls are subjected to high contact stresses in bearing applications, any surface or sub-surface damage will be detrimental. The current experimental study is the first step in aiming to resolve these uncertainties; the second step will be the rolling contact fatigue testing of balls finished at high loads or speeds.

2 EXPERIMENTAL PROCEDURE

2.1 Test machine

The small-scale eccentric lapping machine used in this investigation is shown in Figure 1. This machine was designed by authors and manufactured in-house. The upper plate is stationary and has a flat lapping surface, and the lower plate which has a circular V-groove on it is driven by a micro-inverter controlled AC motor through a belt and pulley system. The diameter of circular V-groove is 65 mm which allows finishing of 13.4 mm diameter ball blanks in small batches of 15 balls. The V-groove angle is 90° and the eccentricity (distance between the centre of circular V-groove and the rotating axis) is 8 mm. A more detailed description of the machine can be found in Reference [3]. Cast iron plates are used for lapping, but steel plates are used for polishing.

Fig. 1 Eccentric lapping machine

2.2 Test samples

Two types of commercial HIPed silicon nitride ball blanks were used for the tests. Diameters of the blanks were 13.2 - 13.4 to give a nominal finished size of ½" (12.700 mm). Material A was used for the load tests. This material is produced by directly HIP process and has a density of 3160 kg/m^3 and a surface hardness (Vickers Hardness Number) of HV 1682. For the high speed tests, a sinter+HIP type of silicon nitride, Material C, was used. This has a density of 3238 kg/m^3 with surface hardness of HV 1532.

2.3 Test programme

The test programme consisted of two series of lapping tests (high load and high speed) and post-test ball surface inspection. In the load tests, 8 tests were carried out with nominal ball loads of 1.85kg/ball, 2.22kg/ball, 2.57kg/ball, 3.16kg/ball, 4.37kg/ball, 6kg/ball, 7.95kg/ball, and 10.87kg/ball. All the other lapping parameters were maintained the same - speed 169 rpm, diamond particle size 45μm and diamond paste concentration 1g:30ml. The second series of tests covered two speeds - 270 rpm and 500 rpm - using a lapping load of 1.85 kg/ball, diamond particle size 15μm and diamond paste concentration 1g:30ml..

Before and after each lapping test, balls and lapping plates were cleaned and the diameter of each ball was measured to 1.0 μm. The diamond paste and lapping fluid were mixed according to paste concentration parameter (1g diamond paste with 30ml lapping fluid) and stirred magnetically to disperse the diamond paste fully in the lapping fluid. 5ml of the mixture were applied to the V-groove of the lower lapping plate before each test. Each lapping test last one hour.

In order to further reveal possible sub-surface damage of the balls under high lapping load, some of the balls lapped at 10.87kg/ball and 4.37kg/ball were polished using a ¼ μm diamond paste and lubricating fluid under a light load and low speed to remove a 5 μm thickness layer from the top surface, a equivalent of ball diameter reduce of 10 μm.

After each test, half of the batch of balls (7 or 8 balls) was processed directly in fluorescent dye penetrant. Because some small cracks or other defects can become filled by lapping fluid and debris particles which hinders penetration by the fluorescent chemical, the other half of the balls in each test batch were treated in a furnace to 500 °C for one hour then cooled in air in order to 'burn out' residues in any small surface defects. Ball surfaces were inspected using optical microscopy at different magnifications under white light and ultra-violet (UV) light.

3 EXPERIMENTAL RESULTS AND DISCUSSION

3.1 The effects of load and speed on lapping rates

The material removal rate, as defined by the decrease in diameter per hour, at first increased with lapping load reaching the maximum, 68μm/hour, at a load of 4.37kg/ball. At higher loads, the material removal rate decreased to 55μm/hour at a load of 6kg/ball, 25 μm/hour at 7.95kg/ball and 20 μm/hour at 10.87kg/ball. At the two highest loads (i.e., 7.95kg/ball and 10.87kg/ball), the lapped ball roundness error appeared much bigger, typically 3~4 μm, some individual ball even 6 μm, in sharp contrast to the ball lapped at load 6kg/ball or less with a typical roundness error only 1~2 μm. It is anticipated that at these extreme high load conditions, the balls being lapped were not rolling freely, so the material removal rate decreased and the ball roundness error increased. This need to be validated both theoretically and experimentally. The material removal rates at lapping speed 270 rpm and 500 rpm were 10 μm/hour and 11 μm/hour respectively.

3.2 The effects of aggressive lapping on ball surface and sub-surface damage

The surface and sub-surface damage effects on balls lapped at the highest load of 10.87kg/ball are clearly revealed after polishing, Figure 2 shows typical examples of surface damage. Fig 2 (a) and (b) are the same surface spall under white light and UV light respectively. The dimensions of this spall are approximately 170μm×100μm with the deepest depth about 35 μm. The white fluorescent regions shown in Fig 2 (b) indicate sub-surface lateral cracks adjacent to the spall. Fig 2 (c) and (d) show a typical surface damage under white light and UV light. Again, there is evidence of sub-surface cracking. These types of damage could be due to the micro-fracture caused by the excessive lapping load. A point or 'star' type feature is shown in Fig 2 (e) and (f). This type of defect is considered to be due to a point overload caused by over-rolling of a large unbroken diamond particle. Radial cracks are formed similar to those that occur during hardness testing with a diamond indenter. Further lapping removes the dent but the cracks and sub-surface damage are revealed with UV light. No significant surface or sub-surface damage was found on balls lapped at loads 4.37kg/ball or less. Fig 3 (a) and (b) show the typical surface appearance of balls lapped at 4.37kg/ball load under white light and UV light.

3.3 The effects of high lapping speed on ball surface and sub-surface damage

Ball surface damage after lapping at a speed of 500 rpm was characterised by spalls of varying sizes. The largest spall observed which was 143μm×214μm in extent is shown in Fig. 4 (a). A typical small spall is shown in Fig 4 (b); this was only 28μm×21μm. The large spalls may caused by ball collisions, while the small spalls are thought to be due to the incursion of diamond particles. No significant ball surface damage was found on balls lapped at the lower speed of 270rpm, except C-cracks which will be discussed next

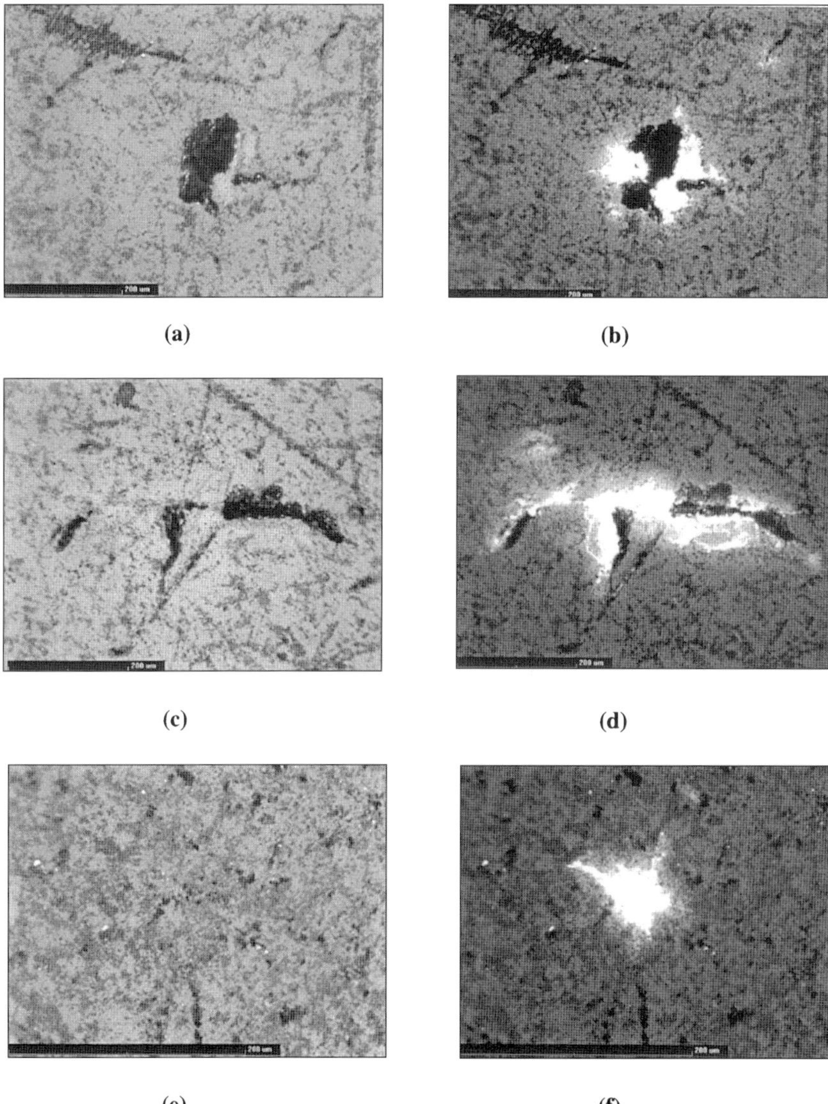

Figure 2 Typical surface and sub-surface damage on balls lapped under a load of 10.87kg/ball

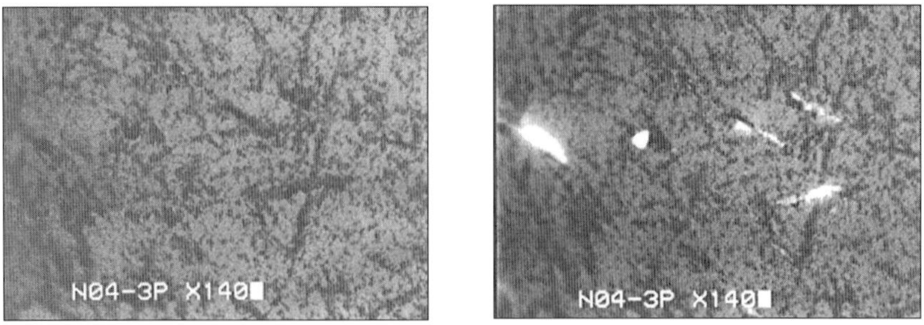

(a) **(b)**

Figure 3 Typical surface appearance of balls lapped under a load of 4.37kg/ball

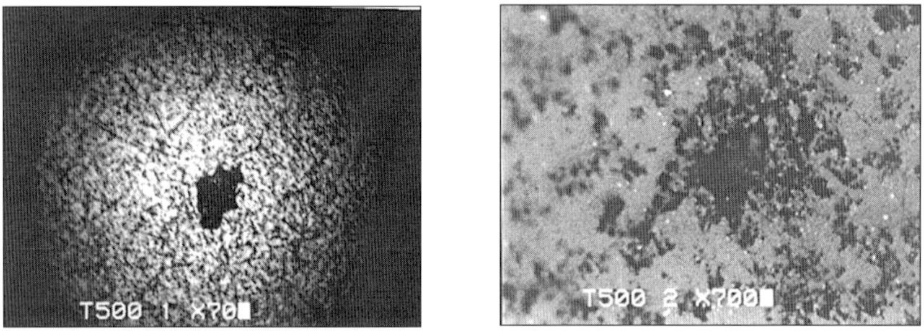

(a) **(b)**

Figure 4 Large and small spalls on balls lapped at a speed of 500 rpm

3.4 C-Cracks

C-cracks are part circular Hertzian cracks that can be present on silicon nitride balls with diameters of 10 mm and above. Large C-cracks are typically formed by impacts of balls caused by improper handling, while smaller C-cracks are thought to be a result of collisions between balls during the lapping processes. Small C-cracks are more easily observed if the balls are heat-treated before fluorescent penetrant inspection. Fig. 5 (a) and (b) show C-cracks found on balls made from material C lapped at a speed of 270 rpm with a load of 1.85kg/ball. The radii of these C-cracks were between 0.4~0.425 mm. Fig 5 (c) and (d) show C-cracks on balls made from material A, lapped at the lower speed of 169 rpm; also at a load of 1.85kg/ball. These smaller C-cracks had radii of 0.25 to 0.275 mm. Fig 5 (e) and (f) show C-cracks present on balls supplied in the finished state by the manufacturer of material C. These C-cracks were smaller in the radius (0.15 to 0.175 mm). However no C-cracks were observed on balls lapped at the highest speed 500rpm. Also, no C-cracks were found on different batches of material A and C balls lapped at speed 169~270 rpm then polished at speed 60~94 rpm. One possibility is that the C-crack formation is closely related to the final lapping (polishing) speed: within a certain speed range, C-cracks are most likely to be formed by collision between balls. Above and below this speed range, C-cracks are unlikely to be formed. Within this speed range, the higher the speed, the larger are the C-cracks formed. If this is the case, the final polishing speed of the grade 5 balls from manufacturer C could be in the range equivalent to the lapping speeds of 94 rpm to 169 rpm on the eccentric lapping machine. This also suggests that if the tangential speed of the rotating lapping plate on the contact points with balls is from 0.4m/s to 1.5m/s, then there is a possibility of generating C-cracks during lapping. The ball motion in an eccentric lapping machine is much more complicated than in a conventional concentric lapping machine, since there are ball accelerations and decelerations, changes in ball spin angle and skidding. More experimental studies and theoretical analysis on ball kinematics are needed before drawing any conclusion.

4 CONCLUSION

The effects of aggressive lapping on HIPed silicon nitride ball surface integrity have been investigated experimentally. Surface spalls and sub-surface cracks were found on balls after lapping under the highest load of 10.87kg/ball. No substantial surface or sub-surface damage was observed on balls lapped under average load 4.37kg/ball. Surface spalls were found on balls after lapping at the highest speed of 500 rpm, but there was no evidence of lapping induced surface damage on balls lapped at the lower speed of 270 rpm except for C-cracks. It appears that C-crack formation is directly related to lapping speed although more extensive experimental study and theoretical analysis are needed in the future.

REFEREENCES

[1] Cundill, R. T., "High-precision silicon nitride balls for bearings", Int. Soc. for Optical Engineering, Vol. 1573. Commercial Applications of Precision Manufacturing at the Sub-Micron Level, Bellingham, WA, USA, 1991, pp75-86.

[2] Cundill, R. T., "Light-weight material for the rolling elements of aircraft bearings", *Ball Bearing Journal,* 1983, Vol. 216, pp. 33-36.

[3] Kang, J. and Hadfield, M., "A study on the lapping of ceramic ball", *Surface treatment for computer methods and experimental measurements*, Computational Mechanics Publications, Southampton, 1999, pp. 389-398.

[4] Kang, J. and Hadfield, M., "Parameter optimization by Taguchi Methods for finishing advanced ceramic balls using a novel eccentric lapping machine" *Journal of Engineering Manufacture, Proceedings of Institution of Mechanical Engineer, Part B,* submitted in October 1999, revised February 2000, accepted April 2000.

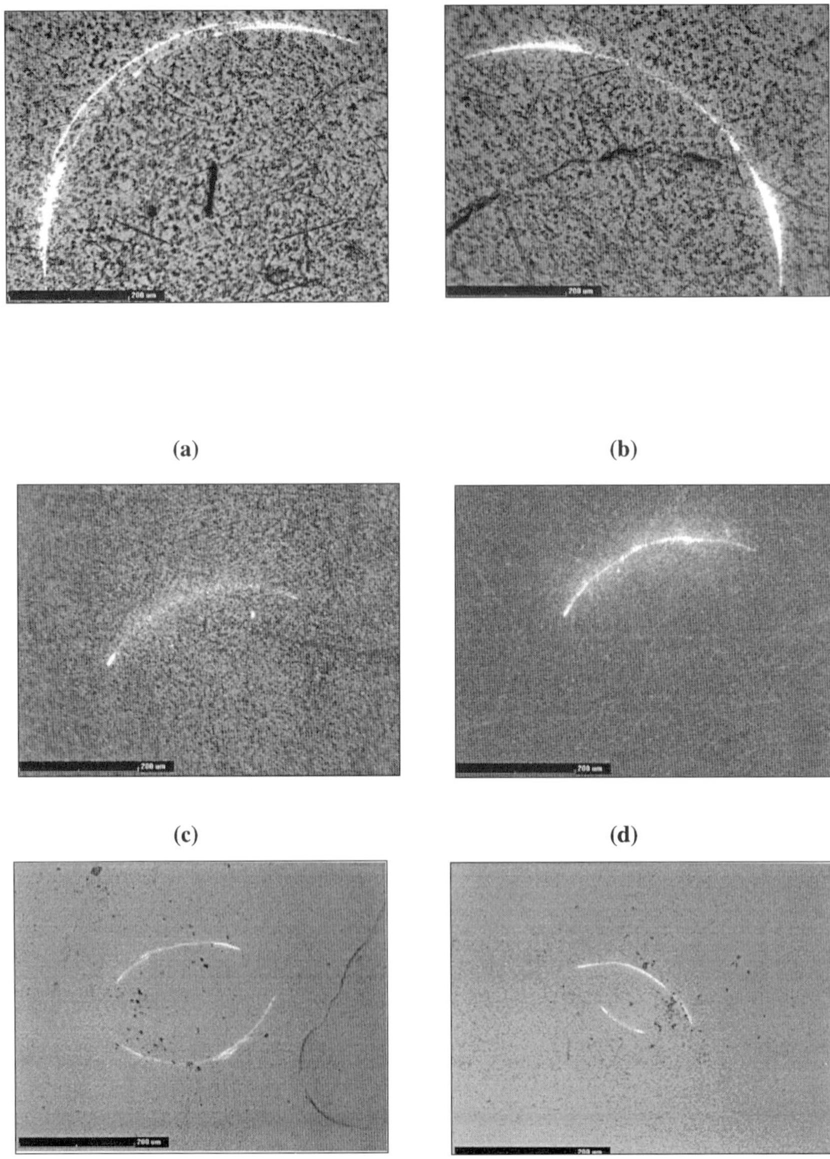

(a) (b)

(c) (d)

(e) (f)

Figure 5 Typical C-cracks found on balls lapped at different speeds

Behaviour of three families of stainless steels subjected to tribological conditions of abrasion and erosion

S JORGE, R GUILLERMO, and **S XENIA**
Centro de Investigaciones Metalúrgicas, La Habana, Cuba
R RIGOBERTO
Facultad de Ingenieria, Universidad de Los Andes, Mérida, Venezuela

ABSTRACT

The present work deals with a comparative tribological study of three families of stainless steels, austenitic, martensitic, and Duplex In the case of Duplex steel, AISI 329, recently available in the market, is used. In this paper, a tribological characterisation of steels in the presence of abrasive and erosive wear mechanisms is achieved, thereby allowing the comparison and definition of the field of application of these steels under the influence of said mechanisms. The relationships between wear rate and contact time during abrasive wear, as well as relationships between impact angle and wear rate during erosion at two particle-impact velocities are shown.

INTRODUCTION

Stainless steels [1] are divided into five families, each one with a unique microstructure, specific alloying elements and specified mechanical properties values which are also specified. Based upon their microstructure, these steels are classified into ferritic, austenitic, two-phase, martensitic, and precipitation-hardened. In terms of number of alloys and applications, austenitic steels is the biggest family of stainless steels. They can not be hardened by heat treatment, are non-magnetic, have a face-centered cubic structure, excellent ductility, and good weldability at cryogenic temperatures. Martensitic stainless steels are similar to carbon steels as they can be heated to the austenitic transformation temperature, tempered and then their toughness and ductility can be increased by annealing. Two-phase stainless steels are Cr-Ni-Mo alloys, which are balanced to form a mixture of austenite and ferrite. The two-phase structure results in an increased resistance to cracking under load.

The selection of a particular type of stainless steel intended for a specific application depends upon factors such as corrosion resistance, mechanical properties, ease of parts construction, and cost. However, in applications such as pumps, valves, fasteners, gas pipes, etc. the tribological behaviour of the material to be used must be taken into account in order to avoid considerable costs due to unexpected failures and the replacement of parts as a consequence of defects that reduce their useful life. The effect of two important tribological processes such as abrasion and erosion should be considered in selecting the adequate steels, type of steel to be used.

Milichenko and Alexandrov [2] in erosive tests of some of the stainless most used, found that 13 % chromium stainless steels showed the smallest resistance to erosion. This steel has ferrite microstructure with a small quantity of annealing martensite. From metallurgical analysis, it was observed that the most extensive wear was in the ferrite phase. It is due to the plastic deformation and the development of microcracks both inside as at the grain boundaries. Also, it was observed that an increase of carbon content in these steels, increases greatly their erosion resistance. Kuklik and et al [3], carried out erosive tests in several steels, (CSN 19.436, CSN 14.209, CSN 42.2931, CSN 17.248, CSN 12.014) subjected to different thermal treatments, with foundry sand of grain 0.8 mm, at different impact angles and two impact speeds, 50 and 90 m/s. They found that the CSN 19.436 and CSN 14.209 steels showed a similar behaviour, having the maximum lost weight to an impact angle of 75°. This was due to their martensite microstructure being hard and fragile. The steel CSN 19.436 showed a better resistance to erosion due to the presence of carbides. At impact angles of 15° and 45°, the steel CSN 42.2921 showed a better erosion resistance than the CSN 17.248 steel. Starting from 45° impact angle, both steels showed a tendency to increase their erosion resistance. This is a characteristic behaviour of the ductile materials. This behaviour was observed with more clarity in the CSN 12.014 steel, which is the softest and more ductile of the studied materials. This steel showed a less erosive resistance than CSN 19.436 and CNS 14.209 steels. These authors concluded that the material microstructure influences the existent relationship between the amount of weight loss by erosion wear and the particle impact angles.

For ferrous materials, abrasion resistance is highly dependent on three metallurgical variables: microstructure, hardness and carbon content. The hard martensitic microstructure is preferable to the soft ferritic and austenitic microstructures, the preference in particular in low stress abrasion. Under high stress conditions, alloys with a high strain-hardening rate show a better abrasion resistance than those of low hardening by cold working [4]. Austenitic stainless steel is more easily hardened by cold working than the other stainless steel families. Therefore in high stress abrasion conditions, martensitic stainless steels are preferably used, as its hardness is obtained by thermal treatment. Austenitic stainless steels with a high strain-hardening rate can be used where the groove formation is a problem and a high tenacity is required. In general, stainless steels are considered under abrasion conditions where the environment is corrosive or when one works at high temperature [5]. Lansdowd and Price [6] in abrasion tests in dry sand found that the austenitic stainless steels have an abrasion resistance smaller than a martensitic one. Also, these authors observed that stainless steel type AISI 410, 0.1% C, has a abrasion resistance similar to the best austenitic alloy, while the abrasion resistance of the AISI 440C is three times superior that of austenitic alloys.

EXPERIMENTAL PROCEDURE

Three stainless steels, representatives of three families of widely industrially applied stainless steels, are studied in this paper. The chemical composition and mechanical properties of stainless steels used are shown in Table 1.

Table 1. Chemical composition in wt % and mechanical properties of materials under study

Material	C	Si	Mn	P	S	Cr	Ni	σ_y (MPa)	σ_u (MPa)	HV
AISI 304	0.060	0.450	1.390	0.035	0.100	18.34	8.350	234.7	573.6	190
AISI 420	0.385	0.660	0.490	0.020	0.060	15.32	0.023	906.9	1410.8	464
AISI 329	0.200	0.0750	1.000	0.040	0.030	25.00	3.500	506.7	647.9	254

The tribological processes of abrasive and erosive wear were then simulated according to the ASTM 665-94 and ASTM 676-95 standards.

Abrasive wear

The abrasive wear test was carried out in a machine intended for these purposes. The experimental regime used was obtained through a modelling [7], which allowed to simulate the working process of materials in the presence of this wear mechanism. The experimental regime parameters were: Frequency of rotation, 63 rpm. Radial forced applied, 40 N. Abrasive, silica sand. Abrasive grain size, 0.45 mm. Contact time, 30, 60, 90, 120 and 150 minutes. Initial surface roughness, 0.4 µm.

The simulation of this wear mechanism made it possible to obtain the wear rate values of the families of stainless steels evaluated. These results are shown in figure 1.

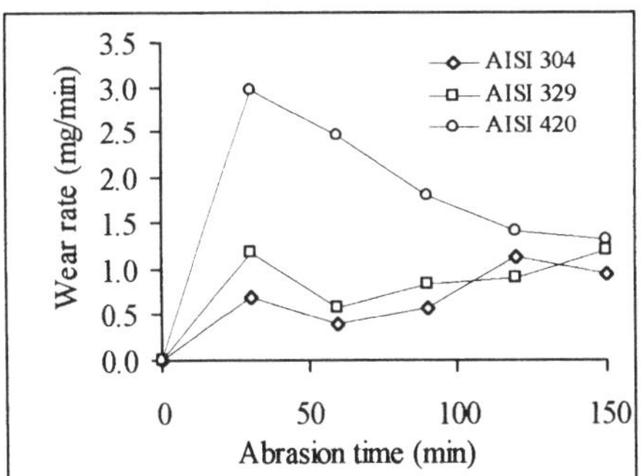

Fig. 1 Behaviour of the wear rate with respect to the contact time of the three stainless steels.

Erosive Wear

The simulation of the erosion wear process of steels investigated was carried out by using centrifugal acceleration equipment intended for these purposes. The simulation allowed the evaluation of two impact velocities of abrasive particles on the material, V = 38 m/s and 76 m/s, as well as four impact angles of abrasive particles (30^0, 45^0, 60^0, and 90^0). An amount of abrasives totalling 5 kg being consumed in a 30-minute working time and with a grain size of 0.045 mm, was used for each run.

The wear rate of each material was obtained as a function of the impact angle of abrasive particles for each impact velocity. The test results are shown in figures 2 and 3.

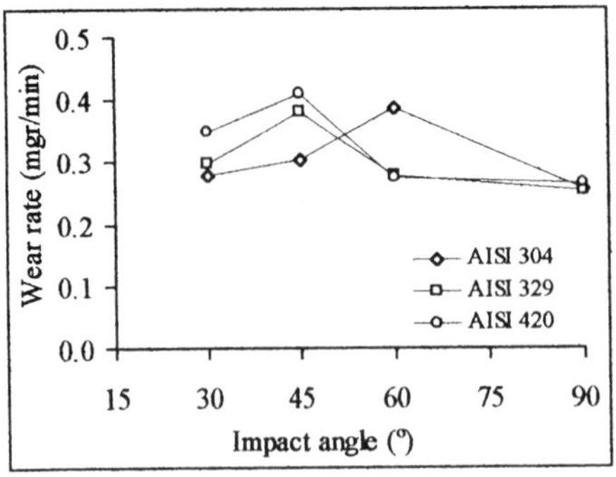

Fig. 2 Behaviour of erosive wear rate as a function of the impact angle for the three stainless steels. V = 38 m/s.

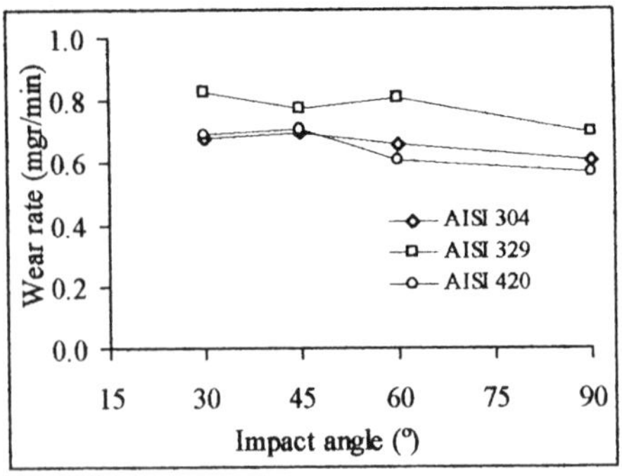

Fig. 3 Behaviour of erosive wear rate as a function of the impact angle for the three stainless steels. V = 76 m/s.

In figure 2, the behaviour of each material as a function of the impact angle of abrasive particles for V = 38 m/s can be observed. In figure 3, the behaviour of wear for each material as a function of the impact angle of abrasive particles, V = 76 m/s, can also be seen. In figures 2 and 3 a comparison of the behaviour of wear rate of the three materials investigated is illustrated for each one of the impact velocities of abrasive particles.

RESULT ANALYSIS

As can be seen in Figure 1, we can assert that for abrasion times of 30, 60, 90, and 120 minutes, the AISI 304 and AISI 329 stainless steels show a similar behaviour, whereas AISI 420 steel exhibits a higher wear rate for the same times. In general, the lowest abrasion wear rate showed by AISI 304 stainless steel at the different abrasion times can be explained in terms of its ability to undergo strain hardening, since during the abrasive wear process the material is subjected to a high plastic deformation. For AISI 304 austenitic stainless steel [1] the strain hardening is caused by an austenite to martensite transformation, which generates an increase in hardness of the worn-out material, and consequently, a decrease in wear rate [8]. The satisfactory abrasive wear rate showed by Duplex (AISI 329) steel might be also due to its ability to get cold deformed, although in a lower proportion than austenitic steel. In general, it can be assumed that the better behaviour of AISI 304 and AISI 329 steels over AISI 420 steel for this wear mechanism is due to their condition of ductile materials. It renders them more capable of absorbing deformation, and impedes a surface cracking wear process, which generates a reduction in abrasion resistance. At t = 60 minutes, both steels reach their minimum abrasive wear rate, i.e., their maximum strain hardness before brittleness, and on continuing the deformation process of abrasive particles on the hardened and worn-out surface, the material is no longer able to absorb the strain energy and initiates a process of macrocracks formation. These microcracks, on propagating with the continuous process of interaction of the particles with the surface, give rise to the fracture of material, generating, in turn, a higher detachment of particles from the surface of the materials.

For the case of erosive wear, figures 2 and 3, it can be asserted that the behaviour exhibited by the three materials at impact angles of 30°, 45°, 60°, 90° can be caused by the following mechanism: The kinetic energy of the abrasive particles when colliding with the material surface, is partly transformed into deformation energy which is absorbed by the material, and the other part is dissipated. The deformation energy absorption capacity of a material depends on its capacity to deform both in the elastic and in the plastic ranges. This property is related to the material hardness, the higher the hardness, the lower the ability to get deformed. Therefore, the lower wear rate shown by AISI 304 and AISI 329 stainless steels at the angles of incidence of 30° and 45°, in comparison with that shown by AISI 420, can be caused by the lower ability of the martensitic steel to deform, wear taking place by a rapid surface fracture in spite of its higher hardness. The atypical behaviour shown by AISI 304 steel at 60° is due to its mechanism of strain hardening and to a loss of material; that is, the loss of material as martensite particles makes the lower hardness austenitic substrate form again a martensitic layer [9]. In the case of AISI 420 and AISI 329 steels, which show a similar wear rate, which is lower than that of AISI 304 steel at an angle of incidence of 45°, the maximum wear corresponding to the classical behaviour of steels at this angle of incidence, takes place. The increased impact velocity of particles generated an increase in wear for each one of the materials. In a greater extent for Duplex steel, which reaches the maximum wear rate at θ = 30°. In the martensitic and austenitic steels, minimum wear rate would be reached at 45°. The close agreement between the wear rate values of AISI 304 and AISI 420 steels may be due to the austenite to martensite transformation occurring more rapidly at 76 m/s than at 38 m/s. On the other hand, the lower wear shown by the martensitic steel is due to the reduction of the tangential component of the deformation energy when increasing the angle of incidence up to its null value at θ = 90° where a purely compressive effect takes place.

In general, it can be asserted that AISI 329 Duplex steel can be used under tribological conditions of abrasion, since its behaviour is similar to that of AISI 304. In the case of erosion something similar happens, most particularly at angles of incidence of 30° and 45°. This assertion is further corroborated when considering excellent anticorrosion properties of this steel, which are also similar to that of AISI 304 steel, and very superior to that of AISI 420 steel.

CONCLUSIONS

The austenitic stainless steel exhibited the lowest abrasive wear rate at all the abrasion times, followed by the two-phase steel. AISI 420 stainless steel showed the highest abrasive wear rate among the steels studied.

For an impact velocity of erosive particles of 38 m/s and at angles of incidence of 30° and 45°, the lowest wear rate was shown by AISI 304 stainless steel, although the difference in relation to that of AISI 329 steel is small. At an impact angle of 90°, the lowest wear was reached by AISI 420 stainless steel.

For an impact velocity of 76 m/s, AISI 329 steel showed the highest wear rate. Therefore its industrial application at high erosive impact velocity in relation to the remainder steels investigated is limited. At this impact velocity, the lowest wear rate corresponded to AISI 304 steel at impact angles of 30° and 45°, and to AISI 420 steel at impact angles of 60° and 90°.

For AISI 329 and AISI 420 stainless steels, the maximum erosive wear rate at an impact velocity of 38 m/s was achieved at an impact angle of 45°, and for AISI 304 steel, at an impact angle of 60°.

The highest erosive wear rate of AISI 304 and AISI 420 steels at an impact velocity of 76 m/s was achieved at an impact angle of 45°, and that of AISI 329 steel, at an angle of 30°.

The important effect of the impact velocity of particles and their impact angle on erosive wear was shown. On increasing impact velocity, the wear rate for the three steels was increased, the opposite being true for the impact angle, which, on increasing, diminished the wear rate.

REFERENCES

[1] Columbier, L., Hochman, J., "Aceros inoxidables y aceros refractarios", Ed. Urmo, España, 1968.

[2] Milichenko, S. L. and Alexandrov, A. G., "Resistencia Hidroabrasiva de los Aceros Inoxidables", Problemas de Fricción y Desgaste, Conf. Cient. Tec. Sobre Tribología, De. Técnica, kiev, 11-115, 1975.

[3] Kuklik, V., Vooel, M. y Suchánek, J., "Vli Struktury na Velikost Erozívního Opotrebení a na Hodnotu Rychlostniho Exponentu", Strojirenstvi 37, cls. 4, 242-245, 1987.

[4] Childs, T. H. C., "The Mapping of Metallic Sliding Wear", Proc. Instn. Mech. Engrs., 202, C6, 1988.

[5] Blaun, P. J., "Friction, Lubrication and Wear Technology", ASM HANDBOOK, 18, ASM International, USA, 1992.

[6] Lansdowd, A. R. and Price, A. L., " Materials to Resist Wear", Pergamon Press, Great Britain, 1986.

[7] Suárez, J., "Modelación y Caracterización de Pares Tribológicos Fabricados y Recuperados". Tesis de Maestría, Universidad Técnica de La Habana, 1997.

[8] Finnie, L., Wolek, J., Kabil, Y.H., "Erosion of Metals by Solid Particles", J. Mater., 2, 1967, pp. 682-702.

[9] Reinoza, R., "Comportamiento de los Aceros Inoxidables ante Solicitudes Mecánico Corrosivas", Dr. C.T. Thesis, Instituto Superior Politécnico J. A. Echeverría, Cuba, 2000.

Advanced Materials – I

Mechanics of metal forming

C DE LA CRUZ GÓMEZ
Departamento de Mecánica Aplicada, Universidad de Castilla-La Mancha, Almaden, Spain

ABSTRACT

The degree of correlation between any theory and the physical situation it attempts to describe depends on how closely the theoretical model can be made to fit the real conditions. It is with this in mind that fairly full descriptive accounts are given of experimental investigations of a number of metalforming processes. Wherever possible the accuracy of any theoretical solution will be checked against experimental work.

In this work the theoretical solutions are developed from an appreciation of homogeneous deformation, without friction between tools and defórming material. Therefore, íf we consider a bar of non-hardening material of initial crosssectional area A_0 deformed to a final eross-sectional area A, the work done per unit volume is [Y • ln • (A_0/ A)] where Y is the yield stress of the material. This is an expression often met with in elementary treatments of metal-forming, though Y is sometimes looked on as a factor to be adjusted to give the best agreement with experiment. In practice, the work per unit volume required for a given defórmation will be greater than that given in the above expression because of work-hardening of the material, friction between tools and material, and because of non-useful work of distortion.

In the fóllowing theoretical treatments of a number of metal-formíng processes, elastic strains are neglected as being small compared with the plastic strains. This assumption of a rigid-plastic material means that residual stresses cannot be studied theoretically, though experimental determination of residual stresses will be discussed.

1.- THE SINKÍNG OF A THIN-WALLED TUBE: GENERAL CONSIDERATIONS.

Hollow sinking or tensile sinking consists in drawing a tube through a die such that its external diameter is reduced, Fig. 1 (a). The operation modifies the mechanical properties of the material, gives a closer tolerance on the outside diameter and improves the degree of surface finish. In plug drawing, Fig. 1(b), a mandrel (or plug) is inserted in the die to goverrí the final thickness of the drawn tube. (See an investigation of the plug drawing process by BLAZYNSM and COLE, 1960, also MOORE and WALLACE, 1967/68, and FLINN, 1969.) The hollow-sinking method generally allows larger reduction and is the process referred to below. In practice, the maxímum reduction in diameter is approximately 30 per cent.

Equipment for the cold drawing of steel tubes is described at length 1 Rozov (1968).
raight.

The drawing of a tube of uniform wall thickness through a straítapered die, Fig. 2(a), will be examined, the tube wall being assurn so thin in comparison with the diameter that the effects of plastic bending annd the variation of stress through the wall thickness can be neglected. As the tube is drawn through the die, each part of the tube passes through exactly the same operation, and if the tube is sufficiently long, a steady state Is attained.

The stresses in the tube are the tensile stress σ_1, parallel to the die face, the compressive stress σ_2 due to contact with the die face and the compressive hoop stress σ_3. If the tube wall is very thin, the compressive stress σ_3 may be great enough to cause the tube to buckle, a fóld forming along the length of the tube, SACHS and BALDWIN, (1946a). For convenience the stresses on the element of tube in Fig. 2(b) are afl shown as tensile, i.e., positive. Theactual direction of the stresses will become apparent on solution.
actua

There is also a frictional force between the die face and the tube wall dependent upon the degree of lubrication and the value of the pressure σ_2. Remembering that σ_2 is, in fact, compressive, the frictional stress is Remembering that $-\mu\sigma_2$ in the direction shown, where μ is the coefficient of friction. Strictly speaking, σ_2 , is not therefore a principal stress. Also σ_1 , on the perpendicular plane must be associated with a non-uniform shear stress, but for normal coefficients of friction this latter stress is negligible.
If a steady state condition has been reached, we can consider the equilibriun, of an element of the tube, Fig. 2(b).

Resolving forces perpendicular to the element of the die,

$$\sigma_2\left(r\Delta\theta\,\frac{dr}{sin\ \alpha}\right) = \alpha_3\left(t\,\frac{dr}{sin\ \alpha}\right)\Delta\theta\cos\alpha$$

(1)

$$\alpha_2 = \alpha_3\,\frac{t}{r}\cos\alpha$$

σ_2 is small in comparison with σ_3 when (t/r) is small, as is assumed in this analysis.

Resolving forces parallel to the element of the die

$$(\sigma_1 + d\sigma_1)(r+dr)\Delta\theta(t+dt) - \sigma_1 tr\Delta\theta \sin\alpha - \sigma_3\left(t\frac{dr}{\sin\alpha}\right)\Delta\theta \sin\alpha - \mu\alpha_2\left(r\Delta\theta\frac{dr}{\sin\alpha}\right) = 0$$

Neglecting second-order small quantities,

$$d(\sigma_1 tr)\Delta\theta - \sigma_3(t\ dr)\Delta\theta - \mu\left(r\Delta\theta\frac{dr}{\sin\alpha}\right)$$

$$\frac{d}{dr}(\alpha_1\ tr) - \alpha_3 t - \mu\frac{\alpha_2 r}{\sin\alpha} = 0 \qquad (2)$$

Substituting the value of a, from equation (1) into equation (2), we obtain

$$\frac{d}{dr}(\alpha_1\ tr) - \alpha_3 t(1 + \mu\cot\alpha) = 0 \qquad (3)$$

It has been shown by SWIFT (1949) that little difference is made to the values of the calculated stresses and the thickness strains if t is assumed constant in the above equation -and therefore the variation of t in the equilibrium equation will be neglected. Hence

$$\frac{d}{dr}(\alpha_1\ tr) - \alpha_3(1 + \mu\cot\alpha) = 0 \qquad (4)$$

A second equation for the stresses is given by the yield criterion of the metal. Under the supposed and known stress conditions $\sigma_1 > 0 > \sigma_3$ and $\sigma_2 = 0$, which substituted into the von Mises´ criterion gives

$$\sigma^2 - \sigma_1\sigma_3 + \sigma_3^2 = Y^2$$

where Y is the uniaxial yield stress.

Under the same stress conditions, the Tresca yield criterion gives

$$\sigma_1 - \sigma_3 = Y$$

remembering that σ_3 has been given a positive direction, Fig. 2(b).

While von Mises's criterion is more correct for non-ferrous metals, Tresca's criterion is mathematically simpler to apply when the magnitude and sense of the stresses are known. As a compromise, therefore, the y¡eld criterion adopted will be

$$\sigma_1 - \sigma_3 = mY$$

where m is a constant obtained by the method of least squares to satisfy von Mises' criterion. The graphícal representation of this modified criterion is shown in Fig. 3 and the value of m is 1•1.

The relations between stresses and strain increments for the assumed.rigidplastic material are given by the Lévy-Mises flow condition

$$\frac{d\,\varepsilon_2}{d\,\varepsilon_3} = \frac{dt/t}{dr/r} = \frac{\sigma_2 - \sigma_m}{\sigma_3 - \sigma_m}$$

where α_{m}. = $(\sigma_1 + \sigma_2 + \sigma_3)/3$, the hydrostatic component of the stress system. Neglecting α_3, in comparison with al α_l and αu,

$$\frac{dt/t}{dr/r} = \frac{\sigma_1 - \sigma_m}{\sigma_1 - 2\sigma_3} \qquad (6)$$

In the following solutions, work-hardening will be neglected and the effect of this will be discussed later.

The final relation required for completing the solution is given by the equation of incompressibility

$$\frac{dt}{t} + \frac{dr}{r} + \frac{dl}{l} = 0 \qquad (7)$$

we where: dl /l is the strain increment in the direction of σ_1.

2.- FRICTIONLESS TUBE-SINKING.

The simplest analytical solution is obtained for the hypothetical case of no friction between the tube and the die face, ie., $\mu = 0$. The equilibrium equation (4) then reduces to

$$\frac{d}{dr}(\sigma_1 r) - \sigma_3 = 0 \qquad (8)$$

Combining this equation with the yield criterion, equation (5),

$$d\sigma_1 = -m \cdot Y \cdot \frac{dr}{r}$$

and, integrating

$$\sigma_1 = -m \cdot Y \cdot \ln r + C$$

where C is a constant ascertainable from the known boundary conditions.The longitudinal stress $\sigma_r = 0$ at the entrance to the die, where $r = r_0$ provides $C = m \bullet Y \bullet . \ln r_0$ and the stresses at any radius r are then

$$\alpha_1 = m \cdot Y \cdot \ln \frac{r_0}{r} \quad \text{and} \quad \sigma_3 = m \cdot Y \cdot \left(\ln \frac{r_0}{r} - 1 \right) \tag{9}$$

In particular, the sinking stress $\sigma_1 = m \bullet Y \bullet \ln (r_0 / r_1)$ where r, is the radius at exít.

The thickness strain is obtained directly by substituting the above stress values in the Lévy-Mises relation, equation (6). Thus,

$$\frac{dt}{t} = \frac{2 \cdot \ln(\frac{r_0}{r}) - 1}{2 - \ln\left(\frac{r_0}{r} \right)} \bullet \frac{dr}{r}$$

Integrating and putting $t = t_0$, the initial wall thickness at entry to the die,

$$\ln \frac{t}{t_0} = 2 \cdot \ln \frac{r_0}{r} + 3 \ln \left[1 - \frac{\ln\left(\frac{r_0}{r} \right)}{2} \right]$$

Finally, the axial strain at any given die radius may be found from the incompressibility condition, equation (7).

3.- TUBE-SINKING WITH WALL FRICTION.

The stress, σ_3, is obtained from the assumed yield criterion (11.5) and substítuted ín the equilibrium equation (11.4), to give

$$\frac{d\sigma_1}{\sigma_1\mu\cdot\cot\alpha - m\cdot Y\cdot(1+\mu\cdot\cot\alpha)} = \frac{dr}{r}$$

Integrating,

$$\frac{1}{\mu\cdot\cot\alpha}\cdot\ln\{\sigma_1\cdot\mu\cdot\cot\alpha - m\cdot Y(1+\mu\cdot\cot\alpha)\} = \ln r\cdot K$$

or

$$\alpha_1\cdot\mu\cdot\cot\alpha - m\cdot Y\cdot(1+\mu\cdot\cot\alpha) = (r\cdot K)^{\mu\cdot\cot\alpha}$$

where K is a constant obtained by knowing that $\sigma_l = 0$ at the entry where $r = r_0$. Hence σ_l is given by

$$\frac{\sigma_1}{m\cdot Y} = \left(1+\frac{1}{\mu\cdot\cot\alpha}\right)\left\{1-\left(\frac{r}{r_0}\right)^{\mu\cdot\cot\alpha}\right\} \qquad (10)$$

By writing $\left(\frac{r}{r_0}\right)^{\mu\cdot\cot\alpha} = e^{\left[\mu\cdot\cot\alpha\cdot\ln\left(\frac{r}{r_0}\right)\cong 1-\mu\cdot\cot\alpha\cdot\ln\frac{r}{r_0}\right]}$ and allowing μ to tend to zero, equation (10) reduces to $\alpha_l = m \bullet Y \bullet \ln(r_0 / r)$ the stress in frictionless drawing.

The variation of sinkíng stress $\frac{\sigma_1}{m\cdot Y}$ with reduction $\frac{r_1}{r_0}$ is shown in Fig. 4 for different values of ($\mu \bullet \cot\alpha$). The load required for hollow sinking is $\frac{2\cdot\pi\cdot r_1\cdot t_1\cdot\sigma_1}{\cos\alpha}$ where t_l is the thickness at the exit side of the die. The thickness strain has already been found when $\mu = 0$ and the procedure is exactly the same when friction is considered.

Putting ($\mu \bullet \cot\alpha$) and $b = \frac{(1+a)}{a}$ the hoop stress σ_3 is, from the yield criterion

$$\frac{\sigma_3}{m\cdot Y} = b\left\{1-\left(\frac{r}{r_0}\right)^{\alpha}\right\}-1$$

Substituting the values of the stresses into the Lévy-Mises equation (6)

$$\frac{dt}{t} = \frac{2 \cdot b \cdot \left\{1 - \left(\frac{r}{r_0}\right)^{\alpha}\right\} - 1}{2 - b \cdot \left\{1 - \left(\frac{r}{r_0}\right)^{\alpha}\right\}} \cdot \frac{dr}{r}$$

Integrating and using the boundary condition ($t = t_0$) where ($r = r_0$)

$$\ln\frac{t}{t_0} = \frac{3}{a-1}\ln\left[\frac{2 \cdot \left(\frac{r}{r_0}\right)^{\alpha}}{2 - b \cdot \left\{1 - \left(\frac{r}{r_0}\right)^{\alpha}\right\}}\right] - 2 \cdot \ln\frac{r}{r_0} \qquad (11)$$

The variation of the thickness ratio (t / t_0) with reduction (r_1 / r_0) is shown in coefficients of friction considered, there is a resultant thíckening of the tube-wall.

The thickness, t_1 , of the tube wall at exit can be found and the load required for hollow sinking evaluated.

It has already been stated that litúe difference is made to the values of the thickness strains if t is assumed constant in the equilibrium equation. It has also been shown again by Swift that strain-hardening has littl effect on the strains; however, it may have a marked effect on sinking loads. If the tube is initially work-hardened, it may be sufficient to assume a mean yield stress, but for a tube that is initially annealed, a stress-strain relationship Must be brought into the solution. (SWIFT, 1949; CHUNG, 1951; HILL, 1950.)

4.- COMPARISON OF THEORETICAL AND EXPERIMENTAL STRESSES IN TUBE-SINKING.

For a comparison of theoretical and experimental sinking stresses, some experimental results published by SACcHs and BALDWIN (1946) are reproduced in Fig.6. In these experiments the ratio (t_0 / r_0) was approximately 0-06, coefficient of friction μ = 0-1, α = 14°; a mean yield stress (m • Y) = 60,000 lbf/in^2 was assumed. It is seen that when the frictional effect is taken into account, the correlation between theoretical and experimental results is good while neglecting the frictional effect may cause errors of up to 20 per cent.

The above analysis sets down the basic equations for the problem of hollow sinking. In industrial practice, the dies are curved and the tubes cannot always be considered as thin-walled (Fig. 1). This means that stresses due to bending and unbending must be taken into account and also the distribution of the stress, a, across the thickness of the tube must be studied. CHUNG and SWIF (1952) have put forward a theory which within the normal industrial range predicts satisfactorily the sinking loads, wall thicknesses and length increments. Descriptively, the me die composed of two oppositely directed radú is as follows, referring to Fig. 1(a).

When an elenient of material in the original tube enters the die, plastic deformation begins at point A, by bending the element to the profile radius since there is no drawing tension at this section, no appreciable alteration in thickness'accompanies the bending. From. A to B, the point where the first profile radius ends, the tensile stress in the tube wall increases continuously owing to "radial" drawing and interfacial die friction; the metal also continues to thícken under the compressive hoop stress. At B the metal unbends and then bends to the reverse curvature of the second part of the die-profile radius, a process involving thinning and a very rapid increase in the sinking stress. From B to C, both the sinking stress and the metal thickness increase continuously. At C unbending, under considerable tensile stress occurs, causing a final decrease in the metal thickness.

The experiments by Chung and Swift consisted in reducing from 1 ¾ to 1 ¼ in diameter well-lubricated steel, brass, and copper tubes of different wall thickrtesses, by drawing them through dies of total vertical conical angle 26° and profile radius 1-226 in, at the rate of 1 in/min. In Fig. 7 the results of thickness-strain /reductio-in-radius determinations, for mild-steel tubes, are compared with those predieted theoretically. The "steps" in the predicted curves correspond to the (theoretically) sudden thinning due to plastic bending at the section contraflexure, B. The stress, σ_2 , in the thickness direction is assumed to have a value equal to half the die face pressure.

A typical. analysis of the constituents of the total sinking stress is given in Fig. 8.

5.- THE DETECTION AND MEASUREMENT OF RESIDUAL STRESSES IN COLD-DRAWN TUBES.

The term residual or internal stress is applied to a stress or a stress system which is induced in an artiele during its manufacture and which does not disappear in the natural relaxation of the article when all external constraints are removed. Thotigh the subject has reccived much attention, the measurement of residual stress has, until recently been only approximate. The object of much experimental work has been to improve the teclinique of measuring residual stresses, particularly as it applies to cold-drawn tubes and other tubular components.

Many methods haye been proposed for detecting and measuring the internal strains in a residually stressed body, but generally these can be classified under tliree main headings:

1. Strain release and measurement by bending deflection.
2. Direct strain release and measurement.
3. X-ray and other non-destructive strain-detection methods.

The basic principles involved in the determination of residual stresses by bending-deflection methods are as follows: For circurriferential stresses, a rin spcimen is cut from a parent tube (the tube under test) and after careful measurement of all the leading dimensions, it is slit longitudinally along any one radial plane. The amount by which the outer diameter of the specimen in the plane perpendicular to that containing the slit alters from its original unstit diameter is a measure of the stress released. By successively reducing the tube-wall thicktiess unilaterally by machining or pickling, and by measuring the current "sprung" diameter at cach stage, the distribution of the residual circurriferential stress throughout the tube wall can be deduced. For longitudinal stresses, a longitudinal tongue is cut from the tube wall and the amount of its deflection and curvature froin the original straight line is a measure of the released stress. Further stress components are released by reducing the thickness of the tongue by pickling, and the longitudinal stress distribution throughout the walls of the tube can then be calculated.

Apart from approximate analyses for the calculation of the primary surface stresses released on slitting a tubular specimen or on cutting a longitudinal strip out of such a specimen, the only two analyses based on the bendingdeflection method of strain release which attempt to provide means for measuring the stress distribution in cylindrical components are those of DAVIDENKOV (1932), and SACHS and ESPEY (1941). The Sachs and Espey analysis is fotind to be more convenient in application than the Davidenkov analysis, but it can be strictly applied otily to relatively thin-walled tubes.

The measurement of residual stresses by direct strain release embodies in principle a reversal of the process by which the strains were produced, i.e., elements are removed from a stressed body and observations are made of the resultant deforroations in the remaining material. The basic work in this field was published by Batier and Heyri, but they considered orily unidirectional stress systems, and in 1927 SACHS published a more general tliree-dimensional stress analysis for what is now widely known as the Sachs boring method. In this method, thin annular layers of material are successively removed from either the inner or the outer surface of a tubular specimen, and the consequent changes in length, diarneter and wall thickness at each stage are measured.

Both bending-deflection and direct-strain-release methods have been considered, and in each case an acid pickling process has been used for removing material from the specimens. The measurement of the strains resulting from. metal removal is generally rnade by micrometers in the bending deflection tests and by electric resistance straingauges in the direct-strain-release experiments.

The residual stresses present in a commercially produced tube are not generally the same at all point around the tube wall, and wide variations in derived-stress curves are obtained frorti identical specimens in the sarne tube that have been slit along different generators.

The residual-stress distribution in hollow-drawn steel tubes is found to be largely tensile in the outer 60 per cent of the be wall (except very near the outer surface) and compressive over the remainder, as Fig.9 shows. The rapidly changing compressive stresses at, and near to, the outer surface of the tube shotild be noted. This surface stress is affected by die friction and is reduced by the use of an efficient lubricant.

The residual stresses which develop in the wall of a hollow-drawn tube do not necessarily increase with the degree of reduction, and in fact, it has been shown experimentally that an inverse relationship may exist.

An extensive review of the status of some of the analyses referred to abóve and analytical analyses of residual stress distribution will be found in a paper by DENTON and ALEXANDER (1963).

6.- IRONING.

Ironing consists, principally, in reducing the wall thickness of a cup by restricting the elearance between punch and die, the reduction in interna diameter of the cup usually being sufficient to allow free entry of the punch only.

In early investigations at Sheffield University, iron ing-load/reduction characteristics for various die angles and states of lubrication were obtained. These have been extended, and the two components of the ironing load, the punch-head load and the frictional or tractional load along the punch, have been measured separately as well as together, by FREEMAN and LEEMING (1953); the punch-head load is of primary importance, since it determines the tension in the cup walls and hence the maximuni reduction possible for a given punch load.

Their experimental apparatus was based on a special 4 in diameter composite ironing punch, having a separate hemispherical head with electric resistance strain-gauges as load-sensitive elements. The recording was electronic and photographic. Their tests were made en hemispherical-headed cups, 0.039 in thick, to determine the ironing-load components for different conditions of wall thinning, die profile, and lubrication, and a typical set of load/reduction curves is shown in Fig.10. A theoretical study of ironing has been made by FUKUI and HANSSON (1970).

7.- SUPERPLASTICITY.

Superplasticity is often taken to refer to large neck-free extensions when deforming certain metals and alloys; the materials which behave superplastically show low strength and high elongation before fracture, of the order of 1000 per cent in some cases. Such behaviour is obviously attractive for use in a forming process, especially as it may well be able to make use of other known pressure-forming techniques, e.g., those of the polymer and glass industries. Aluminium-zinc rods can be drawn into filaments as can glass, and very satisfactory deep-drawing-type products can be formed with superplastic materials, see JOHNSON, AL-NAIB and DUNCAN (1972).
There are two principal classes of superplastic process, (1) transformation plasticity and (2) micrograin plasticity.

7.1.- Transformation Plasticity.

In this class, deformation occtirs during a pliase change whether it be martensitie or diffusion controlled, and because transformation reactions in steeis are well understood many studies have concentrated on ferrous alloys. But allotropic transfórmatícitis in iron, cobalt, titanium and zinc and various eutectoid and precipitation reactions, as well as the martensitic reactions, have also been investigated. It is observed that a threshold stress is necessary to initiate gross deformation; this is attributed to having to move dislocations out froni their sources and to produce pileups at grain boundaries; in some systems the nucleation rate is then increased and thence the transformation rate. Coherency is then lost at the grain boundaries, the dislocations are freed and can move, and thus deformation proceeds. The following features, according to WEISS and KOT (1967), appear to characterize transformation plasticity,

(a) Large neck-free elongation.
(b) A linear relationship between applied stress and strain.
(c) Relative insensitivity to grain size.
(d) Plastic flow duríng a pliase transformation even under external stresses much smaller than that required for normal yield or flow in the material.
(e) A sensitivity to the direction of the transformation and to heating and cooling rates.

GREENWOOD and JOHNSON (1965) have proposed a continuum plasticity model to account for this behaviour which is based on the volume differences that arise during transformation and which then create high internal stresses so that sinall external stresses thereafter can induce larger uniform elongations.

7.2.- Micrograin Plasticity.

Perhaps the best known example of superplasticity is that of the eutectoídd system of approximately 80 per cent aluminium and 20 per cent zinc Extremely high uniform elongations are found with this system at just below the eutectoid temperatures, for relatively small pressures. A lead-tin alloy of eutectic composition also shows superplastic behaviour. The distinguishing feature of these compositions is the stability and fineness of the microstructure.

The behaviour of superplastic alloys is higlily strain-rate sensitive and Fig. 11 is typical, taken from a paper of BACKOFEN, TURNER and AVERY (1964). The strain-rate exponent in $\sigma = B \cdot \dot{\varepsilon}^m$, is usually about 0-2 for many metals but it may reach 0-7 for these special alloys. The consequence of a high value of m is easíly demonstrated thus. In a tensile test, if P denotes the constant tensile force and A the current cross-sectional area then,

$$\sigma = \frac{P}{A} = B \cdot \dot{\varepsilon}^m \quad \text{or} \quad \mathrm{A} = \frac{P}{B} = \cdot\, \dot{\varepsilon}^{-m}$$

$$\text{so that} \qquad -\frac{dA}{dt} = \frac{P}{B} = \cdot\, \dot{\varepsilon}^{-m+1} = \frac{P^{\frac{1}{m}}}{B} \cdot \frac{1}{A^{\frac{(1-m)}{m}}}$$

on eliminating $\dot{\varepsilon}$. If m = 1, then the time rate of decrease of the crosssectional area of the specimen is independent of the cross-sectional area; the load that could be carried would be independent of current cross-sectional area. For a low value of m, say 0-2, the rate of area reduction would. be proportional to A^{-4} and thus it would reduce at an increasingly rapid rate. With this sort of reasoning in mind, it is usually stated that a high value of m inhibits neck development; in the act of fórming a neck, the strain rate increases and thus the material strength, so that load bearing capacity increases.

The characteristics of micrograín plasticity are:

1.Large extensions due to a quasi-viscous behaviour associated with a relatively high strain rate sensitivity index, $m \cong 0\text{-}6$, and a test temperature in excess of 0-5 of the absolute melting point of the alloy or metal.
2. A stable microstructure and a small grain size: in the micron range; the grain boundaries should permit grain boundary sliding. An eutectic or an eutectoid is not essential if a fine grain size can be produced and held.
3. Three mechanisms have been proposed to account for the high strain rate sensitivity, (a) vacancy creep, (b) creep by grain boundary diffusion and (c) grain boundary sliding.

The potentiality of superplasticity is not clear, but its applications are likely to be for non-load bearing (e.g. covers) and decorative components. Certainly the forming process innovations evoked will need to be carefully studied and developed. Forming times anyhow are slow, and there will be a critical need for optimizing forming pressures, stress, strain rate and deflection in sheet-forming. The paper Of JOVANE (1968) is interesting in this respect. Papers of interest concerning superplastic forming will be found in the *International Journal of Mechanical Sciences* pp. 463-524, 1970 and pp. 63-76,1971.

REFERENCES

[1] BACKOFEN, W. A., TURNER, 1. and AVERY, H... "Superplasticity in an Al-Zn Alloy". Trans. A.S.M.. 1966. **57**, p 981

[2] BLAZYNSM, T. Z. and COLE, I. M. "An Investigation of the Plug Drawing Process" Proc. Instn mech. Engrs. 1960. **2,** p 60

[3] CHUNG, S. Y. "Theory of Hollow Sinking of Thin-walled Tubes". Metallurgia. 1951. **43**, p 215

[4] CHUNG, S. Y. and SWIFT, H. W. J. "A Theory of Tube-sinking" Iron Steel Inst. 1952. **170**, p 29

[5] DAVIDENKOV, N.N. "Berechnung der Restsparmungen in kaltgezogenen Rohren" Z. Metallk., 1932, (Feb.)

[6] DENTON, A. A. and ALEXANDER, J. M. "On the Determination of Residual Stresses in. Tubes" J. mech. Engng Sci. 1963, **5**, p 75

[7] FLINN, J. E. "Wall thickness changes-in hollow drawn tubing" Trans. A.S.M.E. 1969, 792

[8] FREEMAN, P. and LEEMING, H. "Ironing of Thin-walled Metal Cups-the Distribution of the Punch Load" B.LS.R.A. Report No. MW/E/46/53. 1953

[9] FUKUI, S. And HANSSON, A. "Analytical Study of Wall Ironing, Considering Work-hardening". Annals of C.I.R.P., 1970, **18**, p 593

[10] GREENWOOD, G. W. and JOHNSON, R. PL. "The Deformation of Metais Under Small Stresses During Phase Transformations". Proc. R. Soc., Series A, 282, **1965,** p 403

[11] HILL, R. "Tube-sinking". The Mathematical Theory of Plasticity 0.U.P. 1950

[12] JOHNSON, W., AL-NAIB, 1972 T. Y. M., and DUNCAN, J. L. "Superplastic Forming Techniques and Strain, Distribution in a Zinc-Aluminium Alloy".' J. Inst. Melais, 1972, **100**, p 45

[13] JOVANE, F. "An Approximate Analysis of the Superplastic Forming of a Thin Circular Diaphragm: Theory and Experiments". Int. J. mech. Sci. 1968, **10**, p 403

[14] MOORE, G.G. and WALLACE, J. F. "Theories and Experiments on Tube Sinking Through Conical Dies". Proe. Instn mech. Engrs, 1967, **182**, p 19
[15] SACHS, G. and BALDWIN, W. M . 'Folding in Tube-sinking'. Trans. A.S.M.E., 1946, **68**, p 647
[16] SWIFT, H. W. "Stresses and Strains in Tube-drawing" Phil. Mag. Ser. 7, 1949, **11**, p 883
[17] WEISS, V. and KOT, R."Superplasticity".Int. Conf. on Manufacturing Methods, A.S.T.M.E., 1967, p. 1031

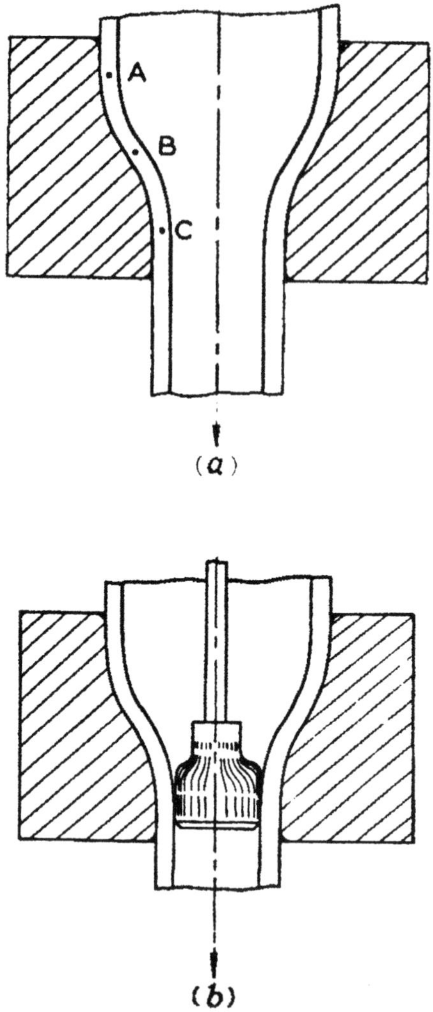

FIGURE 1.-METHODS OF COLD-DRAWING
A) HOLLOW SINKING
B) PLUG DRAWING

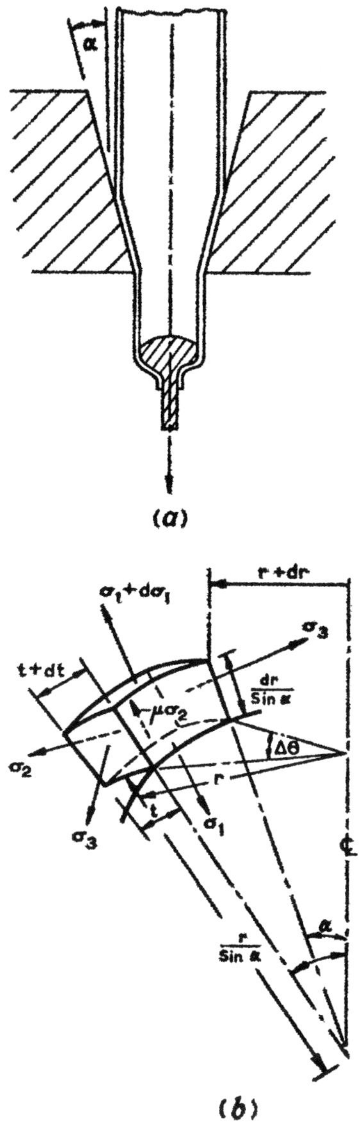

FIGURE 2(a).-HOLLOW SINKING THROUGH STRAIGHT DIE
FIGURE 2(b).- STRESSES ACTING ON SMALL ELEMENT OD THE WALL

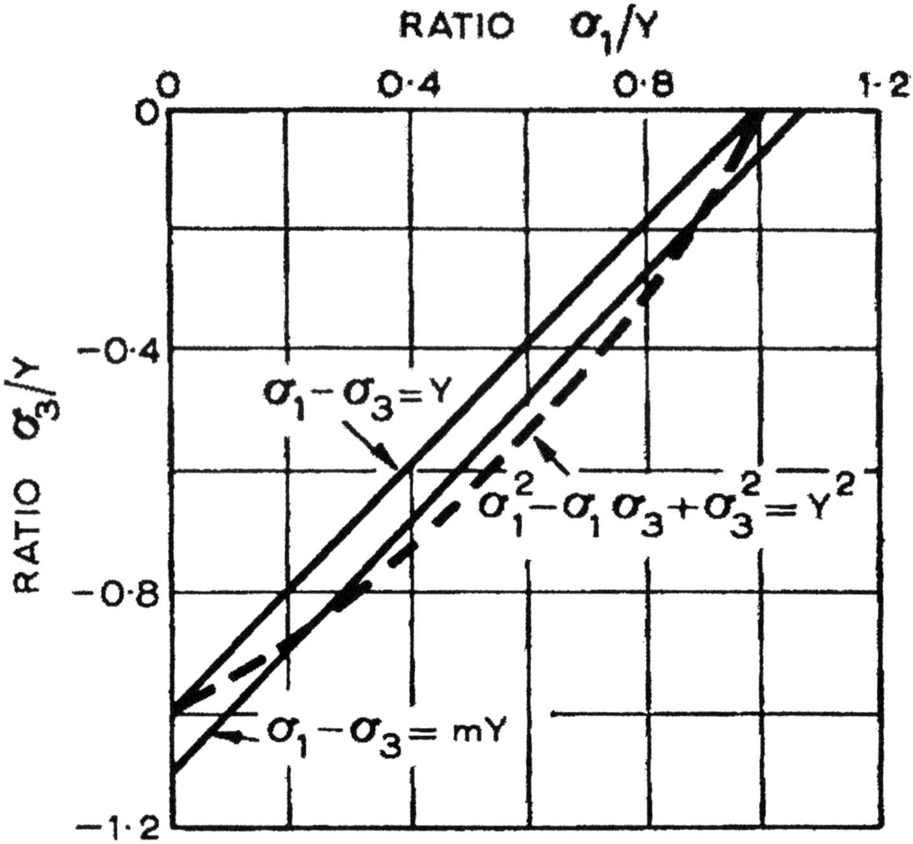

FIGURE 3.-GRAPHICAL REPRESENTATION OF YIELD CRITERIA

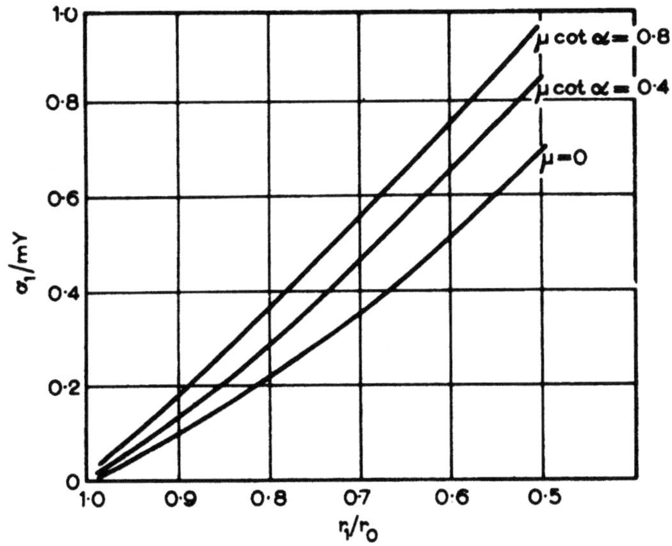

FIGURE 4.- VARIATION OF SINKING STRESS WHIT REDUCTION IN RADIU.

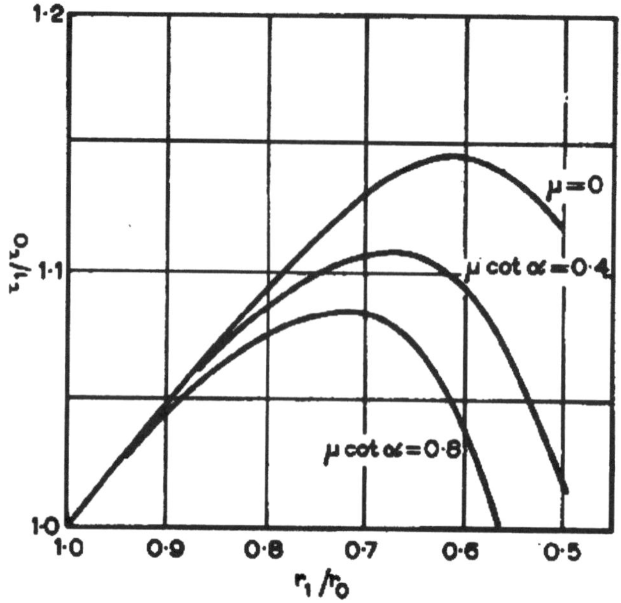

FIGURA 5.- VARIATION OF THICKNESS STRAINT WITH REDUCTION IN RADIUS.

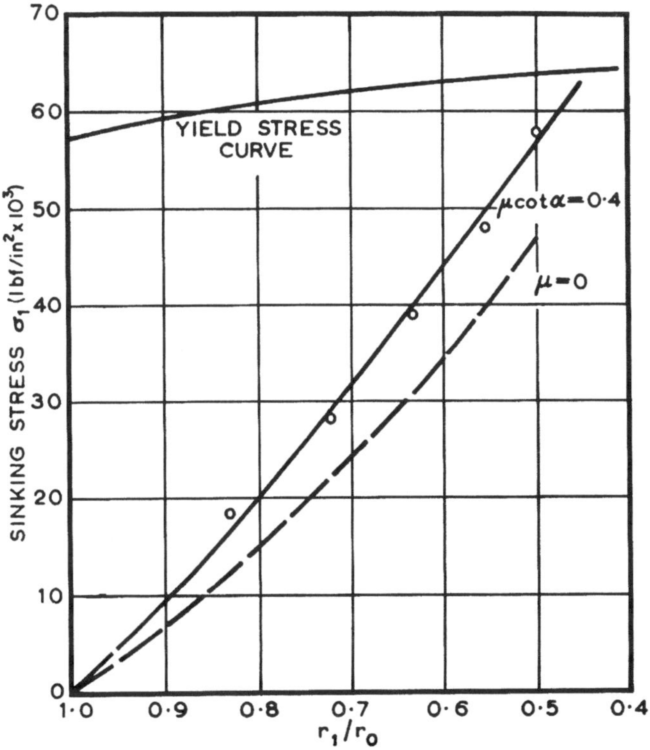

FIGURE 6.- THEORETICAL AND EXPERIMENTAL SINKING STRESS FOR HARD COPPER.

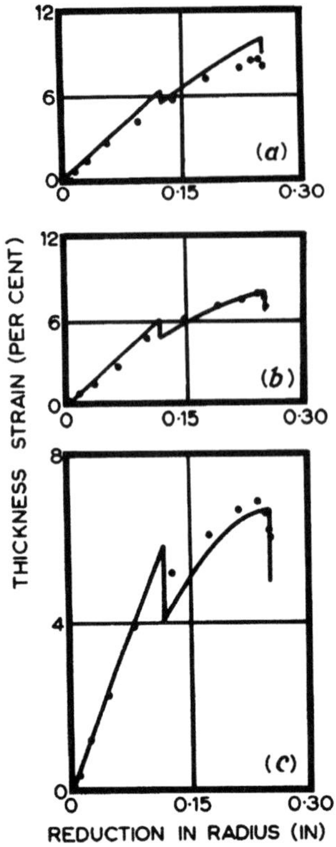

FIGURE 7.- COMPARISION OF THEORETICAL AND EXPERIMENTAL THICKNESS STRAINS DURING THE SINKING OF MILD-STEEL TUBES

(c) 0,0635 in

(d) 0,1052 in

0,1305 in thick tubes

Constituent	Wall Thickness 0.0635 in		Wall Thickness 0,1305	
	lbf/in^2	%	lbf/in^2	%
Pure radial deformation	24,80	75.6	27.320	62.2
Bending and unbending	3920	11.9	8280	20.7
Surface frication	5530	16.8	5970	15.0
Thickness variation	-1400	-4.3	-1560	-3.9
TOTAL	32.850	100	39.990	100

FIGURE 8.- CONSTITUENTS OF TOTAL SINKING STRESS.

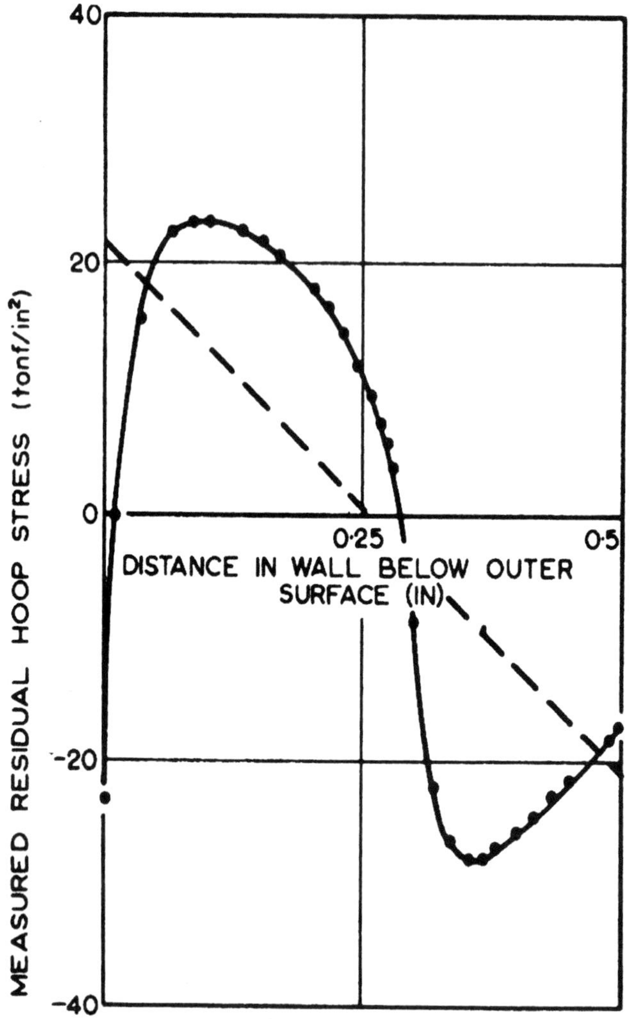

FIGURE 9.- RESIDUAL HOOP-STRESS DISTRIBUTIÓN IN A 3.5 IN OUTSIDE DIAMETER TUBE DRAWN DOWN AT 16 ft/min FROM A TUBE 16 ft/min.WALL TICKNESS OF TUBE = 0.499

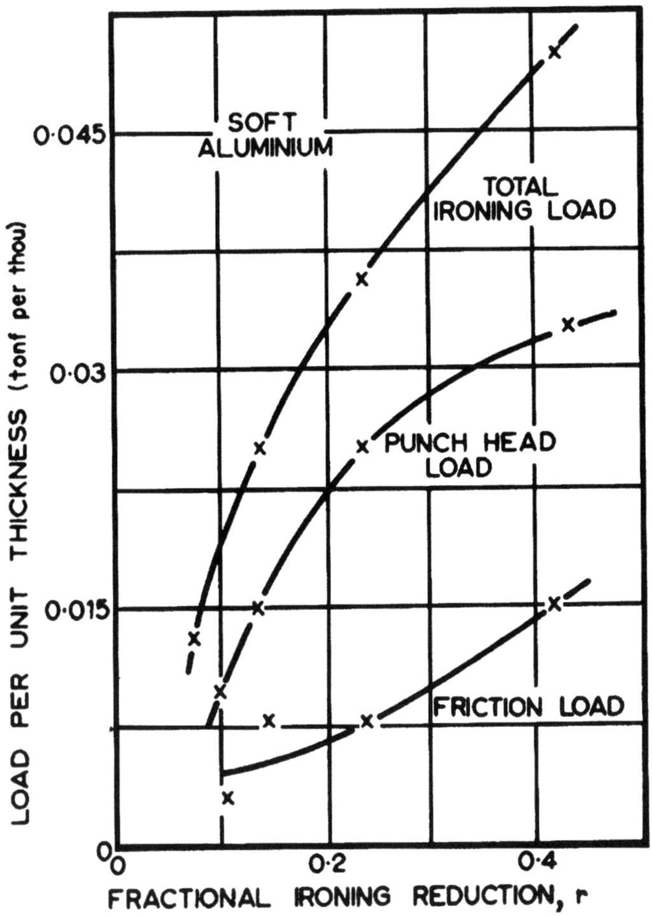

FIGURE 10.- VARIATION OF IRONING-LOAD COMPONÓNTS WITH REDUCTION.

Semi-angle of die = 10°.

Graphíte-in-tallow lubrication.

Elaboration of cast Fe-Cr-Cu-C system composites for sliding friction

V G NOVYTSKYY, V I TIKHONOVICH, and **V P HAVRYLIUK**
New Cast Material Department, Phisico-Technological Institute of Metals and Alloys of NASU, Kyiv, Ukraine

ABSTRACT

The correct selection of materials and determination of wear conditions, when a selective transfer is realized, are essential to wear decrease of metal–to–metal contacts. The stable indication of the selective transfer is a formation of a protective metal film depending on initial lubricating medium. In this case the direct alloying of alloys by elements able to form the protective metal film in contact zone is of an interest. Copper added over its solubility limit in the Fe-Cr-C alloys can serve as such an element. The alloys containing 1.3 wt% carbon, 15 wt% chromium and additionally alloyed by copper, vanadium, titanium were selected as subject of inquiry. On the bases of the studies it has been established the optimum technology parameters for production of Fe-Cr-Cu-C-X-alloys and the mechanism of the phase with high content of copper formation.

1. INTRODUCTION

Investigation and elaboration of new materials are of state interest for most industrial countries [1,2]. The problem to raise the reliability and the longevity of machines and equipment is narrowly connected with both the elaboration of wear resistance of materials and technology of it manufacture. The correct selection of materials and determination of wear conditions , when an selective transfer is realized, are essential to wear decrease of metal-to-metal contacts.

The selective transfer arising from chemical and physical-chemical processes on the friction surfaces favours the wear autocompensation and this way leads to wear minimization. The stable indication of the selective transfer is a formation of a protective metal film depending on initial lubricating medium with different ability to wear decrease. One of the methods to obtain the thin plastic interlayer in contact zone is the application of lubrication containing a powder of film-forming metal (bronze, brass) [3,4]. This plastic interlayer, where the diffusion-vacancy shear mechanism is realized, protected the friction surfaces against intensive wear. However in a number of cases the application of these lubrications is not possible because of the constructive considerations and the wear conditions of the friction units. In this case the direct alloying of alloys by elements able to form the protective metal film in contact zone is of an interest. Copper added over its solubility limit in the Fe-Cr-C alloys can serve as such an element. It leads to appearance of the phase with high content of copper (ε-phase). Unfortunately the manufacture of those cast composites is combined with some technological difficulties due to insufficient investigation of the Fe-Cr-Cu-C alloys in liquid and solid states.

The copper alloying of the Fe-C alloys has been applied to practice for a long time [5,6,7]. Over the last years the copper draws attention as alloying element raising a technological and service properties of wear resistance steels and cast irons [8,9,10,11].

2. EXPERIMENTAL DETAILS

The alloys containing 1.3 wt % carbon, 15 wt % chromium and additionally alloyed by copper, vanadium, titanium were selected as subject inquiry. The compositions of the alloys are given in the Table 1.

Table 1. Identification and composition of the alloys studied

Sample identification	Composition weight %				
	C	Cr	Cu	Ti	V
1	1,33	15,36	9,82	-	-
2	1,29	15,28	9,91	0,31	-
3	1,32	15,41	10,21	-	0,32

To determine the initial data for the melting technology elaboration of experimental Fe-Cr-Cu-C alloys, the critical points of alloys were fixed. They were determined with the help of γ-absorption method allowing to fix the density of alloys in range between 20 and 1700^{0}C [12,13]. The slackening of the γ-radiation current depends on the density of examined alloys. The measuring of the alloys density was 0,022 g/cm^3 accuracy.

The alloys studied were prepared by melting in 2-kg capacity high-frequency induction furnace. Each melt was tapped in to dry sand moulds. The specimens were examined with optical and scanning electron microscopes. The wear tests were performed on pin-on-ring machine with pins 10x10 mm and 15 mm length. The ring was made of a special steel alloy containing 0,2 wt % C, 13 wt % Cr with surface hardness of 38-40 HRC. The nominal contact surface of pin with ring was 100 mm^2. The ratio of contact areas pin to counter-sample (ring) was 1/13. The studied alloys were tested under dry wear (in air), boundary wear (the water was supplied to contact zone), and mixed wear (the oil И20А with kinematic viscosity (17-23)x10^{6}m^{2}s^{-1} by 50 ^{0}C was supplied to contact zone). The sliding velocity and the specific load were 1 m/s and 5 MPa accordingly. The wear measured by the weight loss, using an analytic balance with 0,0001 g accuracy.

3. RESULTS AND DISCUSSION

The density measuring of alloys by melting and solidification allowed to determine the temperature of liquidus, solidus, and homogenization of studied alloys. The density polyterms of 1 and 2 alloys are shown in Fig. 1. One of polyterm branch relates to the stable condition of melt (cooling polyterm), the other relates to metastable condition (heating polyterm) [13].

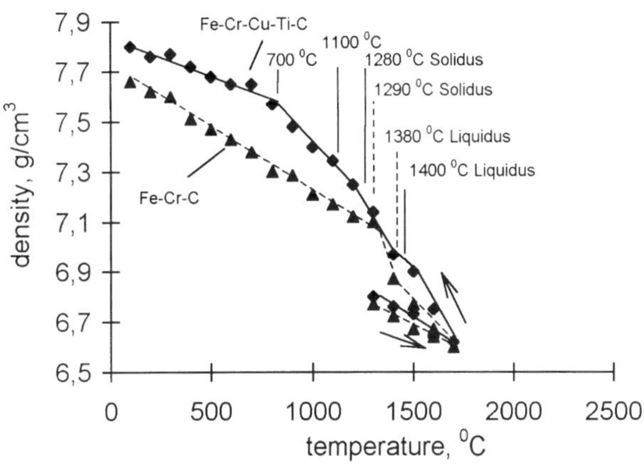

Figure 1. Representative dependencies of alloys density on temperature.

The transition temperature of alloys from the metastable to stable condition is the homogenization temperature of melts.

This point of polyterm branching shows the hysteresis of properties. The analysis of heating polyterm curves (Fig.1) indicates different density of studied alloys. The density of melt 2 is more then density of melt 1, which is attributed by copper content in alloy 2. The density of studied alloys is increased as the result of homogenization process. The comparative analysis of density curves dependency on a temperature (by cooling) shows some difference between the curves. The curve (alloy without copper) has only one sharp inflection in temperature interval ($T_{liquidus}$=1380 ^{0}C and $T_{solidus}$=1290 ^{0}C). The curves of alloy 2 does not have the sharp inflection in the temperature interval ($T_{liquidus}$=1400 ^{0}C and $T_{solidus}$=1280 ^{0}C); however, it has 2 extremes: one extreme at 1050-1100 ^{0}C and the other at 700 ^{0}C.

The first extreme shows the final solidification of precipitates with high copper content (ε-phase). The second extreme shows the ε-phase precipitation out of α-solid solution. Copper has a little effect on the liquidus and solidus temperatures, which are found in the range of 1380-1400 ^{0}C and 1280-1290 ^{0}C, and on homogenization temperature, which is 1650 ^{0}C for studied alloys. The determination of liquidus, solidus temperatures and the date work [13] gave an opportunity to determine a termo-time processing (TTP) conditions of alloy 2. The melt was overheated above liquidus curve by 250-280 ^{0}C, but not less then homogenization temperature. The melt held at this temperature leads to the change in its short range order structure of the parent composition and of complex similar to carbides.

The changes of the complex similar to carbides due to TTP and transition of some carbon as well as weak carbide-forming elements into parent composition favour the decrease of statistic mean distance between the atoms and as a result leads to melt parking which is reflected on the density polyterms (cooling polyterms). After the melt was held at homogenization temperature copper and modifier element were added to cooling. Then the melt was deoxidized by silico-calcium in the range 0.2 wt% of total furnace charge. The casting was made at the temperature above liquidus curve by 80-100 ^{0}C. To regulate a distribution and dimensions of ε-phase precipitations, the alloys were modified by the use of titanium or vanadium.

Titanium generates high temperature carbides and carbonitrides which function as I-kind modifier [14]. The modification effect of titanium may be intensified due to the isomorphism of crystal lattices both of titanium carbides and the ε-phase. In this case the crystallization of the ε-phase will be realized on prepared base under the most advantageous energy situation. Besides that, the high temperature titanium carbide will be primary phase under solidification process around which the liquid volumes with less carbon are formed.

One of the fact that copper almost does not interact with carbon, copper accumulates in these volumes that intensifies the liquid solution separation. The more microvolumes with less carbon in the melt, the more crystallization centers of the ε-phase will be. These microvolumes will favour both the minimization of the ε-phase size and the increase of their quantities. It assumed that eutectic carbides that have less carbon zones around themselves will have the same influence on the ε-phase. The crystallization of such alloys is developed by eutectic transformation scheme and it will be completed with crystallization of eutectic and ε-phase. Vanadium added into alloy up to 2.0 wt% dissolved in metallic matrix and eutectic carbides exerts modification effect on the size of austenite grains and carbides [15]; consequently , on the size and quantity of the ε-phase.

Alloying of Fe-Cr-Cu-C alloy (Fig.2a) by ≈0.3 wt % titanium has an effect on their structure, making ε-phase located in the austenite grains in to the sphere form (Fig. 2b). The ε-phase around eutectic carbides has an unspecified nature; however often repeating the eutectic carbides form. Alloying of Fe-Cr-Cu-C alloy by ≈ 0.3 wt% vanadium has analogous effect to titanium . However in this case the ε-phase precipitated in austenite grains has a bigger size than the size of ε-phase obtained by titanium alloying (Fig. 2c).

The quantitative estimate of both size and volume of ε-phase was carried out with the «Epiquant». Alloying of studied alloys by titanium promotes significant increase in small precipitates (up to 20 μm) of ε-phase (Fig. 3), but a quantity of a greater size precipitates almost does not change. The obtained results accentuate the modification effect of titanium not only on the form of precipitates but also on their quantity. The distribution of alloying elements between studied phases of alloy 2 are shown in Table 2.

The results of wear tests of studied alloys under different friction conditions are shown in Fig.4. It is observed that the wear rate of the alloy containing copper under dry friction is decreased by 11,5 times then that of the alloy without copper. The coefficient of friction of alloy 3 is decreased by 1.7 times. The spectral analyze of the counter-sample surface showed over 2.0 wt % copper on it. In this case the transfer of copper from the pin to the counter-sample leads to the wear decrease according to the rule of positive gradient [16].

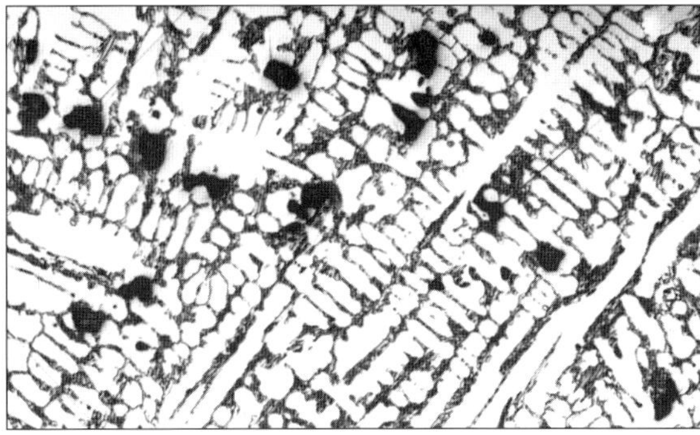

a

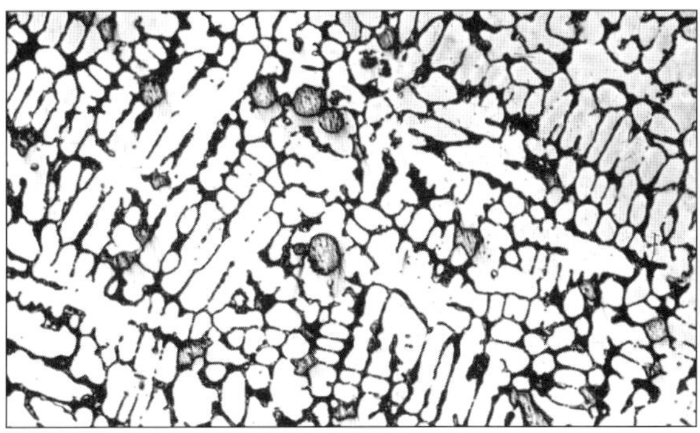

b

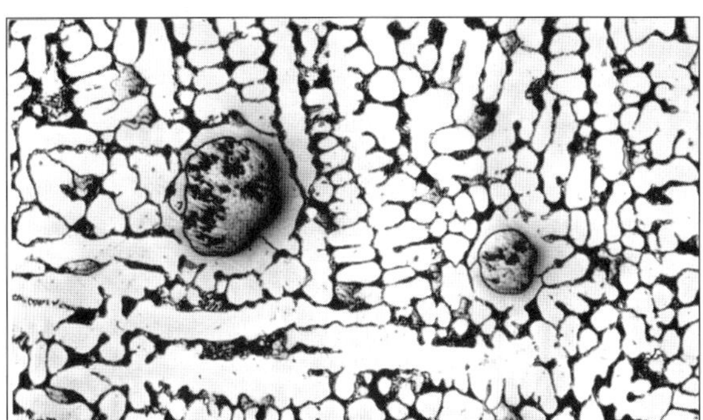

c

Figure 2. Micrographs of initial structure of Fe-Cr-Cu-C (a), Fe-Cr-Cu-Ti-C (b), Fe-Cr-Cu-V-C (c) alloys (x100).

The positive effect of copper is shown under boundary wear when the film with the thickness up to several molecules takes place in contact zone. This film may be interrupted under friction conditions. The wear rate of the pin is decreased by 1.5 times but the friction coefficient does not change. There is up to 3.0 wt % copper on the counter-sample surface that caused the wear decrease of the latter up to 3 times.

Under the mixed wear condition, the interrupted lubrication layer is formed between contact surfaces. In spite of the fact that the kinematic viscosity of lubrication exerts considerable influence on the wear, the antifriction properties of metal-to-metal still are shown. The copper transfer from alloy 3 to the counter-sample surface is up to 0.05 wt %. In this case the wear of the pin isn't fixed, mean while the wear of counter-sample decreases by 10 times.

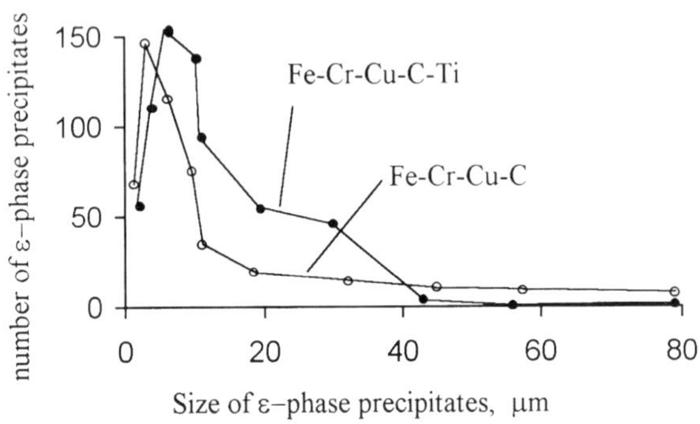

Figure 3. Effect of titanium on quantity and size of precipitated phase (ε-phase) of Fe-Cr-Cu-C steel

Table 2. Identification and composition of the structure components of the Fe-Cr-Cu-C-Ti alloy

Structure components	Composition, %							
	Fe		Cu		Cr		Ti	
	wt %	at %	wt %	at %	wt %	at %	wt %	at %
1	78,6870	69,6435	7,7007	7,0016	13,6723	11,9104	-	-
2	3,5874	2,1372	96,2526	85,0611	0,1567	0,1557	0,0033	0,0037
3	45,6222	24,6232	0,2201	0,1707	53,8404	43,2760	0,0838	0,0668

1- parent austenite; 2 - ε-phase; 3 - eutectic mixture of austenite and Me_7C_3 carbides

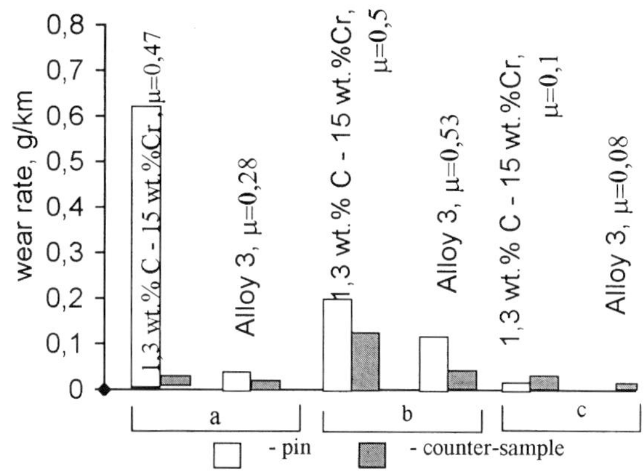

Figure 4. Representative dependencies of wear rate on copper and wear conditions: in air (a), in water (b), in oil (c).

CONCLUSIONS

- The optimum technology parameters for production of Fe-Cr-Cu-C-X alloys are determined.
- The mechanism of the ε-phase formation is proposed. According to it the separation of liquid with formation of zones with less carbon and lattices isomorphism of leading phase under crystallization (austenite or titanium carbides) and high copper contained phase are necessary conditions that facilitate the formation of the ε-phase.
- The copper alloying over its solubility limit in the Fe-Cr-C alloys decreases the wear of contact materials under dry, boundary, and mixed wear conditions.

REFERENCES

[1] Shandrakova I.B., "The future is belonged to the countries creating new technologies", Westnik RAN, 1992, 12, pp. 3-14 (in Russian).

[2] Neale M.I., "New directions" in tribology, First World Tribology Congress, London, 1997, pp. 57-65.

[3] Garkunov D.N., "Tribotekhnika", Mashinostroenie, Moskva, 1985, p.424, (in Russian).

[4] Kragelskiy I.V., "Trenie, iznashivanie i smazka", Spravochnik, 2, Mashinostroenie, Moskva, 1979, p.360 (in Russian).

[5] Gudzov N.T., Mikhailov-Mikheev P.B.," Effect of copper on mild steel", Westnik metalloindustrii", 1928, 3, pp. 81-94 (in Russian).

[6] Lunev A.A.,"The cast wear resistance steels containing copper", Liteynoe proizvodstvo, 1955, 5, pp. 15-18, (in Russian).

[7] Mikhailov P.B., "Chemical endurance and mechanical properties of gray cast iron containing copper", Westnik metalloindustrii, 1926, 9-10, pp. 5-22, (in Russian)

[8] May J, Shetky L.M., "Copper in Iron and Steel", NY, 1982, p.312.
[9] BobroYu.G., Platonova A.A., "Some peculiarities of microstructure of aluminum cast iron containing copper", Novoe w metallovedenii chuguna, 1981, pp.94-99, (in Russian).
[10] Kirievskiy B.A., Izumova T.K., "Improvement of a content, a structure, and properties of chromium cast iron", Liteynoe proizvodstvo, 1992, 9, pp.17-19, (in Russian).
[11] Tikhonovich V.I., Kovalenko O.I., A.Yu.Bobro A.Yu., Novytskyy V.G., "The formation of peculiarities of cast structure of the Fe-Cr-Cu-C alloys", Casting Processes, 1994, 3, pp.23-30, (in Russian).
[12] Kazachkov S.P.,Kochegura N.M., Markovskiy E.A., "Investigation of density of metals and alloys in solid-liquid condition by means of radioactive isotopes", Liteynye splavy, Kyiv, 1973, pp.223-226 (in Russian).
[13] Kochegura N.M., Markovskiy E.A., "New technologies for high quality articles of heat-resistant alloys", Liteynoe proizvodstvo, 1992, 9, 14-16, (in Russian).
[14] Vozdvizenskiy V.M., Grachov B.A., "Liteynye splavy i technologiya vyplavki w mashinostroenii", Metallurgiya, Moskva, 1984, p.432, (in Russian).
[15] Tikhonovich V.I., Gorskiy V.V., Kovalenko O.I., et al., "Carbides and chromium content of 130X15 alloy additionally alloyed by vanadium, niobium, molybdenum, and tungsten", Litye iznosostoykie materialy, Kyiv, 1978, pp. 25-28, (in Russian).
[16] Kragelskiy I.V., Dobychin M.N., Kombalov V.S., "Osnovy raschotov na trenie i iznos", Mashinostroenie, Moskva, 1977, p.528, (in Russian).

Study of the constitutive law for a polycrystal and analysis of rate boundary value problem in finite elastoplasticity

C DE LA CRUZ GÓMEZ
Departamento de Mecánica Aplicada, Universidad de Castilla-La Mancha, Almaden, Spain

ABSTRACT.

The deformation of a single crystal is determined by the elastic energy embedded in the lattice and the plasticity due to slip governed by the Schmid rule or a viscoplastic potential rule. If we refer to the triad of lattice vectors the transformation is described by splitting the velocity gradient into three parts (elastic rate deformation, rotation of the triad of lattice vectors, plastic rate gradient), and the elastic constitutive law is determined by a linear relation between the elastic deformation rate and the objective Jaumann rate of the Cauchy stress in the lattice vectors rotation. With the help of the Hill-Mandel macrohomogeneity hypothesis we can show that the macro velocity gradient has the same decomposition as the micro one and that the stress-strain relation has the same form for a good choice of the triad of vectors which give at each time the orientation of the polycrystal. With this type of description, if the plastic part of the velocity gradient derives from a potential at the micro scale, the macro one derives too from a viscoplastic potential. After this description we analyse the symmetry or non-symmetry of the rate boundary value problem in terms of velocity gradient and rate of internal variables for different yielding surfaces.

1.- INTRODUCTION.

For describing the behaviour of polycrystals we introduce macro-, mechanical quantities, such as stresses, strains and elastic strains, which must be defined from a knowledge of the corresponding micromechanical quantities. To determine the constitutive law forí, polycrystals we use the Hill-Mandel method. We consider the polycrystal as a macrohomogeneous body; in fact it contains several crystals with different shapes and orientations. Inside its volume the stress and strain are not uniform. and are in equilibrium with homogeneous boundary conditions. First we mention the behaviour of an ideal single crystal. After this the hypothesis of macrohomogeneity is formulated and the polycrystal behaviour is analysed. The form. of the macro constitutive law is the same as the micro one and it is necessary to introduce a triad of vectors to ensure uniqueness of the decomposition of the strain rate in the elastic and plastic parts. The evolution of the plastic part is obtained by a yield surface, and for a good choice of plastic potential and its evolution the rate boundary value problem has a symmetric form.

2.- CONSTITUTIVE LAW FOR A SINGLE CRYSTAL

The constitutive law for a single crystal deforming by dislocation glide can be formulated according to the kinematic scheme based largely on the analysis of Mandel (1971) or Hill and Rice (1972), Zarka (1973) and Rice (1971).

The total deformation F can be decomposed as follows: a plastic transformation P given by a set of successively simple shears on the active slip systems which are referred to a fixed lattice, followed by an elastic lattice transformation E which deforms and rotates the material and the lattice together so that $F = E \bullet P$. By unloading the single crystal. and putting its lattice orientation back at its initial position one obtains a unique stress-free state C_K which is defined by P.

It is obvious that the reversible part of the total deformation is determined by the Lagrangian strain tensor $\Delta_K = ½ (E^T E - \ \ )$. Assuming that the hardening has no influence on the elastic properties, we can take on C_K for the free energy the form

$$\Phi\left(\Delta_K, \{\gamma^r\}_r\right) = \frac{1}{2} \Delta_K \Lambda \Delta_K + h\left(\{\gamma^r\}\right) \qquad (1)$$

where h is a function of the internal parameters $\{\gamma^r\}_r$, which are the quantity of slipping γ^r on each slip system (slip in the direction m^r in the plane with normal n^r).
The elastic constitutive law is given by

$$\frac{\Pi_r}{\rho_K} = \frac{\delta\Phi}{\delta\Delta_K}; \quad \frac{\mathrm{A}_r}{\rho_K} = -\frac{\delta\Phi}{\delta\gamma^r} \qquad (2)$$

where Π_K is the Piola-Kirchhoff tensor referred to C_K and A_r, the thermodynamic force associated with the internal parameter γ_r; then the Cauchy stress $\sigma = \rho E^T(\Pi_K / \rho_K)E$ is in equilibrium with the prescribed loading, and the dissipation rate is

$$D = \frac{\sigma}{\rho}\nabla v - \dot{\phi} = \frac{\psi}{\rho_K}\dot{P} \bullet P^{-1} + \frac{A_r}{\rho_K}\dot{\gamma}^r \geq 0 \qquad (3)$$

As the plastic strain rate on C_K is the sum of simple shears on the active slip system, then

$$\dot{P} \times P^{-1} = \sum \dot{\gamma}^r \quad \mathrm{m}_t \otimes \mathrm{n}_r \qquad (4)$$

The expression for D becomes

$$D = \sum_r \left(n_r \times \frac{\psi}{\rho_K} \times m_r + \frac{A_t}{\rho_K} \right) \times \dot{\gamma}^r \qquad (5)$$

with $\psi / \rho_K = E^{-1} x\, \sigma = \rho$ x E.

To determine the evolution of γ^r, one can define by $f^r = n_r \mathrm{x}\, \psi / \rho_K \mathrm{x}\, m_r + A_r / \rho_k - \tau_0$ the yielding function such that $\gamma^r \geq 0$ if $f^r = 0$ and $\gamma^r = 0$ in the other cases.

This is a generalized form of the Schmidt law.

With this definition, the elastic domain is a convex function of V/PK, and we have the law of normality

$$\dot{P} \times P^{-1} = \sum_{r} \dot{\gamma}^{r} \frac{\frac{\delta f^{r}}{\delta \psi}}{\rho_K} \qquad (6)$$

The total plastic gradient rate is determined on C_K; symmetric and antisynunetric parts are given by the plastic potentials f^r.

In the case of small elastic strain (E = SR with S = I + ε and $(\|\varepsilon\| \leq\leq 1)$ the local velocity gradient has the form

$$\nabla \upsilon = d_e + \omega^d + R\dot{P} \times P^{-1} R^{-1} \qquad (7)$$

and the objective stress-strain rate relationship in the actual state is determined by eqn (2) (Mandel, 1971; Halphen, 1975; Stolz, 1982)

$$\frac{D}{Dt}\left(\frac{\sigma}{\rho}\right) = \Lambda d_e \qquad (8)$$

where D(or/p)/Dt is the Jaumann derivative of or/p in the rotation (0 d = k. R-1 of the lattice (Mandel, 1982).

3.-THE MACROHOMOGENEITY HYPOTHESIS.

The polycrystal has a volume V, in which two scales are distinguished: the micro where the material properties vary from point to point l¡ke a highly heterogeneous body and the macro where the properties are those of a homogeneous continuum.

To determine with accuracy the overall properties, it is essential to define the so-called representative volume element (RVE), small enough to allow us to distinguish microscopic heterogeneities and sufficiently large to represent the overall behaviour. For a polycrystalline aggregate it is necessary to have a minimum of information about the geometry of each constituent and to make assumptions of statistical homogeneity or ergodicity in order to define the RVE in a statistical sense (Krúner, 1980; Willis, 1981; Hashin, 1983).

With every microscopic quantity f we can associate its macroscopic value $\bar{f}$ by an averaging process as the RVE:

$$\bar{f} = \frac{1}{\upsilon} \int_{\upsilon} f \, d\upsilon = \langle f \rangle \qquad (9)$$

It is clear that in this way a unique macro state is defined for each micro state.

Conversely, starting with a macro state, the definition of a suitable corresponding micro state requires complementary information or a localization process to determine the nearest possible micro state to the real one. For the choice of a representative volume element suitable boundary conditions must be prescribed and must satisfy some requirements (Francfort et al., 1983; Germain et al., 1983).

In particular, for study of the macroscopic behaviour these boundary conditions can be homogeneous; in this case one prescribes homogeneous stress conditions $T^d = \sum \cdot \vec{n}$ on δV or homogeneous strain conditions (U = E × y for y ε × δV), where Σ and E are two synunetric tensors. For these boundary conditions, a local stress distribution σ and a local deformation ε are developed over the volume V. In the case of homogeneous loading the stress or verifies the equation of equilibrium and the boundary conditions

$$\text{div}\sigma = 0 \ \text{over} V; \qquad \sigma \times \vec{n} = T^d \ \text{on} \ \partial V \qquad (10)$$

In the other case the local deformation ε derives from a local displacement u which verifies in small strain

$$\varepsilon_{ij} = \frac{1}{2}\left(\frac{\partial u_i}{\partial x_i} + \frac{\partial u_j}{\partial x_i}\right) \ \text{over} \, V; \qquad u = U \ \text{on} \ \partial V \qquad (11)$$

We require that the Hill-Mandel macrohomogeneity condition is fulfilled by all fields ε* with $\bar{\sigma}* = \langle \sigma * \rangle$ and by all ε* kinematically admissible so that

$$\bar{\sigma} * \bar{\varepsilon} * = \langle \sigma * \varepsilon * \rangle \qquad (12)$$

With this condition it is obvious that in the case of homogeneous loanding

$$\sum = \bar{\sigma}$$

and the microdisplacement u fluctuates around the homogeneous lacement U such that

$$\int_{\partial V} (U - u) \otimes n \quad da = 0 \qquad (13)$$

Dually, if u = U over δV it is clear that:

$$E = \frac{1}{V} \int_{\partial V} \frac{1}{2} (u \otimes n + n \otimes u) da = \langle \varepsilon \rangle = \bar{\varepsilon} \quad (14)$$

the microstress is such that

$$\int_{\partial V} (\bar{\sigma} n - \sigma n) \otimes y \times da = 0; \quad \Sigma = \langle \sigma \rangle \qquad (15$$

3.1.- Case of a macrohomogeneous body.

It is well known that the macroscopic elastic modulus has not the samevalue when macrohomogeneous strain or stress over V is prescribed. But when the body can be considered as macrohomogeneous in the sense of Hill (1966, 1967) and Mandel (1964) the difference between two estimates is negligible.

So when homogeneous boundary conditions are prescribed in stress ($T^d = K \cdot \vec{n}$ on ∂V; whit a symmetric tensor) the local stress is ined as the solution of an elastic linear problem, and we can prove existence of a localization tensor A_{ijpq} such that

$$\sigma_{ij} = A_{ijpq} \Sigma_{pq} \qquad (16)$$

The tensor A is the elastic localization tensor introduced by Mandel 1964 in the case of a localization process in homogeneous macrostresses.
As $\langle \sigma \rangle = \Sigma$ for all prescribed Σ, we have

$$\langle A_{ijpq} \rangle = \frac{1}{2} (\partial_{ip} \partial_{jq} + \partial_{iq} \partial_{jp}) \qquad (17)$$

For fixed subscripts (p, q) A_{ijpq} satisfies the equilibrium equations

$$A_{ijpq,j} = \frac{1}{2}\left(\partial_{ip}\partial_{jq} + \partial_{iq}\partial_{jp}\right)$$

(18)

$$A_{ijpq} n_j = \frac{1}{2}\left(\partial_{ip}\partial_{jq} + \partial_{iq}\partial_{jp}\right) n_i \quad \text{over } \partial V$$

and the deformation $\varepsilon_{ij} = \Lambda^0_{ijkl} A_{klpq}$ satisfies the condition of compatibility.

3.2.- Global domain of elasticity.

At the local level there exists an elastic domain C defined by $C = \left\{ \frac{\sigma}{f^r(\sigma)} \leq 0, \quad \forall r \right\}$ (we consider here small strain, $\psi = R \cdot \rho \cdot R^T$, and the rotation R is kept constant).

At the macro level there exists an elastic domain **E** defined as follows. Consider two macrostresses Σ and Σ^* by means of a localization process with purely elastic behaviour. It is obvious that $\sigma - \sigma^* = A\ (\Sigma - \Sigma^*)$ where A is the localization tensor introduced in the preceding section (Bui, 1969).

3.3.-On the decomposition of the macroscopic deformation (small strain).

Let Σ be the real macro stress and σ the corresponding micro one. At the initial temperature T_0, the local response in purely elastic behaviour is

$$\sigma_T = A \cdot \Sigma \qquad \textbf{(19)}$$

The stress field $r = \sigma - \sigma_E$ is a self-equilibrated stress. At small strain, the total deformation ε is the sum of the elastic strain ε^e related to σ with the elastic modulus ($\varepsilon^e = \Lambda_0 \cdot \sigma$), and of the plastic strain ε^P. The stress field **r** is such that $\bar{\varepsilon} = \Lambda_0 \cdot r + \varepsilon^P$ satisfies the condition of compatibility. By definition the macroscopic elastic deformation due to the given macrostress Σ is the deformation deduced by a purely elastic unloading (Mandel, 1971). The local deformation $\varepsilon_E = \Lambda_0\, \sigma_E$ is related to E_E by the lemma of macrohomogeneity and $E_E = \langle \varepsilon_E \rangle$. Then for all stresses σ or' in equilibrium with $\Sigma' = \langle \sigma' \rangle$ we have $\Sigma'\, E_E = \langle \sigma' \varepsilon_E \rangle$

For the particular choice of $\sigma' = A\,\Sigma$ we obtain

$$E_E = \overline{{}^t A \Lambda_0 A} \;\; \Sigma = \Lambda \Sigma \qquad (20)$$

which is a definition of the macroscopic modulus Λ.

In the same way, the global deformation $E = \langle \varepsilon \rangle$ satisfies eqn (4), and for the difference $E_P = E - E_E$ we obtain the classical definition of the macroscopic plastic strain

$$E_E = \overline{{}^t A \varepsilon_P}$$

4.- IN FINITE STRAIN.

We must generalize the relation between micro and macro quantities for finite strains. We assume that the volume V is a macrohomogeneous body, that the relation (6) is true with velocity

$$\int_{\partial\Omega} (V - v) \otimes n \quad da \qquad (21)$$

This property is derived from the macrohomogeneity hypothesis for non-symmetric field a* and velocity fields v* and can be rewritten as

$$\nabla_i V_j^* = \langle \nabla_i v_j^* \rangle; \qquad \Sigma^* \nabla\, V^* = \langle \sigma^* \nabla_i v_j^* \rangle; \quad \Sigma^* = \langle \sigma^* \rangle \quad (22)$$

A particular choice of non-symmetric tensor field σ^* is the rate of nominal stresses $\dot{\theta} = \dot{\sigma} - \nabla v \cdot \sigma + \sigma \operatorname{div} v$ applied to the body Ω in theactual state.

This local stress is in equilibrium in Q with boundary homogeneous conditions
$\dot{\theta}^T \vec{n} = \dot{\Theta}^T$ n over ∂V; $\dot{\Theta}^T = \langle \dot{\theta}^T \rangle$.
A particular choice of σ^* in the equation is $\sigma^* = A\Sigma^*$ and we obtain

$$\begin{aligned} D^e &= \langle {}^t A \cdot d^e \rangle \\ D^P &= \langle {}^t A \cdot g^P \rangle = \langle {}^t A \cdot d^P \rangle \end{aligned} \qquad (23)$$

To obtain the antisymmetric parts of ΔV it is necessary to introduce another localization stress tensor H which has the following properties (Mandel, 1982):

$$H_{ijpq,j} = 0 \quad \text{in V}$$

$$H_{ijpq}n_j = \frac{1}{2}\left(\partial_{ip}\partial_{jq} + \partial_{iq}\partial_{jp}\right)\mathrm{n}_j \quad \text{over } \partial \mathrm{V} \tag{24}$$

All $\sigma^* = (A + H)\Sigma^*$ are in equilibrium with $\Sigma^* \bullet \vec{n}$ on $\partial\Omega$; then with this choice we have

$$\nabla V = \left\langle {}^t(A + H)\nabla v \right\rangle \tag{25}$$

and

$$\Delta \mathrm{v} = \mathrm{d}^{\mathrm{e}} + \omega^{\mathrm{d}} + \mathrm{g}^{\mathrm{p}}$$

$$\Delta \mathrm{V} = \mathrm{D}^{\mathrm{e}} + \Omega^{\mathrm{d}} + \mathrm{G}^{\mathrm{p}} \tag{26}$$

with

$$\Omega^d = \left\langle {}^t H\omega^d \right\rangle \quad \text{and}\ \mathrm{G}^{\mathrm{P}}\left\langle {}^t(A+H)g^P \right\rangle \tag{27}$$

Moreover the micro objective stress rate is then

$$\rho\frac{D}{Dt}\left(\frac{\sigma}{\rho}\right) = \frac{DA}{Dt}\Sigma + A\rho\frac{D}{Dt}\left(\frac{\Sigma}{\rho}\right) + \rho\frac{D}{Dt}\left(\frac{r}{\rho}\right)$$

The objective derivation of A is then

$$\frac{DA}{Dt} = \dot{A} - \omega^d A + A\omega^d + \Omega^d\ {}^t A - {}^t A\Omega^d$$

as $\dfrac{D\left(\dfrac{r}{\rho}\right)}{Dt}$ is self-equilibrated and $\rho\dfrac{D\left(\dfrac{\Sigma'}{\rho}\right)}{Dt} = \dot{\Sigma} - \Omega^d\Sigma + \Sigma\Omega^d$

If the hardening has no influence on the elasticity, DA / Dt = 0 and

$$D^e = \langle {}^t A\, d^e \rangle = \left\langle {}^t A \Lambda_0 \frac{D}{Dt}\left(\frac{\sigma}{\rho}\right)\right\rangle = \langle {}^t A\, \Lambda\, A \rangle \frac{D}{Dt}\left(\frac{\Sigma}{\rho}\right) \qquad (28)$$

If $g^P = \frac{\partial \phi}{\partial \sigma}$, it is clear that $\Phi(\Sigma, r) = \langle \phi(\sigma) \rangle$ with $\sigma = \frac{(A+H)}{\Sigma} {}^t r$; ϕ is the viscoplastic macroscopic potential (Mandel, 1982)

$$G^P = \frac{\partial \Phi}{\partial \Sigma} = \left\langle {}^t (A+H) \frac{\partial \phi}{\partial r} \right\rangle = \langle {}^t (A+H) g^P \rangle$$

4.- EVOLUTION LAW AND RATE BOUNDARY VALUE PROBLEM.

The quantity of gliding $\dot{\gamma}^r$ is positive if a loading condition is verified ($f^r = 0 \ \ and \ \ \dot{f}^r = 0, \dot{\gamma} \geq 0$; in other cases $\dot{\gamma}^r = 0$).

The equation $\dot{f} = 0$ can have two expressions (Stolz, 1984):

(a) The normal n^r at the slipping plane is not a material vector; then the evolution of f^r is reduced to

$$\dot{f}^T = n^r \cdot \frac{\dot{\psi}}{\rho_K} \cdot m^r + \frac{\dot{A}}{\rho_K} = 0 \qquad (29)$$

(b) The normal n' is a material vector, and is at each time the normal of the physical slipping plane; in this case $\dot{n}^r + n^r \dot{P} P^{-1} = 0$ and

$$\dot{f}^T = n^r \cdot \left(\frac{\dot{\psi}}{\rho_K} - \dot{P} P^{-1} \frac{\psi}{\rho_K} \right) \cdot m^r + \frac{\dot{A}}{\rho_K} = 0 \qquad (30)$$

In this case ψ is the thermodynamical force associated with P $\left(\frac{\partial \phi_0}{\partial P}\right)$ written on C_K. To prove this is easy: the total Lagrangian strain related to the initial configuration C_0 is

$$\Delta = P^T \Delta_K P + \frac{1}{2}\left(P^T P - I\right) = \frac{1}{2}\left(F^T F - I\right) \qquad (31)$$

The energy becomes a function of Δ, P and $\{\gamma^r\}_r$ such that

$$\Phi_0\left(\Delta; P, \{\gamma^r\}_r\right) = \Phi\left(\Delta_K, \{\gamma^r\}_r\right) \qquad (32)$$

and we have $\psi = -\rho_k P\left(\frac{\partial \phi_0}{\partial \Delta_0}\right)$. For the sake of simplicity we do, not consider the ardening case in the following.

The total potential energy is given by Φ_0, so we have

$$E(u, P, \lambda) = \int_\Omega \rho_0 \varphi_0(\Delta; P) \quad d\omega - \lambda \int_{\delta\Omega} T^d \cdot u \cdot ds$$

where λT^d are the external forces prescribed over $\delta\Omega$. The equations, of equilibrium are $\left(\frac{\partial E}{\partial u}\right)\delta u = 0$; they can be derived to obtain the rate equilibrium equations:

$$\left(\frac{\Pi_0}{\rho_0} = \frac{\partial \phi_0}{\partial \Delta_0}; \qquad \alpha_r = m_r \otimes m_r\right)$$

$$\int_\Omega \dot{\Pi}_0 \cdot F^T \cdot \delta F + \Pi_0 \cdot \dot{F}^T \cdot \delta F \cdot d\omega - \dot{\lambda} \int_{\delta\Omega} T^d \cdot \delta u \cdot ds = 0 \qquad (33)$$

The stress $\dot{\Pi}_0$ is a given function of $\dot{\Delta}$ and $\dot{P}$

$$\dot{\Pi}_0 = \phi''_{\Delta\Delta} \cdot \dot{\Delta} + \phi''_{,\Delta P} \cdot P^T \left(\sum_r \dot{\gamma}^r \cdot \alpha_r\right)$$

where $\dot{\gamma}^r \geq 0$ if $f^r = 0$ and $\dot{f}^r = 0$; then we have

$$\text{(a)} \quad \alpha_r^T : P \cdot \phi''_{,P\Delta} \cdot \Delta + H_{sr} \cdot \dot{\gamma}^s + \alpha_r^T \cdot \Sigma_s \cdot \dot{\gamma}^s \cdot \alpha_s \left(\frac{\psi}{\rho_K} \right) = 0$$

$$\text{(c)} \quad \alpha_r^T : P \cdot \phi''_{,P\Delta} \cdot \Delta + H_{sr} \cdot \dot{\gamma}^s = 0$$

with $H_{rs} = \alpha_r^T \cdot P \cdot \phi''_{,PP} \cdot P^T \cdot \alpha_S$ symmetric in (r, s).

It is clear that case (b) is a symmetric problem and that case (a) is symmetric only for a simple plastic potential, as in the analysis due to Halphen (1975). Case (b) corresponds to the rate boundary value problem in the theory proposed by Nguyen to study the stability of structure in plasticity and rupture (Nguyen, 1984).

REFERENCES.

[1] Bui, H. D. (1969). Etude de Pévolution de la frontiére du domaine élastique avec l'écrouissage et relation de comportement élastoplastique des métaux cubiques. Doctoral Thesis. Paris.

[2] Francfort, G., Nguyen, Q. S. and Suquet, P. (1983). Thermodynamique et lois de comportement thermomécanique. Compt. Rend. Acad. Sci. (Paris), II, **296**, 1007-10.

[3] Germain, P., Nguyen, Q. S. and Suquet, P. (1983). Appl. Mech., **50**,1010-20.

[4] Halphen, B. (1975). Sur le champ des vitesses en thermoplasticité finie. Int. J. Solids Struct., **11**, 947-60.

[5] Hashin, Z. (1983). Analysis of composite materials. J. Appl. Mech., **50**, 481-505.

[6] Hill, R. (1966). Generalised constitutive relations for incremental deformation of metal crystals by multislip. J. Mech. Phys. Solids, **14**, 95-102.

[7] Hill, R. (1967). The essential structure of constitutive laws of metal composites and polycrystals. J. Mech. Phys. Solids, **15**, 79-95.

[8] Hill, R. and Rice, J. (1972). Constitutive analysis of elastic-plastic crystals at arbitrary strain. J. Mech. Phys. Solids, **20**, 401-13.

[9] Krόner, E. (1980). Linear properties of random media. Proc. 15th Colloq. Groupe Frangás de Rhéologie.

[10] Mandel, J. (1964). Contribution Théorique á l'Etude de l'Ecrouissage et des Lois de l'Ecoulement Plastique. (Proc. 11th Int. Congr. Appl. Mech.), Springer-Verlag, Berlin, pp 502-9.

[11] Mandel, J. (1971). Plasticité Classique et Viscoplasticité. (CISM, Udine), Springer-Verlag.

[12] Mandel, J. (1982). Sur la définition d'un repére privilégié pour l'étude des transformations anélastiques d'un polycristal. J. Méc. Théor. Appl., **1**, No. 1.

[13] Nguyen, Q. S. (1984). Bifurcation et stabilité des systémes irréversibles au principe de dissipation maximale. J. Méc. Théor. Appl., **3**, No. 7, 41-61.

[14] Rice, J. (1971). Inelastic constitutive relations for solids: an internal variable theory and its application to crystal plasticity. J. Mech. Phys. Solids, **19**, 413-35.

[15] Stolz, C. (1982). Contribution á l'étude des grandes transformations en élastoplasticité. Doctoral Thesis, ENPC, Paris.

[16] Stol, C. (1984). Etude des milieux á configuration physique et application, Compt. Rend. Acad. Sci. (Paris), 11, **299**, 1153-5.

[17] Willis, J. R. (1981). Overall properties of composites. In Advances in Applied Mechanics. Vol. 21, C. S. Yih (Ed.), Academic Press, New York.

[18] Zarka, J. (1973). Etude du comportement des monocristaux métalliques. J. Mécan.,**12**, No. 2.

Analytical Studies

Porous metal bearing temperature problem

A MARINKOVIC, T MANESKI, and **V MILOSEVIC**
Department of Mechanical Engineering, University of Belgrade, Yugoslavia

SYNOPSIS

By specific sort of bearing such as porous metal bearing, the temperature problem is very important and interesting to analysing with aim to develop theoretical investigation in this field and so for exploitation this sort of bearing. Working temperature of bearing represents the temperature value which become constant after some time from start.
This paper analyses two aspects of temperature problem, which are dependence of temperature upon a time and phenomena of nonuniform temperature distribution on volume of porous metal bearing. The temperature distribution problem of porous metal bearing is studied based on hydrodynamics lubrication theory, used finite elements analysis with appropriate boundaring restriction. The temperature values presented in this paper are compared with results of experimental investigations.

NOTATION

F	*bearing load*
r	*bearing radius*
c	*operating clearance*
p	*specific bearing load*
v	*sliding velocity*
n	*rotational shaft speed*
μ	*coefficient of friction*
η	*dynamic viscosity of oil for lubrication*
ε	*relative excentricity*
k	*heat transfer coefficient (convection)*
T_2	*temperature value on outside surface of bearing*
T_0	*outside air temperature value.*
$\dot{q}_i$	*termic flucks through elementary bearing area*
$\dot{q}_A$	*termic flucks of heat exchange between surface of bearing and outside air*
Ψ	*constructive parameter of porous metal bearing*

1 INTRODUCTION

By specific sort of bearing such as porous metal bearing, the temperature problem is very important and interesting to analysing with aim to develop theoretical investigation in this field and so for exploitation this sort of bearing. Working temperature of bearing represents the temperature value which become constant after some time from start. Friction on contact surface produce heat which is taking off only through housing surface because there is a small amount of oil in porous bearing material. That's a reason of higher work temperature values for this sort of bearing in working life.

Two different temperature problem aspects of porous metal bearing are to be analysed here. As a first it's dependence of operating temperature upon a working time and after that complex phenomena of nonuniform temperature distribution on volume of porous metal bearing.

2 COEFFICIENT OF FRICTION AND WORKING TEMPERATURE

Mechanism of self-lubricating in porous metal bearings makes lubrication better, but value of coefficient of friction is still in wide interval. That can be understand if we know that in bearing life it works in regimes from boundary to hydrodynamics lubrication [1]. So all parameters which have influenced to friction, define lubrication quality regime and in same time coefficient of friction value. Coefficient of friction value also depends of contact surface quality. That is a reason of great care for bearing calibration and fine shaft scraping.

Working temperature is very important and it represents the temperature value which become constant after some time from start. Friction produced heat is taking off only through housing surface because there is a small amount of oil, at work temperature too, in porous material. That is the reason of higher work temperature value for this sort of bearing in working life.

Exploitation this sort of bearing and many experiments show that working temperature become constant after certain time from start and it's change in that time can be approximate with polynome:

$$T(t) = C_0 + C_1 t + C_2 t^2 + C_3 t^3 + C_4 t^4$$

This relation is based on experiments and will be explained and completed with graphics in a chapter of experimental investigations (Fig. 1).

3 TEMPERATURE FIELD OF POROUS METAL BEARING

The working temperature of sliding bearing is the temperature value which is measured on the contact surfaces of bearing and shaft, after certain period of working time necessary to reach stationary state. But, it is very interesting and useful to know the temperature values all around the porous bearing.

When we have mathematical model of porous metal bearing, differential equation for temperature distribution of isotropic and homogeneous material in cylindrical coordinates is [2]:

$$\frac{\partial^2 t}{\partial r^2} + \frac{1}{r}\frac{\partial t}{\partial r} + \frac{1}{r^2}\frac{\partial^2 t}{\partial \theta^2} + \frac{\partial^2 t}{\partial z^2} = 0$$

If we try to find an analytical solution of this equation, it will have form:

$T = R(r) \cdot \Theta(\theta) \cdot Z(z)$

where R,Θ,Z are two differentiable functions.

After a few transformations we have differential equations in this form:

$$Z(z) = A_1 \cos\omega_n z + A_2 \sin\omega_n z$$

$$R(r) = A_3 J_0(\omega_n r) + A_4 Y_0(\omega_n r)$$

$$\Theta(\theta) = A_5 \cos b\theta + A_6 \sin b\theta$$

3.1 Boundary conditions

To solve this problem there are some very important boundary conditions:

3.1.1 Inside surface of the bearing

On inside surface of the bearing we have boundary conditions of second sort, while termic flucks through elementary area is known as a function [1]:

$$\dot{q}_i = p\,v\mu = \frac{3v^2\eta}{r\,c^2}\left[\frac{\varepsilon\sin\theta}{(1+\varepsilon\cos\theta)^3 + 12\Psi}\right]\left[\frac{b^2}{4} - z^2\right]\mu + p_b\,v\mu$$

3.1.2 Outside surface of the bearing

On outside surface of the bearing ($r = r_2$) we have boundary conditions of first sort, while we can suppose that temperature values is constant around the bearing, what means:

$$\frac{\partial t}{\partial \theta} = 0 \quad \text{and} \quad \frac{\partial t}{\partial z} = 0$$

3.1.3 Condition of periodically

For $\theta = 0$ we have a condition of periodically: $\theta\,(0, r) = \theta\,(2\pi, r)$

3.1.4 Condition of simetricity

For $z = 0$ we have a condition of simetricity: $\dfrac{\partial t}{\partial z} = 0$

3.1.5 Heat exchange between surface of bearing and outside air

Boundary conditions of third sort are on outside (noncylindrical) surfaces of the bearing, while we have a value of outside air temperature (T_0) and relation for heat exchange between surface of bearing and outside air:

$$\dot{q}_A = \pm k\,[T_2 - T_0]$$

4 EXPERIMENTAL INVESTIGATION

This paper presents the temperature distribution on porous metal bearing volume, based on result of experiments and will be explained in next chapter.
All experimental investigations are carried out in laboratory on testing equipment USL1 [3], with the porous metal bearings of sintered bronze (CuSn10) produced in Yugoslavia. For all samples have been done full controls and laboratory analyses. Obtain physical characteristics for this material are shown in Table 1.

Table1 Physical characteristics of tested bearings

Material		CuSn10
density,	g/cm^3	6,4.......6,5
open porosity,	%	22,45.....23,17
radial fracture force,	F	3500.......3600
hardness,	HB	33,26.......36,24

Porous metal bearings, which were tested, have dimensions Φ20 / Φ30 × 20 mm. Before testing of temperature and coefficient of friction change, all bearings were worked out.
The dependence of temperature upon a time has been followed on working regime with parameters: radial load of, for example $F = 170$ N and rotation speed of $n = 4780$ rpm (for boundary value of *pv* characteristics) [3]. Temperature values where measured every 2,5 minutes from start, until this parameters become constant and result of this measurement is shown on the Fig. 1.

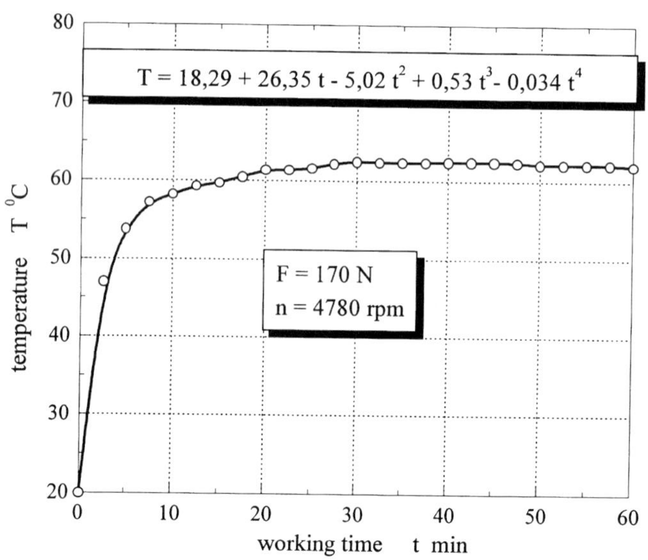

Fig. 1 The dependence of bearing temperature upon a operating time

The temperature distribution on bearing volume have been done on working regime determinate with radial load $F = 550$ N and rotation speed of $n = 1350$ rpm. In eight different points outside around the bearing are measured temperature values with angle of 45^0 between two neighbour measurement points (Fig. 2).

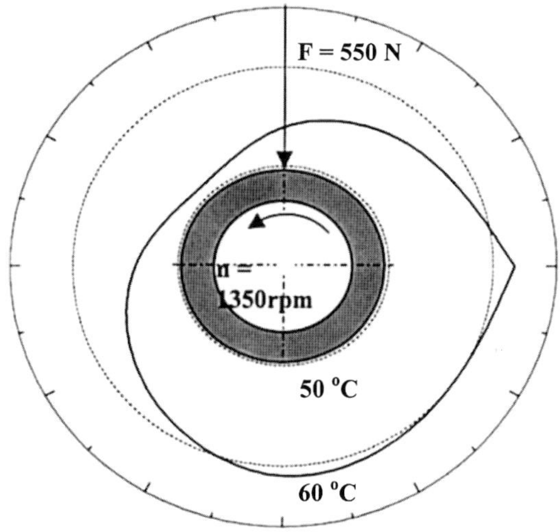

Fig. 2 The temperature distribution on bearing volume

5 CONCLUSION

This investigations presented that temperature of porous metal bearing become constant after (30...50) minutes from working start and it's dependence from time can be best polynomial approximate (Fig.1). The results presented in this paper also shows that temperature distribution around porous metal bearing (Fig. 2), made by experiments This distribution is very similar as the pressure distribution given by hydrodynamic lubrication theory. That means that maximum temperature values are in the area near load and rotating direction and for minimum situation is completely opposite. Actually, temperature analysis of this problem is analysing by finite elements method, using selfdeveloped computer program. Results of that analyse with results presented in this paper will complete the temperature analysis of porous metal bearing.

REFERENCES

[1] Rouleau, W.T., 1963., "Hydrodynamic Lubrication of Narrow Press-Fitted Porous Metal Bearings", Journal of Basic Engineering, march 1963, pp.123-128.

[2] Marinković A., Janković M.,"Experimental determination Dependence of Temperature and Coefficient of friction in time for Porous Metal Bearings", International Paper Tribology in Industry, Vol.18, Kragujevac June 1996., pp.61-64.

[3] Janković M., "The investigation of porosity influence on porous metal bearings working performances", Ph.D. thesis, Faculty of mechanical engineering, Belgrade 1988.

Boundary element analysis of surface ring crack in silicon nitride bearing elements

Y WANG and **M HADFIELD**
Tribology Design Research Unit, Bournemouth University, UK

SYNOPSIS

The boundary element method was used for a fracture mechanics analysis of surface ring crack found on silicon nitride ball surfaces. Hertzian point loading with normal pressure and tangential traction was employed to simulate rolling contact loading. The stress intensity factor calculation was based on a three-dimensional crack propagation model. The crack geometry adopted in the model was determined by the experimental observations, which was defined as a conic shape with a cured line as the crack front. Numerical results were presented for a ring crack with a radius of 0.21 millimetre, crack angles of 50^0 and 40^0, maximum crack depth of 0.05 millimetre, and an arc length equals to ¼ of the circumference of the ring crack circle. The maximum contact pressure p_0 is assigned to the values of 4.0 GPa and 5.6 GPa. Life predictions from the present calculation are consistent with the experimental observations.

1 INTRODUCTION

The use of ceramics such as silicon nitride in rolling bearing applications can have environmental benefits during the in-use phase of the product life cycle. Properties such as low friction will result in low in-use energy consumption and low contact acoustic noise are typical examples. The reliability of the product during in-use is critical to its widespread application. The reliability of silicon nitride products is considerably influenced by surface defects, such as porosity, inclusions and surface cracks. Surface ring crack defects are often found on the ball surfaces even though they are difficult to detect. These ring cracks decrease the fatigue performance considerably. Hence it is important to assess the influences of ring cracks on fatigue life.

Fracture mechanics has been utilised by many investigators [1-4] to study surface crack growth behaviour. The geometry and loading encountered in three-dimensional crack problems are too complex for the stress intensity factor (SIF) to be solved analytically, and numerical analysis has often been adopted in stress intensity factor calculations. A two-dimensional model has been utilised by several investigators [5, 6] and a more realistic three-dimensional analysis has also been proposed [3, 4, 7, 8]. However, the SIF calculation is further complicated because it is a function of the position along the crack front, the crack size and shape, type of loading, and geometry of the structure. Some assumptions were made by investigators to carry out the analysis. Therefore, the simple assumptions in the crack geometry and loading configuration caused the calculated results not to match the real experimental observations.

In this study, a three-dimensional boundary element model is employed to perform the fracture mechanics analysis of 'natural ring crack' found on ceramic ball surfaces. The SIF solutions are compared with the fracture toughness K_{IC} and crack growth threshold value ΔK_{th} of silicon nitride to determine the possibility of the crack growth. Life estimates from the present calculation are in line with previous experimental observations.

2 MODELLING

The rolling contact load is simulated by a moving Hertzian point load with normal pressure $p(x,y)$ and tangential traction $q(x,y)$. Assume that normal pressure $p(x,y)$ and tangential traction $q(x,y)$ within the loading region are expressed by following equations [9]

$$p(x, y) = p_0(1 - x^2 / a^2 - y^2 / a^2)^{1/2} \quad (1)$$

$$q(x, y) = fp(x, y) \quad (2)$$

where p_0 is maximum normal pressure, a is radius of contact circle and f is friction coefficient. Friction coefficient is determined in terms of the lubrication regime. Due to lubricated rolling contact, friction coefficient is assigned to a value of 0.05. Fig. 1 illustrates the ring crack subjected to this simulation loading. The coordinate x measures the distance from the center of contact circle to the center of ring crack circle.

The geometry of a ring crack is very important to fracture mechanics analysis and will be considered carefully. The position of an element on the crack face can be described by the following equations

$$x = (R + l \cos \alpha \sqrt{1 - (\phi / \phi_0)^2}) \cos \phi \quad (3)$$

$$y = R \sin \phi \quad (4)$$

$$z = -l \cos \alpha \sqrt{1 - (\phi / \phi_0)^2} \tan \alpha \quad (5)$$

where R = radius of ring crack
l = crack depth, $0 < l <=$ maximum crack depth ($P_0P_{0'}$)
α = crack angle, $0 < \alpha < \pi/2$
ϕ = angle of an element position on the crack face, $-\phi_0 < \phi < \phi_0$
ϕ_0 = half angle of ring crack arc P_1P_2, $0 < \phi_0 < \pi/2$.

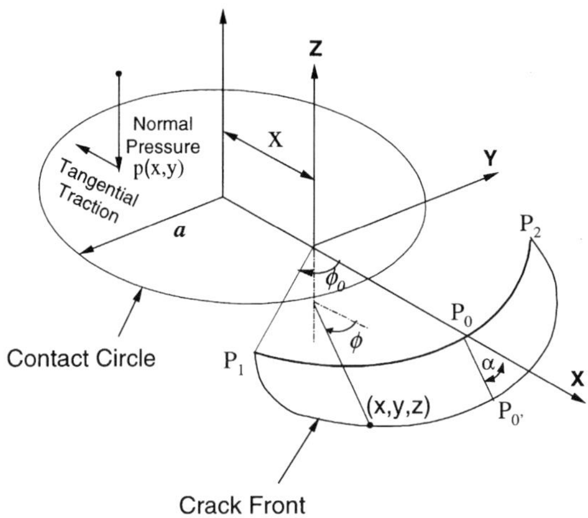

Fig. 1 Geometry and coordinate system of a ring crack subject to rolling contact

The maximum contact pressure p_0 is assigned to the values of 4.0 GPa and 5.6 GPa. Crack geometry is defined in terms of experimental observations. Two crack angles are selected to compare the stress intensity factors, which are 40^0 and 50^0, respectively. The maximum crack depth and ϕ_0 are 0.05 millimetre and 45^0 respectively. These parameters are applied to the SIFs calculation throughout this study.

3 NUMERICAL ANALYSIS

3.1 Boundary element modelling

The boundary element method is utilised in the present study. There are several ways to model a crack in boundary element modelling. The dual boundary element is used for the crack growth analysis in the model. The dual boundary element method is well suited to the analysis of crack growth and growth direction. The dual boundary element method considers two independent equations, the displacement and traction boundary integral equations, within the same integration path for each pair of coincident source point. The crack is represented by two elements occupying the same physical location, each element representing a face of the crack. The displacement and traction boundary integral equations are on the crack surface. The use of dual elements for 3-D crack growth means the only one surface of the crack needs to be defined. Fig. 2 shows a detailed view of the crack mesh.

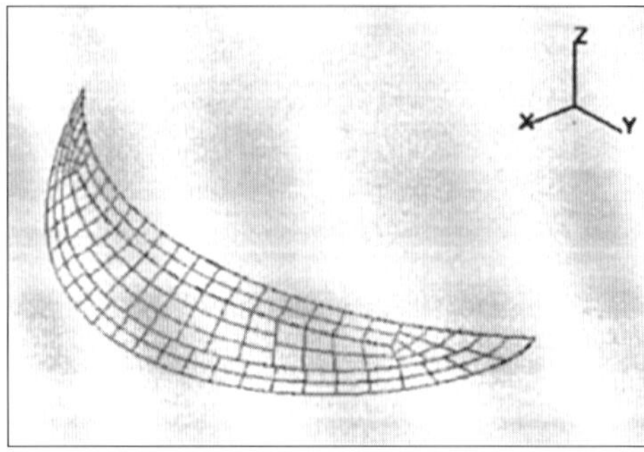

Fig. 2 Detailed view of the crack mesh

The stress intensity factors (SIF) are computed using the crack opening displacement method. The SIF values along the crack front are obtained by the following equations:

$$K_I = \frac{E}{4(1-\nu^2)}\sqrt{\frac{\pi}{2r}}(u_n|_{\theta=\pi} - u_n|_{\theta=-\pi}), \quad (6)$$

$$K_{II} = \frac{E}{4(1-\nu^2)}\sqrt{\frac{\pi}{2r}}(u_t|_{\theta=\pi} - u_t|_{\theta=-\pi}), \quad (7)$$

$$K_{III} = \frac{E}{4(1-\nu^2)}\sqrt{\frac{\pi}{2r}}(u_u|_{\theta=\pi} - u_u|_{\theta=-\pi}), \quad (8)$$

where u is the displacement of the crack faces and the subscripts n, t and u denote the normal, first tangential and second tangential directions respectively.

4 RESULTS AND DISCUSSION

Fig. 3 shows a plot of stress intensity factors (K_I, K_{II}, K_{III}) versus ϕ at the maximum contact pressure of 5.6GPa. As can be seen from the plot, K_I reaches a maximum value at $\phi = 45^0$ and the maximum value is 4.7 MPa $m^{½}$. K_I is higher than both K_{II} and K_{III}, but less than the fracture toughness K_{IC} of silicon nitride bearing materials (6.0 MPa $m^{½}$). It indicates there is no possibility of unstable fracture under the calculation condition. However, crack propagation may happen because the crack propagation threshold ΔK_{th} for silicon nitride ceramics is about 3 MPa $m^{½}$ [11]. The K_{II} and K_{III} also vary in line with ϕ.

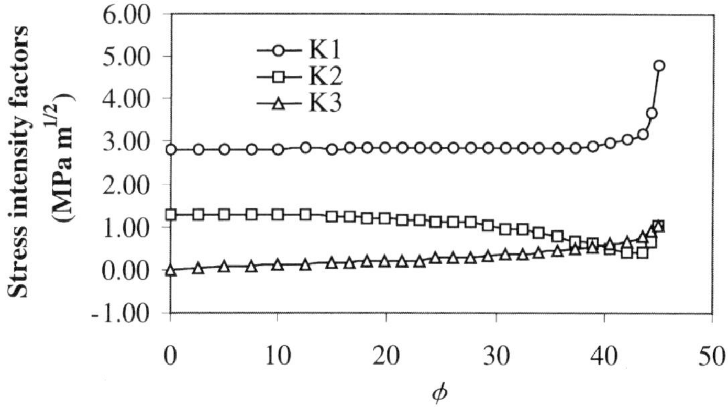

Fig. 3 Stress intensity factors against ϕ at the maximum contact pressure of 5.6 GPa and the crack angle = 50^0.

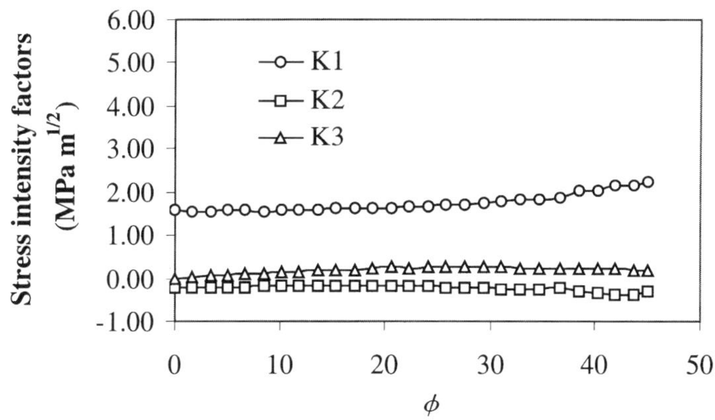

Fig. 4 Stress intensity factors against ϕ at the maximum contact pressure of 4.0 GPa and the crack angle = 40^0.

Fig. 4 is the plot for the maximum contact pressure of 4.0 GPa. K_I is higher than K_{II} and K_{III} at all locations. K_{II} goes down, but K_{III} goes up slightly with ϕ. The K_I value is around 1.7 MPa $m^{1/2}$ and it is less than ΔK_{th}. Therefore, the crack front will not be able to propagate if the maximum contact stress is under 4.0 GPa. The results indicate that it is hard to cause silicon nitride ball failure in a fatigue spalling mode.

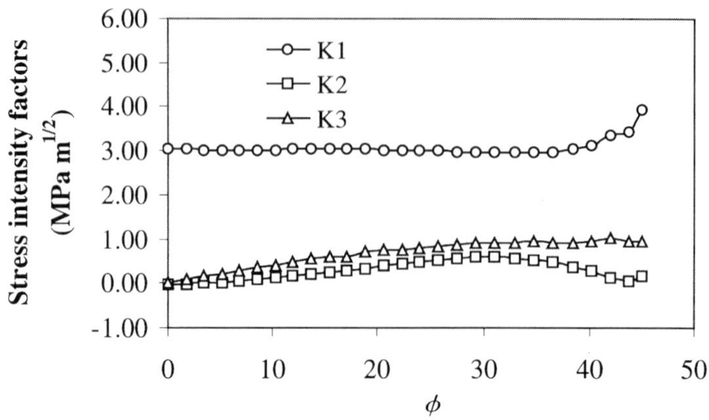

Fig. 5 Stress intensity factors against ϕ at the maximum contact pressure of 5.6 GPa and the crack angle = 40^0.

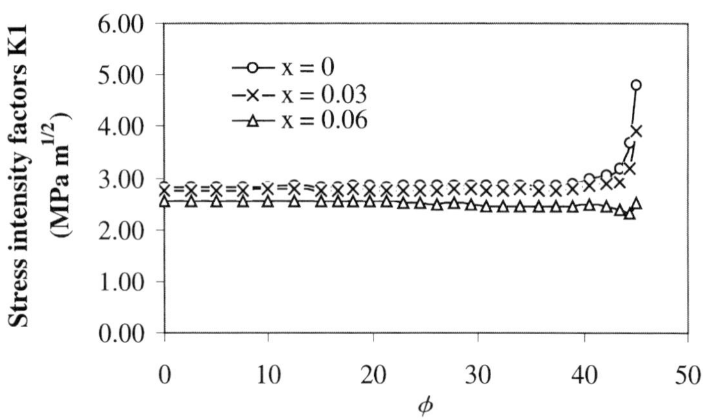

Fig. 6 Stress intensity factor K_I versus ϕ for the crack angle of 50^0 at different x coordinate.

Fig. 5 illustrates the stress intensity variation for the crack angle of 40^0 at the maximum contact pressure of 5.6GPa. As in above cases, the stress intensity factors will change in the similar way. K_I increases about 0.3 MPa $m^{1/2}$ at $\phi = 45^0$. K_{III} has little change. There is no significance change in the K value for the calculated crack geometry.

Figs. 6 shows K_I variation at different co-ordinate ($x = 0, 0.03, 0.06$). This plot demonstrates that the K_I value decreases as the x increases. The K_I distribution along the crack front also changes due to the characteristic of the contact stress field.

The possibility of crack growth has been determined in terms of the K_I value. If $\Delta K_{th} < K_I < K_{IC}$, the crack front may propagate. If $K_I < \Delta K_{th}$, the crack front will not propagate. In other words, if there is a condition for a crack growth, the crack will grow and a fatigue spall may be formed. If no crack growth condition, no fatigue spalling. Experimental verification shows that the present life predication is consistent with the experimental results [12].

5 CONCLUSION

The boundary element method is used for a fracture mechanics analysis of a surface ring crack on a silicon nitride ceramic ball. The stress intensity factor calculations are based on a three-dimensional model of ring crack growth. The crack geometry adopted in the model is determined by the experimental observations of the surface ring crack. Stress intensity factors on the crack front provide the necessary information to determine whether the crack front propagates or not in concentrated contact. Rolling contact fatigue failure of cracked silicon nitride bearing elements can be predicted by a simulation analysis through comparing SIF on the crack front to K_{IC} and ΔK_{th}.

References

[1] J. R. Fleming and N. P. Suh, "Mechanics of crack propagation in delamination wear", Wear, 1977, 44 pp.39-56.

[2] L. M. Keer, M. D. Bryant, and G. K. Haritos, "Subsurface and surface cracking due to Hertzian contact", Journal of Lubrication Technology-Transactions of the ASME, 1982, 104, pp. 347-351.

[3] M. Kaneta, M. Suetsugu, and Y. Murakami, "Mechanism of surface crack growth in lubricated rolling sliding spherical contact", Transacations of ASME, 1986, 53, pp. 354-360.

[4] M. T. Hanson and L. M. Keer, "An analytical life prediction model for the crack propagation occuring in contact fatigue failure", ASLE Trans., 1992 35, pp. 451-461.

[5] L. M. Keer and M. D. Bryant, "A pitting model for rolling-contact fatigue", Journal of Lubrication Technology-Transactions of the ASME, 1983, 105, pp. 198-205.

[6] A. F. Bower, "The influence of crack face friction and trapped fluid on surface initiated rolling-contact fatigue cracks", Journal of Tribology-Transactions of the ASME, 1988, 110, pp. 704-711.

[7] Y. Murakami, M. Kaneta, and H. Yatsuzuka, "Analysis of surface crack-propagation in lubricated rolling-contact", ASLE Transactions, 1985, 28, pp. 60-68.

[8] Y. Murakami, C. Sakae, and K. Inchimaru, "Three-dimensional fracture mechanics analysis of pit formation mechanism under lubricated rolling sliding contact loading", ASLE Trans., 1994, 37, pp. 445-454.

[9] K. L. Johnson, Contact Mechanics, Cambridge University Press, Cambridge, 1985.
[10] Y. Wang and M. Hadfield, "The influence of ring crack location on the rolling contact fatigue failure of lubricated silicon nitride: fracture mechanics analysis", accepted for publication in *WEAR*.
[11] I. Bar-On and J. T. Beals, "Fatigue in silicon nitride bend bars, Fatigue 90 International conference on fatigue and fatigue thresholds", 1990.
[12] Y. Wang and M. Hadfield, "The influence of ring crack location on the rolling contact fatigue failure of lubricated silicon nitride: experimental studies", accepted for publication in *WEAR*.

Wear of worm gears – theory and experiment

P KRAL and **S PAUSCH**
Institute for Machine Elements, Vienna University of Technology, Austria

SYNOPSIS

This article describes a computational procedure for calculating wear of worm gears on the basis of physical fundamentals. For this purpose, the elasto-hydrodynamic theory of lubrication (EHD-Theory), the molecular-mechanical theory of wear and energetic theory of friction is applied to the specific geometrical and kinematic peculiarities of worm gears. The results are compared with test bed experiments as well as with a calculation method based on empirical data according to NEUPERT.

SYMBOLS

Symbol	Unit	Meaning
a_H	m	Half of the Hertzian oblateness width of two cylinders pressed on each other
A_a	m^2	nominal contact surface
E	N/m^2	Young's modulus
h_{fikt}	m	Small distance which the worm is moved in axial direction in order to keep the lubricating gap in its non-deformed state
h_m	m	average lubricant film thickness in 'EHD-Theory'
$h_{min,isoth}$	m	minimal lubricant film thickness in 'Isothermic EHD-Theory'
h_0	m	lubricant gap in non-deformed state
h_v	m	Height of wear removal
$h_{v,sum}$	m	Total height of wear removal
I_h	–	Linear intensity of wear in the 'Molecular-mechanical Theory'
L_{th}	–	thermic load factor according to MURCH and WILSON

M_{OUT}	N·m	Given output torque of the worm gear
n_B	–	Number of the contact lines occurring at the considered time simultaneously.
N_{v0}	–	Number of stress cycles calculated in one arithmetic step
q	N/m	Line load
r	m	Radius of curvature
v^+	m/s	average circumferential speed in 'EHD-Theory'
$v_n^{\ +}$	m/s	average relative velocity of the substitute-cylinder of worm and worm wheel relatively to the contact point projected onto perpendicular surface of the contact line
V_v	m^3	Volume of wear
x, y, z	m	Coordinates
α	m^2/N	Pressure-viscosity index
α_T	1/K	Temperature-viscosity coefficient of the lubricant
η_0	Pa·s	Dynamic viscosity of the lubricant at a given mass temperature and under atmospheric pressure
ϑ	K	Temperature
λ	W/(mK)	Thermal conductivity
ρ	kg/m^3	Density

SYMBOLS

Index	Meaning
B	relatively to contact point
EHD	Elasto-hydrodynamic theory of lubrication
n	Projection onto normal surface
WHEEL	Worm wheel
WORM	Worm
x, y, z	Coordinates
1	Worm
2	Worm wheel
+	Substitutional value to describe worm and worm wheel

1. MATHEMATICAL MODEL DESCRIBING THE WEAR OF WORM GEARS

1.1 Introduction

Within the course of a dissertation [9] at the Institute for Machine Elements and Machine Design a theoretical model of wear, based on the theory of wear according to KRAGELSKI [5, 6] was developed. With this program it is possible to calculate wear-volume and/or wear-mass of worm gears with known geometry, known material and given load and temperature characteristics.

1.2 Model requirements

Between the flanks mixed friction is dominating. Therefore both solid state friction and liquid friction have to be considered.

1.2.1 Solid state friction

A statistical model of the surface is used [9]. Roughness is described by the four parameters R_t, R_p, R_q and R_z, waviness is neglected [3]. The solid-state friction-forces are computed with the add of the energetic friction theory [13].

1.2.2 Liquid friction

Elasto hydrodynamic theory of lubricant (EHD-Theory) comes to application. The minimum lubricant film thickness is determined by the equations founded by DOWSON and HIGGINSON [2] with the modifications from MURCH and WILSON. From the minimum lubricant film thickness, the middle lubricant film thickness is calculated according to DOWSON and TOJODA. The friction force is determined by using 'effective viscosity' according to PREDKI [10].

1.3 EHD-Theory for worm gears

Starting point is the REYNOLD'S equation, which is used to describe the connection between the pressure p in the lubricant, its dynamic viscosity η, density ρ, the lubricant gap height h and the occurring velocities v (see equation 1).

$$\begin{aligned}&\frac{\partial}{\partial x}\left(\frac{\rho\cdot h^3}{12\cdot\eta}\cdot\frac{\partial p}{\partial x}\right)+\frac{\partial}{\partial z}\left(\frac{\rho\cdot h^3}{12\cdot\eta}\cdot\frac{\partial p}{\partial z}\right)=\frac{\partial}{\partial x}\left(\frac{\rho\cdot h\cdot(v_{x1}+v_{x2})}{2}\right)+\\&+\frac{\partial}{\partial z}\left(\frac{\rho\cdot h\cdot(v_{z1}+v_{z2})}{2}\right)-\rho\cdot v_{x2}\cdot\frac{\partial h}{\partial x}-\rho\cdot v_{z2}\cdot\frac{\partial h}{\partial z}+\rho\cdot(v_{y2}-v_{y1})\end{aligned} \tag{1}$$

With the following simplifications: [Dowson/Higginson]

- Lubricant of constant density (incompressible fluid)
- Variations of the velocity are only considered in direction of the lubricant gap height (y-coordinate)
- A convenient choice of the coordinate system makes it possible that the lubricant gap does not vary in z-direction ($\frac{\partial h}{\partial z}=0$)

equation 2 follows.

$$\frac{\partial}{\partial x}\left(\frac{h^3}{12\cdot\eta}\cdot\frac{\partial p}{\partial x}\right)+\frac{\partial}{\partial z}\left(\frac{h^3}{12\cdot\eta}\cdot\frac{\partial p}{\partial z}\right)=\frac{\partial h}{\partial x}\cdot\frac{v_{x1}-v_{x2}}{2}\cdot v^+ \tag{2}$$

with

$$v^+=\frac{v_{x1}+v_{x2}}{2} \tag{3}$$

Using the following methods the EHD-Theory could be applied to the conditions of worm gears:

- The contact lines are bent curves, therefore they must be subdivided finely and the circumstances have to be considered in the respective point of contact.
- The determining values as radii and velocities,... must be projected onto the normal face of the contact line in the considered point of contact (index n).
- Velocities are viewed relatively to the point of contact (index B)

$$\frac{1}{E^+} = \frac{1}{2} \cdot \left(\frac{1-\nu_{WORM}^2}{E_{WORM}} + \frac{1-\nu_{WHEEL}^2}{E_{WHEEL}} \right)$$

$$v_n^+ = \frac{v_{WORM,Bn} + v_{WHEEL,Bn}}{2} \tag{4}$$

$$r_n^+ = \frac{r_{WORM,n} \cdot r_{WHEEL,n}}{r_{WORM,n} + r_{WHEEL,n}}$$

$$h_{min,isoth} = \frac{2.65 \cdot \alpha^{0.54} \cdot \left(\eta_0(\vartheta) \cdot v_n^+\right)^{0.7} \cdot r_n^{0.43}}{E^{+0.03} \cdot q_{EHD}^{0.13}} \tag{5}$$

$$L_{th} = \eta_0(\vartheta) \cdot \frac{4 \cdot \alpha_T \cdot v_n^{+2}}{\lambda} \tag{6}$$

The required values for equations 1-6 are calculated (see 1.5). These are:

- Geometry of worm and worm wheel
- Conditions of the velocities
- Lubricating film thickness and line load
- Torque and efficiency of the gearing

1.4 Estimation of the expected wear

Since worm gears in general run in the mixed friction field, wear occurs in practice. In this case, the determining value is the lubricant film thickness. It specifies, how deep the roughness peaks of the harder body penetrate the softer one and therefore causes wear. In order to be able to apply the model of surface by KRAGELSKI with justifiable expenditure the assumption was made that the harder wear partner is the worm (hardened steel) which penetrates the worm wheel (bronze). This assumption is supported by measurements in practice.

1.4.1 Problems in the calculation of the wear behaviour of worm gears

- The oil temperature in the case of worm gearings with splash lubrication increases while testing the gears. This leads to lower oil viscosity and therefore to a smaller lubricant film thickness which increases the wear. This can be considered by measuring the oil temperature and using these values as input variables for calculation.
- The geometry of the wheel flank is changed insignificantly by wear. This is considered by summing the height of wear h_v in the viewed point of the contact line of the corresponding lubricant gap h_0 at non-deformed state. The change of the radii of curvature however can not be included with reasonable effort.

1.4.2 Simulation of wear while running worm gears

The above-mentioned facts show clearly that a calculation of wear which considers all this possible facts could only be done in the form of a simulation. In this case, the initial state is either 'ideal', or a section already modified by wear, which could be taken from a preceding wear simulation for a gear with identical geometric data.

For calculation of the resulting wear, the knowledge of the glide path L is required. With known glide path and calculated linear intensity of wear I_h, the height of wear removal h_v that results during the turn of the wheel around $\Delta\varphi$, can be computed with the following equation 7:

$$h_v = I_h \cdot L \tag{7}$$

The wear volume V_v on a piece of the wheel flank at a viewed point of contact is calculated by multiplication of the height of wear removal h_v with the nominal contact surface A_a:

$$V_v = h_v \cdot A_a \tag{8}$$

A_a is a rectangle; its side length is on the one hand the length of the viewed contact line Δl and on the other hand the Hertzian oblateness width $2 \cdot a_H$ of the substitute-cylinder under the line load q_{EHD}:

$$A_a = 2 \cdot a_H \cdot \Delta l = 2 \cdot \sqrt{\frac{8 \cdot q_{EHD} \cdot r_n^+}{E^+ \cdot \pi}} \cdot \Delta l \tag{9}$$

In order to receive wear along the current contact lines (specified by z_0), integration along the contact lines has to be done. With the discretization made above, integration can be replaced by summation, therefore:

$$V_v(z_0) = \sum_{j=1}^{j=n_B} \sum_{i=1}^{i=n_y} V_v\left(x_{B,j}(y_i, z_0), y_i\right) \tag{10}$$

This is the wear volume which results if one moves the worm starting from the situation z_0 along Δp in axial direction and/or turns the wheel around $\Delta\varphi$. The wear volume in the case of a revolution of the worm wheel is:

$$V_{v,1U} = z_2 \cdot \sum_{k=0}^{k=n_p-1} V_v(z_{0,k}) = z_2 \cdot \sum_{k=0}^{k=n_p-1} \sum_{j=1}^{j=n_B} \sum_{i=0}^{i=n_y} V_v\left(x_{B,j}(y_i, z_{0,k}), y_i\right) \tag{11}$$

1.5 Carrying out of an individual calculation step:
A single calculation step consists of the computation of the wear volume $V_v(x_{B,j}(y_i, z_{0,k}), y_i, l)$.

1.5.1 Determination of the current contact lines considering $z_{0,k}$

$$\Rightarrow \begin{cases} n_B(z_{0,k}) \\ x_{B,j}(y_i, z_{0,k}) \\ z_{B,j}(y_i, z_{0,k}) \end{cases} \tag{12}$$

1.5.2 Determination of the radii of curvature relevant for the EHD-Theory along the contact line

$$\Rightarrow \begin{cases} r_{WORM,n}\left(x_{B,j}(y_i, z_{0,k}), y_i\right) \\ r_{WHEEL,n}\left(x_{B,j}(y_i, z_{0,k}), y_i\right) \end{cases} \tag{13}$$

1.5.3 Determination of the velocities relevant for the EHD-Theory along the contact line

$$\Rightarrow \begin{cases} v_{WORM,Bn}\left(x_{B,j}\left(y_i, z_{0,k}\right), y_i\right) \\ v_{WHEEL,Bn}\left(x_{B,j}\left(y_i, z_{0,k}\right), y_i\right) \end{cases} \tag{14}$$

1.5.4 Iterative determination of h_{fict} and η_z

Supposing that the given torque M_{OUT} is transmitted onto the wheel. For this purpose, equation 15 must be solved:

$$h_m\left(q_{EHD,y}\right) - h_0\left(y, h_{fict}, h_{v,sum}\right) - \delta\left(q_{EHD}, y\right) = 0 \tag{15}$$

M_{WHEEL} and the gearing efficiency η_z can then be determined out of this.

$$\Rightarrow \begin{cases} h_{fict}\left(z_{0,k}, l\right) \\ h_m\left(x_{B,j}\left(y_i, z_{0,k}\right), y_i, l\right) \\ \eta_z\left(z_{0,k}, l\right) \end{cases}$$

1.5.5 Linear Intensity of wear I_h, and known average lubricant film thickness h_m

$$\Rightarrow \left\{I_h\left(x_{B,j}\left(y_i, z_{0,k}\right), y_i, l\right)\right.$$

1.5.6 Height of wear removal h_v and wear volume V_v

$$\Rightarrow \begin{cases} h_v\left(x_{B,j}\left(y_i, z_{0,k}\right), y_i, l\right) \\ h_{v,sum}\left(x_{B,j}\left(y_i, z_{0,k}\right), y_i, l\right) = \sum_{m=0}^{m=l} h_v\left(x_{B,j}\left(y_i, z_{0,k}\right), y_i, m\right) \cdot N_{v0} \\ V_v\left(x_{B,j}\left(y_i, z_{0,k}\right), y_i, l\right) \end{cases}$$

2 MEASUREMENTS AT THE TEST BED

For measurement of wear of worm gears, a test bed was built up at the Institute for Machine Elements and Machine Design. At this test bed, two worm gears with different flank forms were tested.

2.1 Tested worm gears

The two worm gears have different tooth shapes. Test gear 1 has a worm with involute flanks, test gear 2 one with hollow flanks. Moreover, the materials of worm and worm wheel are different. The copper content, modulus and gear rim width only differ slightly.
The most important technical data of both worm gears are shown in table 1.

2.2 Used lubricating oil

In order to be able to determine the influence of lubricating oil on wear two different lubricating oils were used. A mineral oil (Shell Spirax 90) SAE90 with a viscosity of 170 [mm^2/s] as well as a synthetic oil with 467 (Shell Tivella) [mm^2/s], each at a temperature of 40 degrees Celsius.

Table 1 Technical data of the tested gears

			Testgear 1	Testgear 2
Nominal power	P_N	kW	6.0	6.0
Maximal power	P_{max}	kW	15.0	15.0
Rated speed	n_1	min^{-1}	1000	1000
Centre distance	a	mm	160	160
Number of teeth	z_1	—	2	2
	z_2	—	49	49
Modulus	m	mm	5.33	5.27
Worm shape			ZI	ZH
Gear rim width	z_{b1}	mm	43.87	42.00
Material/worm			SAE8620	case hardening steel
Material/worm wheel			BS1400, 1985 Grade PB2-C	GZ-CuSn12Ni
Copper content		%	88-89	87

2.3 Test bed

The test bed for measuring wear of worm gears is shown in Figure 1. The gearbox (1) is mounted on a stable steel frame. The three-phase alternator motor (2) is directly flanged on the gear to guarantee the same configuration as in the actual escalator. A chain gear (3) (speed ratio of 3:1) between gear and pendulum generator (4) reduces the output-torque and makes it possible to use the available brake generator. The equilibrium of the pendulum generator is achieved by regulation (5) of its excitation current and therefore of its excitation field. The energy produced by the generator is converted into warmth by a resistor (6). To adjust the output-torque $\mathbf{T_{out}}$ exactly the generator is loaded by using weights (7). The effective power of the drive motor is measured with the 2-Wattmeter method (8).

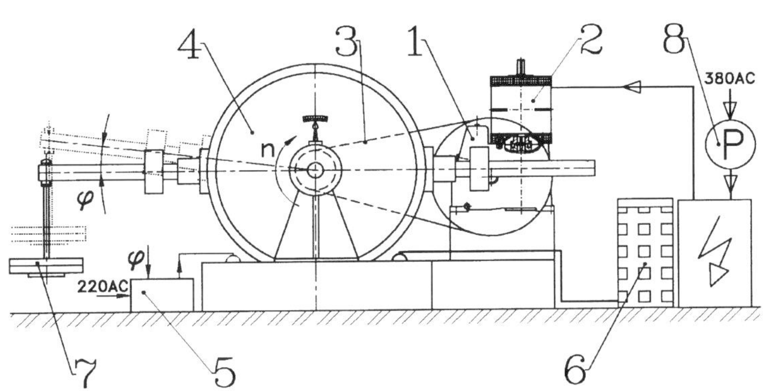

Fig. 1 Test bed for measuring the wear of worm gears

2.4 Measurement of wear

The measurement of wear [4, 11, 12] is made by filtering the applied lubricating oil and subsequent chemical investigation with the aid of the atomic absorption spectrometry (AAS) [1, 7]. For this purpose the lubricating oil is filtered by nitrocellulose filters at 0.8 µm pore size after passing through the full load collective. Then the filters have to be degreased, the wear components are dissolved into nitric acid and afterwards analyzed with AAS. A quite considerable part of wear occurs solved in oil and therefore can not be filtered out, so the remaining wear is determined by AAS of an oil sample. With that method, the copper content in oil is determined and the bronze wear could be calculated out of this (copper content of the bronze is known). The total wear of the worm gear is the sum of the two wear components (in the filter and in oil) which occur during the duration of one load collective cycle.

3 COMPARISON BETWEEN THEORETICAL CALCULATION AND MEASUREMENT

Subsequently, a comparison of the calculated results of the theoretical model of wear as explained in section 1, with the results of wear measurements at the two test gears (see Table 1) and two different oil types (see 2.2) is made. These results are additionally compared with empirical calculations according to NEUPERT [8].

3.1 Lubrication with mineral lubrication oil

The average specific wear depending on the output torque of the gears is shown in illustration 2. It shows, that the measurements correspond quite well with the empirical computation model according to NEUPERT, however the theoretical model of wear overestimates especially the wear at test gear 2 (worm with hollow flanks). The reason could probably be found in the a little worse surface quality of the worm of test gear 2. The surface quality is obvious overrated in the mathematical model of wear, because the wear values measured in the experiment show no major differences.

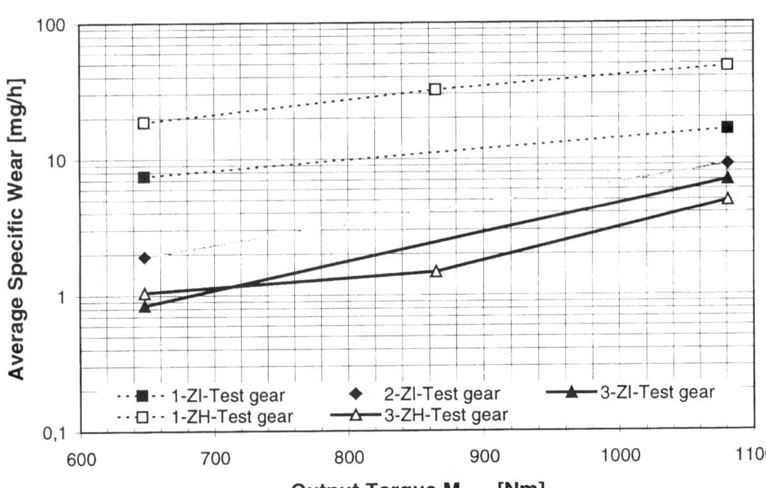

Fig. 2 Average specific wear depending on the output torque, lubrication with mineral lubrication oil, 1-theoretical model of wear; 2-NEUPERT; 3-Measurement

3.2 Lubrication with synthetic lubrication oil

Figure 3 shows that the theoretical calculation model comes up with higher specific wear than the practical experiments. On the other hand, the empirical approach according to NEUPERT shows lower specific wear compared to the test bed measurements.

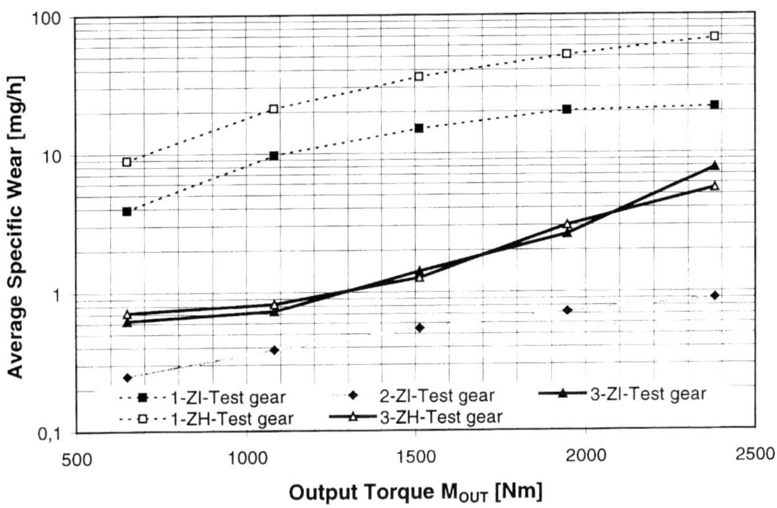

Fig. 3 Average specific wear depending on the output torque, lubrication with synthetic lubrication oil, 1-theoretical model of wear; 2-NEUPERT; 3-Measurement

REFERENCES

[1] Brown, J. R., Saba C. S., Rhine, W. E., Eisentraut K. J., "Particle Size Independent Spectrometric Determination of Wear Metals in Aircraft Lubricating Oils", Analytical Chemistry, Vol. 52, No. 14, December 1980.

[2] Dowson, D., Higginson, G. R., "Elasto-hydrodynamic lubrication", Pergamon Press, Oxford, 1977.

[3] Effendi, J., "Die numerische Lösung der elastohydrodynamischen Kontaktprobleme unter Berücksichtigung der Oberflächenrauheiten", Dissertation, Technische Hochschule Aachen, 1987.

[4] Kovar, W., "Verschleiß- und Wirkungsgraduntersuchungen an einem Schneckengetriebe", VEB Verlag Technik, Berlin, 1971.

[5] Kragelski, I. W., "Reibung und Verschleiß", VEB Verlag Technik, Berlin, 1971.

[6] Kragelski, I. W., Dobyčin, M. N., Kombalov, V. S., "Grundlagen der Berechnung von Reibung und Verschleiß", VEB Verlag Technik, Berlin, 1982.

[7] Kriss, R. H., Bartels T. T., "Improved Atomic Absorption Techniques for Measuring Iron Particles in Lubricating Oil", Atomic Absorption Newsletter, Vol. 9, No. 3, May-June 1970.

[8] Neupert, K., "Verschleißtragfähigkeit und Wirkungsgrad von Zylinder-Schneckengetrieben", Dissertation, Technische Universität München, 1990.

[9] Pausch S., "Verschleiß von Schneckengetrieben–Theorie und Experiment", Dissertation, Technische Universität Wien, 1998.

[10]Predki, W., "HERTZ'sche Drücke, Schmierspalthöhen und Wirkungsgrade von Schneckengetrieben", Dissertation, Ruhr Universität Bochum, 1982.
[11]Rinder, L., "Tragfähigkeitsuntersuchungen an Schneckenrädern aus der Al-Zn-Legierung Alzen 501", Konstruktion, 28, 8, 1976.
[12]Steingröver, K., "Untersuchungen zu Verschleiß, Verlustgrad und Fressen bei Zylinder-Schneckengetrieben", Dissertation, Technische Universität München, 1993.
[13]Weinhauer, D., "Reibung und verschleiß einer hochbeanspruchten Reibpaarung unter Mischreibungsbedingungen", Dissertation, Otto-von-Guericke-Universität, Magdeburg 1996.

Advanced Materials – II

Wear mechanisms in a graphitic metal matrix composite A356 Al-10%SiC-4%Gr

A R RIAHI and **A T ALPAS**
Mechanical, Automotive, and Materials Engineering, University of Windsor, Ontario, Canada

ABSTRACT

An experimental wear map has been constructed for the graphitic cast aluminum matrix composite A356 Al- 10%SiC- 4%Gr in order to delineate load and velocity conditions under which wear transitions occurred. Wear tests were performed using a block- on- ring wear machine within a load range of 0.4- 420 N and a sliding speed range of 0.2- 3.0 m/s against a hardened bearing steel. It was shown that the composite A356 Al- 10%SiC- 4%Gr displayed a transition from mild to severe wear at load and sliding speed combinations which are considerably higher than those of the A356 aluminum alloy and the non graphitic A356 Al- 20%SiC composite. In the ultra mild wear regime wear resistance was primarily due to the hard particles and graphite did not play a significant role. At nearly all sliding speeds and loads in the mild wear regime, a protective tribolayer was formed, which by increasing the speed and load covered a larger proportion of the contact surface and became more compact and smoother. The topmost part of the tribolayer consisted of iron rich layers whose hardness increased with the applied load and the speed. The rest of the tribolayer consisted of fractured SiC and Al_3Ni particles and thin graphite films which were elongated in the direction of sliding thus reducing shear stresses transmitted to the subsurface regions.

1. INTRODUCTION

Cast aluminum alloy engine blocks provide cost effective light-weight material alternatives to the traditional cast iron engines. However, sliding action of a piston ring against an unprotected 319 or 356Al- Si based cylinder bore leads to rapid wear and may cause damage during the cold start. The wear resistance of cast aluminum alloy automotive engine blocks can be improved by protecting the cylinder bores with composite cylinder liners reinforced with hard particles of SiC and Al_2O_3. Scuffing and seizure problems that may occur during the oil starvation periods can be reduced by incorporating solid lubricants namely graphite in the composition of composites [1]. According to Ames and Alpas [2], under severe loading and sliding speed conditions, graphite serves as a temporary safeguard by keeping the initial temperature rise low, therefore allowing the contact surfaces enough time to generate a hard compacted tribolayer with smooth surface and sufficient thickness

(>10μm) to protect the material underneath from excessive subsurface damage. Currently A356 Al - SiC - Graphite, and A356 Al - Al_2O_3 - Graphite composites are being developed on this basis for wear resistant automotive components including cylinder liners, pistons, various types of brakes, and bushings.

The links between the wear resistance and the wear mechanisms can be stablished by constructing wear maps such as the one produced by Lim and Ashby [3] for steels in dry sliding contact against themselves. Liu et al. [4] constructed a wear mechanism map for the aluminum alloys using wear data collected from the literature. Rohatgi et al. [5] developed wear maps for the Al alloys containing solid lubricant particles such as graphite and hard particles such as zirconia. Zhang and Alpas [6] studied the wear behaviour of a wrought aluminum alloy 6061 Al as a function of applied load and sliding velocity and developed an experimental wear transition map for this alloy sliding against SAE 52100 steel. The observation that the transition from mild to severe wear arose when the bulk surface temperature exceeded a critical temperature provided a practical means to predict the onset of severe wear using a single bulk temperature measurement. Wear mechanism maps for an A356 Al alloy and an A356 Al- 20% SiC composite have also been published [7].

The present paper focuses on the wear behaviour of a graphitic metal matrix composite A356 Al- 10%SiC- 4%Gr. Different wear mechanisms were identified and these were plotted on load versus sliding speed axes in a wear map form. The formation of tribolayers and their roles in controlling the wear transitions in the graphitic composites are discussed.

2. EXPERIMENTAL METHODS

Dry sliding wear tests were performed on samples made of A356 Al- 10%SiC- 4%Gr composites (GrANi ®10S:4G). The nominal composition of the A356 Al matrix alloy (in wt pct) was 7 % Si, 0.2 % Mg, 0.11 %Fe, 0.2 %Ti, 0.05 % Mn, 0.05 % Zn, and the balance aluminum. The composite was produced through a molten metal mixing technique (INCO Ltd., Mississauga, ON, Canada). The graphite particles were coated with nickel, to improve surface wetting during casting [8]. The microstructure of the A356 Al- 10%SiC- 4%Gr composite shown in Fig. 1 contains coarse graphite nodules of 138 $\pm$ 25 μm in length and 80 $\pm$ 15 μm in width. The SiC particles have an average diameter of 16 $\pm$ 2 μm. The composite also contains an Al_3Ni intermetallic phase formed as a result of the reaction between the nickel coating around the graphite particles and the molten aluminum. The Vickers hardness of the composite measured at 1 kg load was 90 $\pm$ 2 kg/mm^2, and it's density was 2.75 Mg/m^3.

Wear tests were performed using a block- on- ring type wear machine within a load range of 0.4 - 420 N and a sliding velocity range of 0.2- 3 m/s. The samples were machined from the cast liners into rectangular blocks of 5 mm x 10 mm x 10 mm size. The counterface ring was made of SAE (AISI) 52100 type bearing steel and had an outer diameter of 38 mm. The surface Vickers hardness of the SAE (AISI) 52100 bearing steel ring was (1kg load) 950 $\pm$ 5 kg/mm^2. The sliding distance for each test was normally 6000 m for all sliding speeds and loads except for the severe wear conditions and at the onset of seizure.

A scanning electron microscope (SEM) equipped with energy dispersive X - Ray spectroscope (EDS) was used to characterize the composition and morphology of worn surfaces and debris particles.

Figure 1. Microstructure of A356 Al- 10%SiC- 4%Gr (back scattered SEM) showing the graphite particles (dark constituents), the SiC particulates and the Al_3Ni phase (light constituents) in the A356 Al matrix.

3. RESULTS AND DISCUSSION

3.1 Wear Mechanism Map

The wear rates of the A356 Al- 10%SiC- 4%Gr composite and the main wear mechanisms in each wear regime are presented in the map form on load versus sliding speed axis in Fig. 2. The three major wear regimes, namely, ultra mild, mild and severe wear regimes are depicted on the map. The map also shows sub-regimes such as a scuffing (localized material transfer to the counterface) region and a transient wear region between the mild and severe wear.

The ultra mild regime occurs at very low loads and speeds, with the wear rates ranging between 1×10^{-7} and 9×10^{-6} mm^3/m. In the mild wear regime, the wear rates are an order of magnitude higher (i.e., in the range of 1×10^{-5} to 27×10^{-3} mm^3/m). When the severe wear arises at the other extreme of load and sliding speed conditions, wear rates show a drastic increase and eventually the composite seizes against the steel counterface. The transition to the severe wear occurs at 300 N for the tests at 0.8 m/s, 250 N at 1.2 m/s, 200 N at 2.0 m/s and 120 N at 3 m/s. At the relatively low sliding speeds (less than 1.2 m/s) but high loads (above 350N) in the severe wear regime, wear occurs by the extrusion of material from the sliding interface. At other conditions (sliding speed of 1.2 to 2 m/s and loads of 270 N to 200 N), specimens experience a period of surface melting prior to seizure. Tribolayers are the most characteristic features of mild wear regime and play a significant role in extending the realm of the mild wear to higher speeds and loads compared to non-graphitic composites. Their removal cause transition to the severe wear which is soon followed by the seizure. It should be emphasized that the wear map in Fig. 2, is not only based on the wear rate information but also on the metallographic and compositional studies that were used to delineate microscopic aspects of the wear mechanisms. These will be discussed in the following sections.

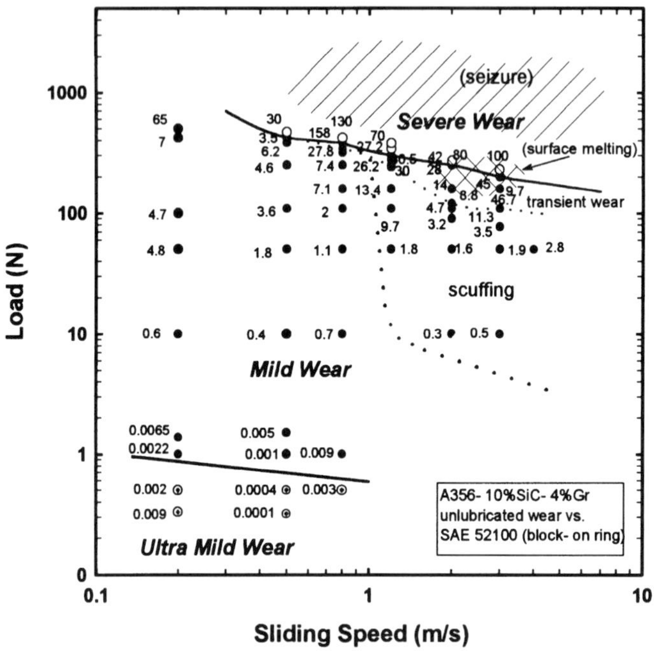

Figure 2. Wear mechanism map for the A356 Al- 10%SiC- 4%Gr composite worn against the SAE 52100 steel counterface. Three major wear regimes, namely, ultra mild, mild and severe wear regimes are shown on the map. A scuffing sub-regime and a transient wear region between the mild and severe wear are also shown. Multiply the wear rates shown by 10^{-3} to find the actual wear rates in mm^3/m.

3.2 Ultra Mild Wear Regime

Zhang and Alpas [9] have shown that the ultra mild wear is due to the load supporting effect of ceramic reinforcing particles at very low loads. The worn surface of the A356 Al-10%SiC- 4%Gr composite was covered by a continuous layer of reddish iron oxide possibly $Fe_2O_3.H_2O$. The debris particles were typically equiaxed and ranged in size from 0.2 μm to approximately 3.0 μm in diameter. Equiaxed particles were mixed with occasionally elongated chips of 10 μm in length. The EDS analysis of the fine equiaxed particles indicated that they were comprised of Fe and O and thus the loose debris generation was the result of the oxidation of the steel counterface during wear rather than that of the surface of the composite. The worn surface of the steel counterface ring exhibited continuous parallel grooves that are indicative of the abrasive effect of the SiC particles. Similar wear processes were observed in the ultra mild wear regimes of non graphitic Al-Al_2O_3 and Al-SiC composites [7, 9] implying that the graphite does not play a significant role in this regime. There is no significant difference between the wear rates (nor the transition loads- to the mild wear) of the graphitic and non-graphitic versions of the A356- SiC composites.

3.3 Mild Wear Regime

Mild wear regime extends in a wide range of loads (5 N to 420 N) and sliding speeds (0.2 m/s to 3.0m/s). The boundaries between the mild wear and the others could easily be determined by visual examination of the changes in the morphologies of worn surfaces and wear debris during the wear tests. Detailed metallographic examinations supported by compositional analyses indicated that the mild wear in the A356 Al- 10%SiC- 4%Gr was primarily controlled by the formation of tribolayers (mechanically mixed and oxidized surface layers) on the contact surfaces. The tribolayers contained fractured SiC, $NiAl_3$ and Si particles mixed with aluminum. The top surfaces of the tribolayers were covered with the iron oxide films. Formation of lamellar graphite films was another aspect of the tribolayers that played an important role in delaying the mild to severe wear transition. Metallographic details of the tribolayers are given in section. 3.3.1. Another wear process that had significant effect on the counterface material (but does not influence mild wear rates of the composite) was the scuffing phenomenon which will be examined in section 3.3.2.

3.3.1 Tribolayer Formation

Mechanically mixed and oxidized tribolayers were formed as soon as the transition to mild wear was realized. However, when the applied loads and sliding velocities were modest the tribolayers were thin (about 5μm) and not continuous throughout the wear tracks as shown in Fig. 3.a, which is from a sample tested at 0.5 m/s and 10 N. Under these conditions approximately 40% of the contact area was covered with the tribolayers. Increasing the sliding speed and the load increased the proportion of the contact surface area covered by the tribolayers. For example at a sliding speed of 2 m/s and a load of 10 N, the wear track was entirely covered by the tribolayers with smooth surfaces (Fig 3.b).

Tribolayers incorporated fragments of fractured SiC and $NiAl_3$ particles mixed with iron and aluminum oxides as is shown in Fig. 3.c. This is a cross sectional SEM microstructure of a tribolayer (and the bulk material underneath the tribolayers) in a sample tested at a sliding speed of 0.5 m/s and a load of 10 N. The mechanically mixed tribolayer had an average thickness of about 5- 15 μm in this section. The iron transferred from the counterface was concentrated on the topmost layers of the tribolayer as iron oxide.

The average thickness of the tribolayers increased with the load and speed but more importantly the top layers adjacent to the contact surfaces became richer in iron oxide as shown in Fig 3.d. (275 N and 2m/s). The hardness of the tribolayers, even when the applied load and the sliding speed were modest was considerably higher than that of the bulk material. (HV_{1000g}= 83 kg/mm^2). At 10N and 2.0 m/s the hardness of the tribolayer was 380 kg/mm^2 (measured on the contact surface). By increasing the applied load to 275 N at the same speed

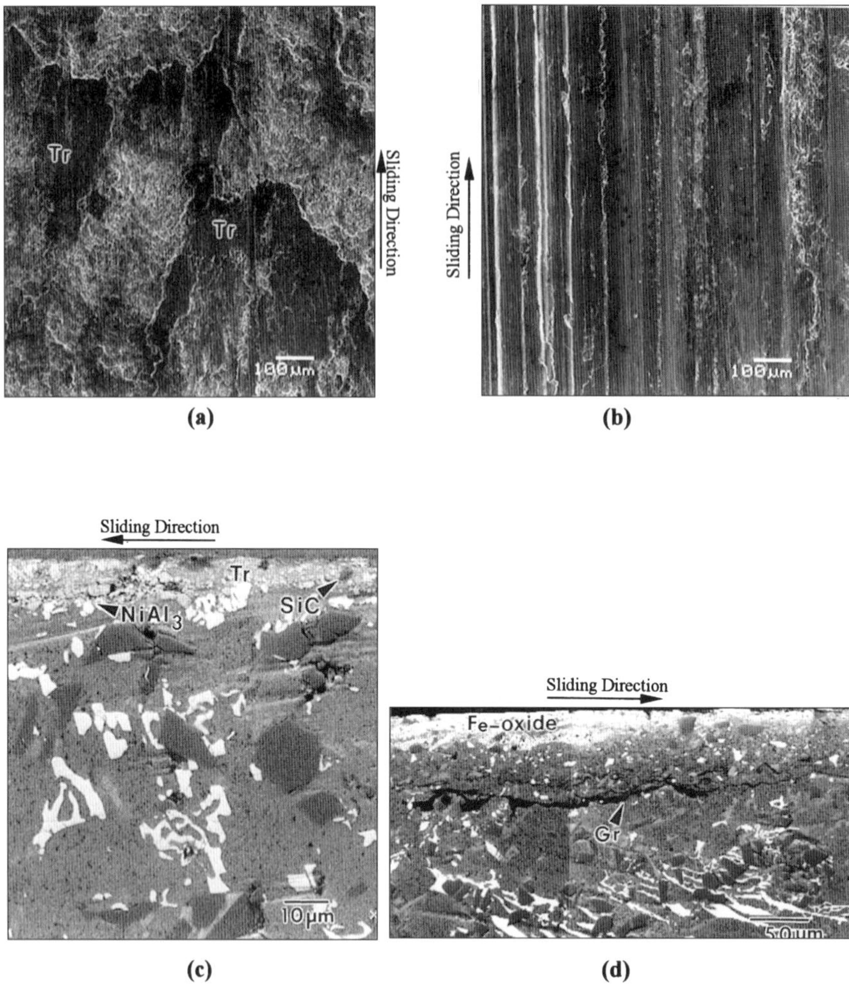

Figure 3. (a) Secondary electron SEM micrograph of the worn surface at sliding speed of 0.5 m/s and load of 10 N. The darker area indicated as "Tr" is the tribolayer. (b) Secondary electron SEM micograph of the worn surface at sliding speed of 2.0 m/s and load of 10 N. The wear track is entirely covered by the tribolayer. (c) Cross sectional SEM microstructure of the tribolayer "Tr" and the bulk material underneath the tribolayer in a sample tested at sliding speed of 0.5 m/s and load of 10 N. The tribolayer incorporates fragments of fractured SiC and $NiAl_3$ particles mixed with iron and aluminum oxides. d) Back scattered SEM micrograph at a sliding speed of 2.0 m/s and a load of275 N. The top layer adjacent to the contact surface is rich in iron oxide. The dark constituent is the graphite layer.

the hardness of the tribolayer reached 780 kg/mm^2. Mechanically mixed layers containing fractured hard particles were found on the surfaces of other (non graphitic) aluminum matrix composites rubbed against the bearing steel [7]. However in the composites without graphite the layers were completely removed at loads lower than those formed on the surfaces of the A356 Al- 10%SiC- 4%Gr, hence causing faster transition to severe wear.

Two distinguishing features of the tribolayers in graphitic composites were i) the presence of iron oxide layers at the contact surfaces that were significantly thicker and more compact than those formed in non-graphitic composites tested under the same conditions; ii) the formation of lamellar graphite films that were peculiar to the graphitic composites. The morphology of the graphite films can be seen in Fig 3.d. These thin films appear to have been formed as a result of shearing of graphite nodules located just below the contact surfaces. They were in the form of thin films of a few microns in thickness and spread long distances (typically 0.5-1.0 mm). It can be suggested that these graphite layers reduce the shear stresses transferred to the matrix. Occasionally they cause delamination of the tribolayers but in this case the exposed graphite provides a low friction surface.

3.3.2 Scuffing

At sliding speeds above 1.2 m/s, the steel counterface experienced local material transfer from the composite, a phenomenon known as scuffing (see [10] for the definition of scuffing). Scuffing could readily be identified from the appearance of the worn steel counterface ring and manifested itself in the form of dark spots, elongated in the sliding direction and protruded (up to 100 μm) from the ring surface. Scuffing occurs as the result of localized heating of the surfaces during sliding and in the case of composites it is facilitated by an abrasive wear process initiated by the tribolayers. At high sliding speeds, friction induced heating is high enough to increase the temperature of the asperities to a temperature necessary to initiate local welding of aluminum to steel [11]. Hard particles of SiC or fragments of iron oxide detached from the tribolayers in the A356 Al- 10%SiC- 4%Gr may act as either two body or three body abrasives against the steel counterface. The hard particles cut longitudinal grooves on the contact surface of the counterface and may occasionally become entrapped at the front end of a groove. The entrapped SiC particles act as abrasive elements to the aluminum matrix of the composite. The contact surface temperatures easily reach 300° C at sliding speeds above 1.2 m/s greatly reducing the shear strength of aluminum [12] and thus facilitating shear fracture and transfer to the counterface. Additionally, the rapid local heating and cooling cycles experienced during sliding wear may cause hardening of the edges of the grooves, which may also act as a hard asperity. Welsh [13] observed the local hardening effect during dry sliding of steel and attributed scuffing to the formation of the hard martensitic spots.

Fig. 4.a is a backscattered SEM micrograph that shows the initial stage of aluminum transfer (which is confirmed by the EDS analysis) from the A356 Al- 10%SiC- 4%Gr to the grooves formed on the contact surface of the steel ring. The transferred aluminum particles first accumulated in the leading front of the hard asperity, then spread backwards as their width increased (Fig. 4.b) and formed transferred material spots whose lengths exceeded 100μm.

3.3.3 Severe Wear Regime

Transition to severe wear was marked by a significant decrease in the hardnesses of the worn surfaces of the A356 Al- 10%SiC- 4%Gr composite compared to those reached in the mild wear, where the tribolayers were present . As discussed in section 3.3.1, the hardness of tribolayer was almost ten times higher than the original bulk hardness of the unworn surfaces at high speed and load conditions in the mild wear regime. In the severe wear, the hardness of the worn surface was still two times higher than the original hardness (200 kg/mm^2 at 300 N, 2m/s) but clearly, the surfaces have lost the hardness provided by the tribolayers. Any tribolayer that might be formed was quickly removed from the contact surfaces in the form of long (2-3mm) fragments of about 0.1mm thick by extrusion in the sliding direction. In regime marked as "transient wear" in the wear map (Fig. 2) tribolayers that were initially formed during sliding were delaminated after a certain period of time, therefore transient wear demarcates conditions under which a time dependent mild to severe wear transition occurs in the A356 Al- 10%SiC- 4%Gr.

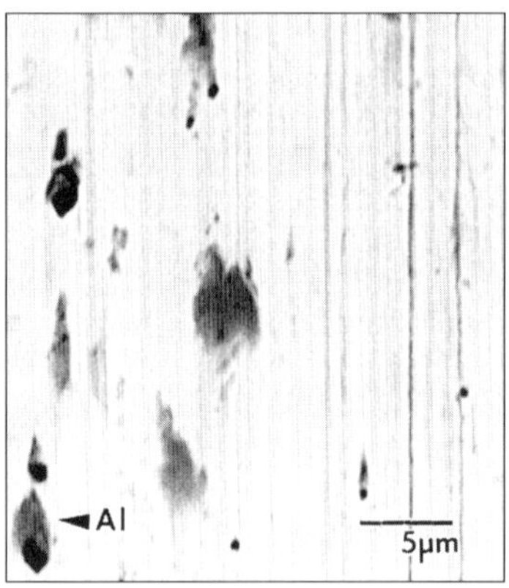

(a)

(b)

Figure 4. (a) Back scattered SEM micrograph of the worn 52100 steel counterface at 50 N and 4.0 m/s showing the initial stage of aluminum transfer from the A356 Al-10%SiC- 4%Gr to the grooves formed on the contact surface of the steel ring. (b) Secondary electron SEM micrograph of the worn 52100 steel counterface at sliding speed of 3.0 m/s and load of 10 N. The transferred aluminum particles first accumulate in the leading front of the hard asperity, then spread backwards as their width increases and form transferred material spots whose lengths exceeds 100 μm.

It is well established that in aluminum alloys the transition to severe wear is controlled by a critical temperature criterion [2, 7, 14, and 15]. The transition temperature from the mild to severe wear was measured as 125° C for an unreinforced A356 alloy [7]. In the A356 Al-20% SiC alloy a much higher surface temperature of 328° C was needed for the transition to severe wear to occur. The critical temperature for the A356 Al- 10%SiC- 4%Gr was 310° C, therefore similar to that of the non-graphitic composite. Although the critical transition temperatures for both composites were similar, the transition to severe wear occurred at higher loads and speeds in the A356 Al- 10%SiC- 4%Gr. In Fig. 5 the mild to severe wear transition boundary of the A356 Al- 10%SiC- 4%Gr composite is compared with those of the A356 Al- 20% SiC composite and the unreinforced A356 alloy. The graphitic composite is clearly superior to the others in delaying the transition to severe wear and hence provides the best seizure resistance against the steel. The graphite films in the tribolayers appear to be responsible for the improved seizure resistance. The reduction in shear stresses transmitted to the aluminum layer and reduction in the heat generated during sliding by the exposed graphite layers are among the plausible causes that can be cited for this effect, but further investigations are currently under way to delineate the exact role played by graphite in delaying the seizure. It can be concluded that the graphitic composites are well suited to the applications where seizure resistance is important such as cylinder liners in the internal combustion engines.

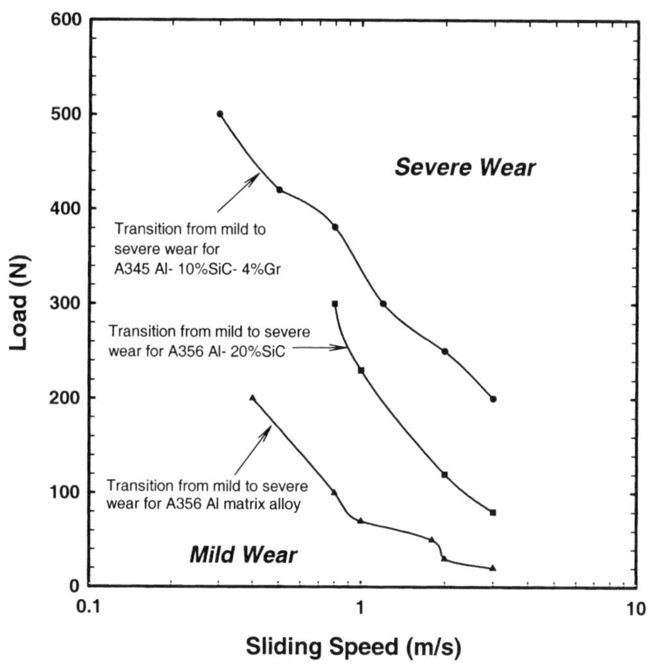

Figure 5. Comparison of the mild to severe wear transition boundary of the A356 Al- 10%SiC- 4%Gr composite with those of the A356 Al- 20%SiC composite and the unreinforced A356 alloy. The graphitic composite is superior to the others in delaying the transition to severe wear.

4. SUMMARY AND CONCLUSIONS

Sliding wear of the graphitic aluminum matrix composite A356 Al- 10%SiC- 4%Gr worn against the SAE 52100 type bearing steel has been investigated under a broad range of loading and sliding speed conditions. The wear rates have been measured and micromechanisms responsible for them have been identified. The results were summarized in the form of a wear mechanism map. The wear behaviour of A356 Al- 10%SiC- 4%Gr was compared with a non-graphitic composite with the same matrix A356 Al- 20% SiC and an unreinforced A356 Al alloy. The main results can be summarized as follows

1) Mild wear in the A356 Al- 10%SiC- 4%Gr is primarily controlled by the formation of the tribolayers or mechanically mixed and oxidized surface layers on the contact surfaces. The mechanically mixed layers contain fractured SiC, $NiAl_3$ and Si particles mixed with aluminum. Increasing the sliding speed and the load increases the proportion of the contact surface area covered by the tribolayers.
2) Two important aspects of the tribolayers in A356 Al- 10%SiC- 4%Gr composites are i) the presence of iron oxide layers at the contact surfaces that are significantly thicker and more compact than those formed in non graphitic composites tested under the same conditions; ii) the formation of lamellar graphite films within the tribolayers.
3) At sliding speeds above 1.2 m/s, the steel counterface is subjected to scuffing or local material transfer from the composite. This is mainly attributed to the roughening of the surface of the counterface by the abrasive action of the counterface and frictional heating induced softening of aluminum.
4) In the severe wear regime the tribolayers are removed and the hardness of the worn surfaces is greatly reduced. However, A356 Al- 10%SiC- 4%Gr displayed a transition from mild to severe wear at load and sliding speed combinations which are considerably higher than those of the unreinforced A356 aluminum alloy and the non graphitic A356 Al- 20%SiC composite.

Acknowledgements

This work is supported by Materials and Manufacturing Ontario and INCO Technical Services Ltd. through a collaborative research grant.

References

[1] Gibson, P.R., Clegg, A.J. and Das, A.A., Production and Evaluation of Squeeze-cast Graphitic Al-Si Alloys, Mater. Sci. & Techn., 1 (1985) 559- 567

[2] Ames, W. and Alpas, A.T., Wear Mechanisms in Hybrid Composites of Graphite-20 Pct SiC in A356 Aluminum Alloy (Al-7 Pct Si-0.3 Pct Mg), Metall. Mater. Trans. A, 26A (1995) 85- 98

[3] Lim, S. , Ashby, M. B., Wear-Mechanism Maps, Acta Metallurgica, 35 (1987) 1- 24

[4] Liu, Y. , Asthana, R., Rohatgi, P.K., A Map for Wear Mechanisms in Aluminum Alloys, J. Materials Sci. 26 (1991) 99- 102

[5] Rohatgi, P.K., Liu, Y., Asthana, R. in Rohatgi, P.K., Blau, P.J., Yust, C.S. (Eds.), Some Issues in the Construction of Wear Mechanism Maps for Metal Matrix Composites, Tribology of Composite Materials, ASM, 1990, pp. 69- 80

[6] Zhang, J. and Alpas, A.T., Transition Between Mild and Severe Wear in Aluminum Alloys, Acta Mater. 45, (1997) 513- 528

[7] Wilson, S. and Alpas, A.T., Wear Mechanism Maps for Metal Matrix Composites, Wear 212 (1997) 41- 49

[8] Stephenson, T.F. and Bell, J.A.E.: Die Casting Reinforced with Nickel Coated Fiber, Development & Applications of Ceramics & New Metal Alloys, R.A.L. Drew and H. Mostaghaci, eds., CIM, Montreal,. (1993) 51- 60

[9] Zhang, J. and Alpas, A.T., Wear Regimes and Transitions in Al_2O_3 Particulate-Reinforced Aluminum Alloys, Mater. Sci. Eng. A161 (1993) 273- 284

[10] Ludema, K.C., A Review of Scuffing and Running-in of Lubricated Surface, with Asperities and Oxides in Perspective, Wear, 100 (1984) 315- 331

[11] Wilson, S. and Alpas, A.T., Thermal Effects on Mild Wear Transition in Dry Sliding of an Aluminum Alloy, Wear 00 (1999) pp. 1- 10

[12] Singh, J. and Alpas, A.T., High Temperature Wear and Deformation Processes in Metal Matrix Composites, Metall. Mater. Trans 27A (1996) 3135- 3148

[13] Welsh, N.C., Frictional Heating and its Influence on the Wear of Steel, J. of Appl. Phys., Vol. 28, No. 9, (1957) 960- 968

[14] Martin, A., Martinez, M.A. and llorca, J., Wear of SiC-Reinforced Al-Matrix Composites in the Temperature Range 20 °C to 200 °C, J. Wear 193 (1996) 169- 179

[15] Reddy, A.S., Pramila Bia, B.N., Murthy, K.S.S and Biswas, S.K., Wear and Seizure of Binery Al-Si Alloys, Wear, 171 (1994) 115-127

Effective-cluster simulation of polycrystal plasticity

C DE LA CRUZ GÓMEZ
Departamento de Mecánica Aplicada, Universidad de Castilla-La Mancha, Almaden, Spain

Abstract. A model is proposed in which each grain, representative of a certain orientation and strength, is embedded in a thin shell of plastic effective medium; this assembly is called an "effective cluster". For each grain the plastic strain increment is chosen such that the yield surface of the effective cluster gives the macroscopic stress. The results differ from uniform-strain theories only in some cases, such as in the compression of fcc polycrystals, where "grain curling" occurs.

1.- THE MODEL.

The most widely used theories of polycrystal plasticity, despite many modifications, all start with the assumption of Taylor (1938) that: the plastic strains are uniform throughout the body. It is well known that this is not true in fact, and cannot be true if equilibrium and yield conditions are to be satisfied everywhere (Bishop and Hill, 1951). For some of the known and expected nonuniformities, plausible arguments can be given for why they might not influence the macroscopic average behavior significantly (Kocks, 1970); for others this is not so evident. In this work we propose a model to deal with one of these.

A useful picture of the elemental process of deformation in a polycrystal is this (Kröner, 1958): consider one grain and the "hole" in which it was at the start; now consider separately the deformation of the hole and of the grain and how they influence each other. The Taylor assumption is equivalent to saying that the hole deforms flke the bulk and the grain follows this deformation exactly. When the particular grain chosen is, for example, stronger than the average, internal stresses are set up which he1p the grain to meet the demanded performance, but they are assumed not to influence the deformation of the hole. In the so-called "self-consistent" elastic-plastic treatments (Kröner, 1958; Budiansky and Wu, 1962) the elastic deformation of the hole is adjusted, but it is not verified whether the total stress in the vicinity of the hole might exceed the yield condition there; in this respect the model is not "self-consistent". This problem has been recognized (Hill, 1965; Berveiller and Zaoui, 1979), but not solved, in the sense that no large-strain computer code has been suggested that incorporates a plastic accommodation of heterogeneities in strength.

We here propose a model in which the "immediate environment" of a grain is allowed to accommodate the individuality of the considered grain in a plastic response; the grain and its immediate environment are together considered as a "cluster" which itself is now embedded in an elastic medium. This serves to average out the effect of especially different grains on a local basis. The aim is to provide an approximate solution rather than a rigorously self-consistent one.

A case of particular interest is when the grains are not really different in "'strength" (in the sense of the size of their yield surface) but have a yield surface of very anisotropic shape. Then to achieve certain components of strain (expressed in crystal coordinates) an especially high stress is required; in sample coordinates these "hard" components will vary from grain to grain. This could be particularly important in materials of hexagonal crystal structure (Kocks and Westlake, 1967; MacEwen et al., 1983; Tomé and Kocks, 1985). In cubic materials there are certain situations in which plane-strain deformation is considerably easier than uniaxial deformation; then, even though uniaxial deformation be macroscopically prescribed, different grains may choose different directions of plane-strain deformation, leading to the phenomenon of "grain curling" (Hosford, 1964).

One of the problems to be addressed is the fact that such a grain is surrounded by only a dozen or so other grains which are not enough to be considered representative of the average. Of course each part of a grain feels one or two of the specific neighboring grains more strongly than it does the others. To solve this problem exactly one would have to do a finite-element calculation. However, an elegant concept inherent in the "self-consistent" treatments (Kröner, 1958) is to assume that a *representative* grain, which occurs many times within a certain interval of orientation and strength, behaves as if all the various environments were averaged and thus constitute an "effective medium". Thus the nonuniformity within each grain (due to the proximity of different specific neighbors) and that from grain to grain (due to different total environments) are ignored. We shall retain the concept of a representative grain of a certain orientation (or strength, or hard component) and apply the effective-medium concept to the "immediate environment"; the combination is what we call an "effective cluster".

2.- THE ALGORITHM.

The model has been used (Tomé et al., 1986) to modify an established Computer code that simulates polycrystal plasticity (Tomé et al., 1984). Each of N grains is imagined to be surrounded by n of the other grains (picked in a systematic-random way); we have so far tried N = 800, n = 12 and N = 200, n = 50; the latter case corresponds in some way to an "effective environmen"; in the former an equivalent averaging occurs through treating more grains. The yield surface of this cluster is obtained by weighting the stresses in the following way:

$$\sigma_c = \left(1 - \frac{V_g}{V_c} \right) \sigma_{.c} + \frac{V_g}{V_c} \sigma_g \qquad (1)$$

where σ_c, is the average stress in the cluster, σ_e, is that in the environment, and σ_g is that in the (central) grain. All are in responsse to a postulated direction $\varepsilon_g{}^0$ of the strain-increment, which is initially taken to be the macroscopic one, $\bar{\varepsilon}^0$, but is then iterated such as to attain

$$\sigma_c(\varepsilon_g{}^0) = \bar{\sigma}(\varepsilon^0) \qquad (2)$$

where $\bar{\sigma}$ fulfills macroscopic equilibrium conditions.

In equation (1), V_g is the volume of the grain, Vc , that of the cluster; the ratio is not equal to 1/n, but rather an independent parameter which characterizes the range of plastic acconunodation around the grain. If V_g were infinitesimally small the influence of the grain would be nil; it conforms completely to the environment and $\varepsilon_g{}^0 \cong \bar{\varepsilon}^0$ (Taylor, 1938). On the other hand, in the limit V_g /Vc = 1, eqn (1) becomes equivalent to the Sachs (1928) model. We have found a value of 1/12 to be essentially equivalent to uniform strain, and 1/4 to give differences of the kind and magnitude that agree qualitatively with experimentally observed deviations from Taylor behavior. V_g /Vc = 1/4 corresponds roughly to a cluster diameter of 1-3 grain diameters.

Equations (1) and (2) were written in terms of the complete stres and strain-increment tensors. However, some of the stress and strain-increment components were exactly prt, the major imposed strain componens were exactly prescribed. Firstly there is the majort imposed strain component, which serves as a scaling factor. (On the other hand, the major stress component was left to come out as it would). In addition we have used the "relaxed contraints" (Honneff and Mecking, 1978; Kocks and Canova, 1981; Tomé et al., 1984). Here the developing anisotropy of grain shape is taken into account by relaxing compatibility conditions on some strain components but, on the other hand, enforcing the corresponding stress components. (Van Houtte et al., 1984, have relaxed some additional components without, however, considering the interaction with the environment).

The iteration in ε_g^0 led to rapid fulfillment of eqn (2) in many cases; it was immaterial what tolerance was allowed for the approximate relation (2). In other cases no convergence could be obtained even after many iterations. We considered this to be a physically meaningful occurrence: sometimes the environment will not be able to accomniodate the hard component of its center grain, and the latter has to follow Taylor conditions.

So far we have only used vertex states of the single-crystal yield surface for calculating σ_g. Thus a change in ε_g^0, would often not give a change in σ_g (although it: would change the slip system activity). In general one should consider grain stress states along the fourth-order or third-order edges of the single-crystal yield surface; a local deviation is expected to occur especially when fewer slip systems can attain a deformation similar (though not identical) to the imposed one.

To resolve ambiguities in slip system activity at vertices we have used a vertex-rounding procedure based on the assumption of some rate sensitivity (Canova and Kocks, 1984). Improvements to this method are still in progress (Canova et al., 1985); they have no significant effect in the limit of relaxed constraints, where fewer cases of nonuniqueness occur.

3.- SOME RESULTS.

Figure 1 shows the results of three different simulations of compression of a polycrystal of fcc lattice structure, in terms of the preferred direction of the normal to the compression plane in a stereographic triangle, after a true strain of 1-0, starting with a random texture. The three methods used were: (a) classical Taylor/Bishop-Hill "full constraints" (FC) theory, using an averaging procedure to resolve ambiguities (Bishop, 1954); (b) relaxed-constraints (RC) theory, based on the evolution in grain shape with deformation and attaining tri-slip conditions in the later stages (Tomé et al., 1984); and (c) "effective cluster" (EC) theory, as proposed here. It is seen that the last, as compared to the first, gives a wider scatter of orientations, as may well be expected, but the general shape of the direction distribution is rather similar. The RC theory (in the form used) gives a sharper texture in which, importantly, the experimentally observed main component, {llO}, is missing; this has been a major problem of the RC theory and led, in fact, to the present new developments.

Figure 2 shows the development of the average Taylor factor, $\bar{M}$, with strain for the three cases. In the FC mode it continues to increase, corresponding to the fact that M is larger than average in the vicinity of {110}, the developing texture component. In the RC model, FC conditions obtain initially, until the aspect ratio of the grains has become sufficiently large to relax the constraints; M then decreases since fewer slip systems are required. In the EC calculation it decreases right from the start; this is due to the fact that an increasing number of grains elect to deform. in the plane-strain mode (which ideally requires only 2/3 of the stress) and are acconimodated by the effective environment.

Since, in this model, there is a direct correlation between the strength of a grain and its behavior, we needed to assume a hardening law in the effective-cluster simulation. We have, assumed it to be of the Taylor (1938) type in that the flow stress depends only on the algebraic sum of shears; the scalar hardening law uses an exponential approach to a final low (but not níl) linear hardening rate, as we have found to be appropriate for compression in copper in Tomé et al. (1984).

4.- SUMMARY.

A further modification of the Taylor theory of polycrystal plasticity is proposed. The difference between the varíous modeis may be expressed in terms of the boundary conditions experienced by a grain. In the true Taylor ("full constraints") theory all components of strain increment are rigorously prescribed. In the 'relaxed constraints' models only some of the strain-increment components are rigorously prescribed; on the complementary components the stresses are rigorous1y prescribed. In the current model there are two units: the grain and the "effective cluster"; the grain follows rigorously the (full or relaxed) strain-increments of the cluster, but the cluster follows approximately the macroscopic stress conditions. Only the grain is used to obtain the macroscopic average properties, but its strain may deviate from the average, the more so the harder ¡t is for the grain to conform and the easier ¡t is for the cluster to accommodate.

REFERENCES

[1] Berveiller, M. and Zaoui, A. (1979). J. Mech. Phys. Solids, 26, 325

[2] Bishop, J. F. W. (1954). J. Mech. Phys. Solids, 3, 130.

[3] Bishop, J. F. W. and Hill, R. (1951). Phil. Mag., 42, 1298.

[4] Budiansky, B. and Wu, T. T. (1962). Proc. 4th Congr. Appl. Mech., p. 1175.

[5] Canova, G. R. and Kocks, U. F. (1984). In 7th Int. Conf. on Textures in Materials, C. M. Brakman, P. Jongenburger and E. J. Mittemeijer. (Eds), Netherlands Society for Materiais Science, Amsterdam.

[6] Canova, G. R., Molínari, A., Kocks, U. F. and Fressengeas, C. (1986).

[7] Hill, R. (1965). J. Mech. Phys. Solids, 13, 89.

[8] Honneff, H. and Mecking, H. (1978). In Textures of Materials, Vol. 1, G. Goustein and K. Lücke (Eds), Springer, p. 265.

[9] Hosford, W. F. (1964). Trans. Met. Soc. AIME, 230,12.

[10] Houtte, P. van, Wenk, H. R. and Wagner, F. (1984). In 7th Int. Conf. on Textures in Materials, C. M. Brakinan, P. Jongenburger and E. J. Mittemeijer (Eds), Netherlands Society for Materials Science, Amsterdam.

[10] Kocks, U. F. (1970). Metall. Trans., 1, 1121.

[12] Kocks, U. F. and Canova, G. R. (1981). In Deformation of Polycrystals, N. Hansen et al. (Eds), Riso National Laboratory, Denmark, p. 135.

[13] Kocks, U. F. and Wesfiake, D. G. (1967). Trans. Met. Soc. AIME, 239,1107. Króner, E. (1958). Z. Physik, 151, 504.

[14] MacEwen, S. R., Faber, J. and Turner, A. P. L. (1983). Acta Metall., 31, 657Sachs, G. (1928). Z. Verein. Deut. Ing., 72, 734.

[15]Taylor, G. 1. (1938). J. Inst. Metals, 62, 307.

[16] Tomé, C. and Kocks, U. F. (1985). Acta Metall., 33, 603.
[17] Tomé, C., Canova, G. R., Kocks, U. F., Christodoulou, N. and Jonas, J. J. (1984). Acta Metall., 32, 1637.

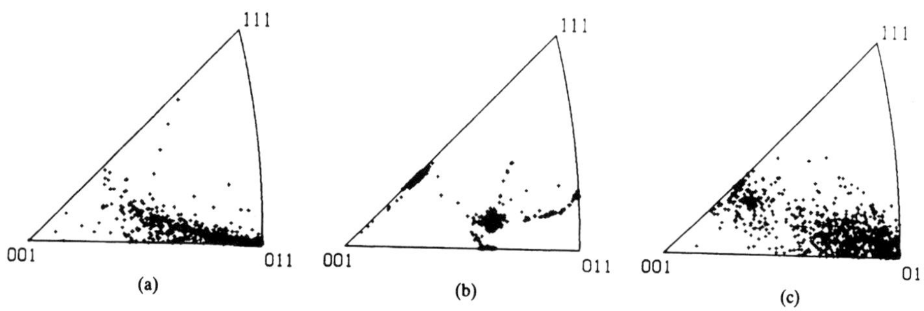

FIGURE. 1. INVERSE POLE FIGURES FOR THE COMPRESSION AXIS IN AN FCC POLYCRYSTAL AFTER A TRUE STRAIN OF 1-0, ACCORDING TO CALCULATION BY (A) FC, (B) RC, (C) EC METHODS.

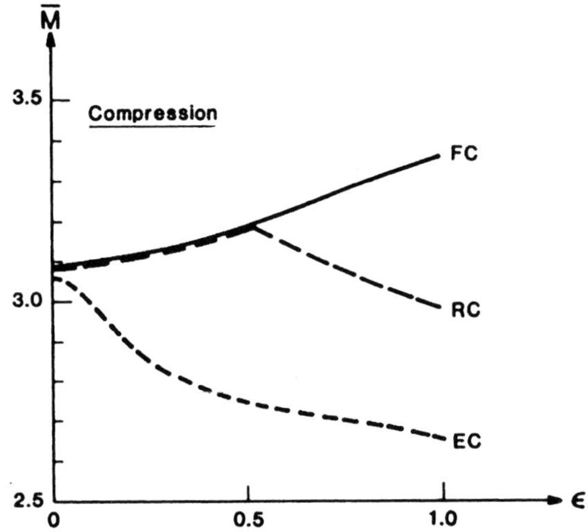

FIGURE 2. DEVELOPMENT OF THE AVERAGE TAYLOR FACTOR $\bar{M}$ WITH COMPRESSIVE STRAIN.

Influence of microstructure on contact fatigue of silicon nitride

B LEGIN EYZOP and **S KARLSSON**
Swedish Ceramic Institute, Gothenburg, Sweden

ABSTRACT

Ceramics are interesting in various tribological applications, such as roller bearings, cam roller followers and valves as it means improved performance and an increased lifetime. The objective of this work was to study the influence of microstructure and test environment on cyclic contact fatigue of silicon nitride. With a coarse microstructure the indentations are deep and large but a more bimodal microstructure has more cracks. A high content of glass phase makes the material sensitive to water due to stress corrosion. Humidity promotes stress corrosion resulting in deep and large indentations with cracks.

1 INTRODUCTION

Thanks to its mechanical and physical properties silicon nitride is interesting to use in several tribological applications, such as cutting tools, cam roller followers and ball bearings. In the early 1960s bearing designers started working on ceramic bearings, but now the main interest is focused on hybrid bearings that consist of silicon nitride balls in steel races. With ceramics in bearings there are performance improvements such as reduced heat generation, good dry-running properties and better corrosion resistance resulting in an increased lifetime by up to ten times. Other advantages are increased speed, increased operating temperature, electrical insulation, less vibration and reduced fuel consumption. When selecting materials for bearing applications several requirements must be considered. The most important ones are low porosity, i.e. less than 0.1 vol%, a uniform microstructure with fine-equiaxed grains, a uniform distribution and minimal content of a second phase, high hardness and high fracture toughness, which result in a better surface finish, and a high resistance to damage from foreign particles. Low thermal expansion in order to reduce the sensitivity to temperature differences, a moderate elastic modulus and spalling failure mode are other important factors. The elastic modulus is higher than for steel, which decreases the vibration, but if the elastic modulus is too high it causes stress concentration. [1-3]

The properties of a silicon nitride material depend on its microstructure, grain boundary phase characteristics and the additives used for densification. A silicon nitride with fine interlocking grains has the best impact properties [4]. The micropore size is considered to have a greater effect on the mechanical properties than the ratio of the α/β silicon nitride phases [5]. According to Tikare and Choi [6] the toughness is higher for silicon nitrides with elongated and coarser grains due to bridging and crack deflection during crack propagation; this enhances long-crack resistance but not short-crack resistance [7]. Mechanical properties such as strength and toughness affect the indentation damage, which is a lifetime limiting factor for ceramics in many engineering applications [8].

Basic features in cyclic fatigue of silicon nitride are an enhancement of the damage growth rate by up to three times compared with static loading, a lack of fatigue striations, an environmental sensitivity and a strong dependence on maximum applied stress intensity [9]. Cyclic contact fatigue tests give information about the short-crack resistance that determines the performance of the material, whereas traditional cyclic fatigue tests are conducted in the long-crack region [10].

Environmental effects on the wear of ceramics are caused by the formation of adsorption layers, tribochemical reactions and a change of the mechanical properties by adsorption, due to stress-corrosion cracking and chemo-mechanical effects [11]. Water and water vapour increase cyclic fatigue damage by promoting tribo-oxidation and by inducing stress corrosion, which leads to microfracture and wear at bridging sites [12,13]. If the grain boundary phase contains yttria and alumina the silicon nitride matrix is more prone to fatigue under hydrothermal conditions [9]. Also the asperity of the crack surfaces affects the cyclic damage in ceramic materials [5].

The main purpose of the work described in this paper was to study cyclic contact fatigue, which is an important wear mechanism in ceramic hybrid bearing systems. A spherical indenter closely reproduces the loading conditions in such applications as roller bearings. The environmental influence on the fatigue strength was also investigated by tests performed in air, ethylene glycol and distilled water. Similar cyclic contact fatigue tests in oil have been conducted by Chen *et al.* [14].

2 EXPERIMENTAL PROCEDURE

Four silicon nitrides were selected as test materials; two commercial silicon nitrides for bearing applications designated A and B, one commercial silicon nitride for valves designated C, and a non-commercial silicon nitride manufactured at the Swedish Ceramic Institute, material D. Materials A and B are HIPed whereas materials C and D are sintered by gas pressure sintering (GPS). The three commercial materials are fine grained, less than 5 μm, and dense; A and C have some rod-like grains longer than 10 μm, whereas B has a more homogeneous but slightly coarser microstructure. Material C has the highest glass phase content. The non-commercial material has a bimodal microstructure with an inhomogeneous glass phase distribution and a density of 99.7 % of the theoretical. The microstructures of the four materials are shown in Figure 1. Material A also contains magnesium oxide, alumina and iron oxide whereas material B contains alumina, titanium oxide and yttria. In materials C and D_2 alumina and yttria are used as sintering additives, but only yttria in D_1.

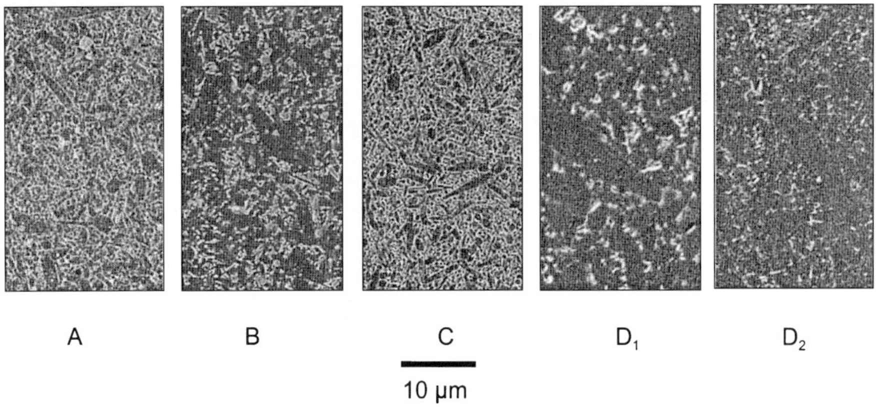

Fig. 1 Microstructures of materials A, B, C and D (plasma etched).

Disc-shaped specimens, thickness 4–6 mm, were prepared from bearing balls, from valves and from laboratory specimens. The test surfaces were ground and polished down to 3 µm diamond paste and finally with a 40 nm colloidal silica suspension in a Struers RotoPol-22.

Hertzian stresses were induced in the silicon nitride discs with a rigid indenter to simulate the loading conditions in silicon nitride hybrid ball bearing systems. The indenter consists of a tungsten carbide ball with 6% cobalt and a diameter of 6.0 mm. The maximum compressive contact pressure (p_0), obtained in the middle of the contact area, was calculated from

$$p_0 = \left(\frac{6W\left(E^*\right)^2}{\pi^3 R^2} \right)^{1/3} \tag{1}$$

with
$$\frac{1}{E^*} = \frac{1-\nu_1^2}{E_1} + \frac{1-\nu_2^2}{E_2} \tag{2}$$

where W is the applied load, R the radius of the tungsten carbide indenter, $E_{1,2}$ elastic moduli of the disc (295-310 GPa) and the indenter (600 GPa), respectively, and $\nu_{1,2}$ Poisson's ratios of the disc (0.27) and indenter (0.26), respectively [15].

The tests were conducted in an MTS 880 servo hydraulic test machine using load control mode. Cyclic loading was applied in the form of a half sine wave with a preload of 5–10 N. The experiments were run at room temperature under normal laboratory conditions. To study the environmental efffect three test media were used, air, ethylene glycol and distilled water. The frequency of the cycles was 10 Hz for all the tests. In hybrid ball bearings the maximum contact pressure is usually 2–5 GPa; 5 GPa was therefore chosen for these tests to avoid machine instability and to minimise test duration [16]. To reduce the number of cycles the higher maximum contact pressures of 10, 12 and 15 GPa were chosen. A schematic of the test equipment is shown in Figure 2. Two series of experiments were conducted. The first one consisted of 9 samples each of materials A and B [17] and the second series consisted of 20 samples each of materials A and B, 3 of material C and 4 of material D.

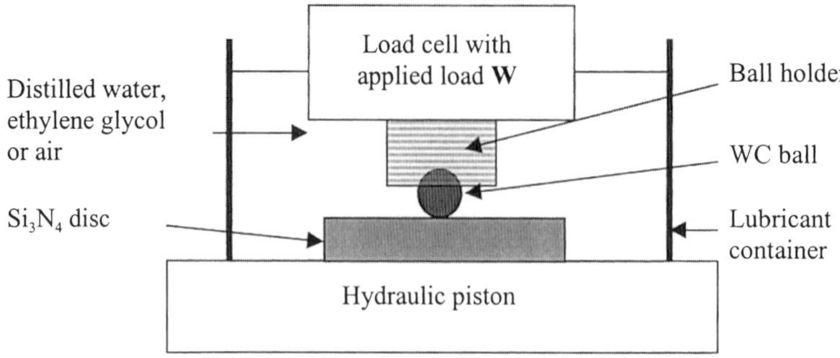

Fig. 2 Schematic of test equipment.

To characterise the materials the fracture toughness was determined by an indentation method and calculated according to Niihara *et al.* [18] and Laugier [19] with an applied load of 98.1N. Vickers hardness was measured with a load of 1kg in accordance with the ENV 843-4 standard.

The contact areas were evaluated using a light optical microscope (LOM), a scanning electron microscope (SEM) and an interference microscope (IM). The topography of the indentation damage, the crack length and type, and the grain pull-outs were investigated in a SEM. The IM was used for measurements of area and depth/height of the damage.

3 RESULTS AND DISCUSSION

Table 1 shows the Vickers hardness and the fracture toughness of the silicon nitride materials. These values are in good agreement with other similar commercial silicon nitride materials.

Table 1 Mechanical properties of the silicon nitride materials.

Material	Hardness (1 kg) (GPa)	Fracture toughness ($MPam^{1/2}$)	
		Laugier	*Niihara et al.*
A	16.4 ± 0.5	4.8 ± 0.3	6.8 ± 0.5
B	16.2 ± 0.7	5.0 ± 0.6	7.3 ± 0.9
C	14.7 ± 0.4	5.8 ± 0.6	8.1 ± 0.9
D_1	11.4 ± 0.3 *	4.0 ± 1.0	5.9 ± 1.4
D_2	11.4 ± 0.4 †	4.3 ± 0.5	6.3 ± 0.8

*) Long cracks at load 1 kg, hardness 18.4 ± 0.5 GPa at load 0.5 kg.
†) Long cracks at load 1 kg, hardness 18.7 ± 0.5 GPa at load 0.5 kg.

The wear process in air starts by accumulating wear debris thereby forming a layer on the silicon nitride disc, which is later on removed revealing the indentation. In ethylene glycol and distilled water the wear debris is continously removed from the indentation in the silicon nitride disc. The cyclic contact fatigue damage starts later for material A than for material B, but when the damage is initiated the damage propagation is faster in material A which results in deeper and larger indentations, see Table 2.

Table 2 Examples of damage.

Medium	Maximum contact pressure (GPa)	# Cycles (k)	Area of damaged zone (mm^2)			
			A	B	C	D
Air	5	2,000	0.00	---	---	---
	10	2,000	0.15	0.15	---	---
		3,000	0.23	0.17	0.12	0.18
	12	200	0.00	0.56	---	---
	15	200	1.20	---	---	---
Ethylene glycol	10	2,000	0.00	0.11	---	---
		3,000	0.21	0.13	0.09	---
	12	200	0.00	0.24	---	---
	15	200	---	1.00	---	---
Distilled water	5	700	0.05	0.05	---	---
		1,500	0.04	0.12	0.61	---
	10	2,000	0.22	0.50	---	---
	15	200	---	1.70	---	---

---) not tested

The environmental effect was studied by comparing the results of the test carried out in air, ethylene glycol and distilled water. Ethylene glycol was chosen as an inert and anhydrous lubricant. A comparison between air and distilled water points out the severe effect of moisture. In distilled water the indentations are deeper and larger and there are more cracks than in air and ethylene glycol. According to Gomes *et al.* [13] this may be explained by a promotion of the tribo-oxidation and by a weakening of the adhesion of the wear debris in water. Kishimoto *et al.* [12] claimed that water enhances the effect of cyclic fatigue by stress corrosion, which causes microfracture and wear at bridging sites. Thus there is an environmental factor influencing the damage process, but the mechanical forces are predominant [10].

When comparing the different materials the medium must also be considered. For the three commercial materials the wear scar is largest in A followed by B and C in both air and ethylene glycol, but in distilled water the order is reversed, see Table 2. In all three media material D is more damaged than the commercial materials. Figure 3 shows the wear scar after 3 million cycles at 10 GPa in air. The silicon nitrides with the most uniform microstructure, materials B and C, have the best performance in air and ethylene glycol, i.e. environments with low moisture content. According to Wani *et al.* [20] a softening the of grain-boundary glass phase at increased contact temperatures helps preventing microfracture and smoothes asperities, thereby reducing the wear, which is shown by the smallest damage in material C. In distilled water material A, a fine-grained silicon nitride with some elongated grains, has the best performance due to crack deflection by the elongated grains, whereas material C, with a high glass-phase content, has the largest wear scars, which according to Kishimoto *et al.* [12], is a result of stress corrosion. Material D with a coarser, bimodal microstructure has deep and large wear scars. The inhomogeneous glass phase distribution increases the sensitivity to water.

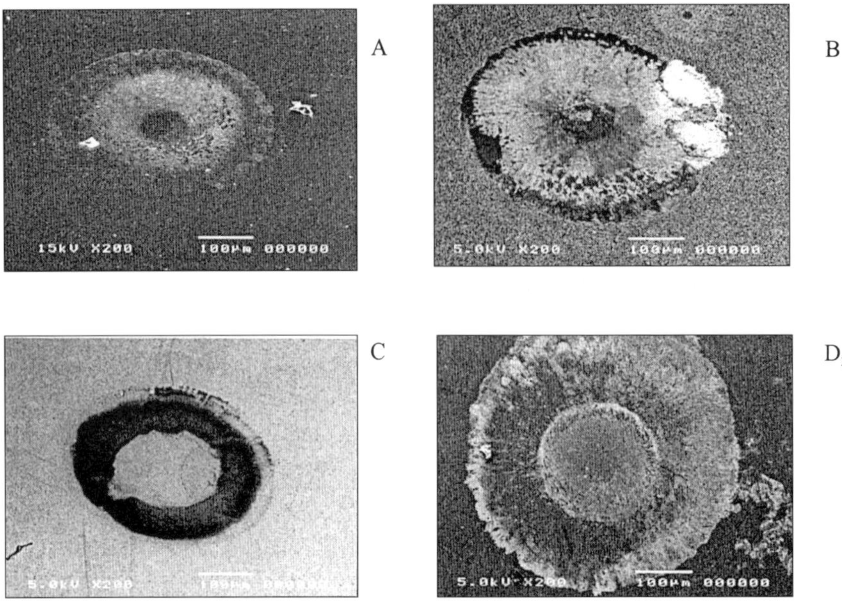

Fig. 3 Wear scar at 10 GPa in air after 3 million cycles for materials A, B, C, and D (plasma etched).

Crack propagation was investigated for the three media. In air cracks appear at 15 GPa only but in distilled water at 5 GPa and higher. Cracks were not observed in ethylene glycol even after 5 million cycles at 10 GPa. The cracks propagating from the indentations are intergranular with grain pull-outs along the longer cracks. Figure 4 shows a long crack in material A tested in air at 15 GPa for 200,000 cycles. This difference in crack behaviour between the three media can be explained by the moisture content. In ethylene glycol the humidity is very low and therefore there is no stress corrosion, whereas the stress corrosion promotes crack propagation in water. There is also a difference in crack behaviour between the silicon nitrides. In water material A shows cracks at 5 GPa whereas cracks first appear at 10 GPa for material B. For material C there are no cracks at all and material D has cracks at 10 GPa in water, but for these two materials the number of tests are insufficient for further analysis.

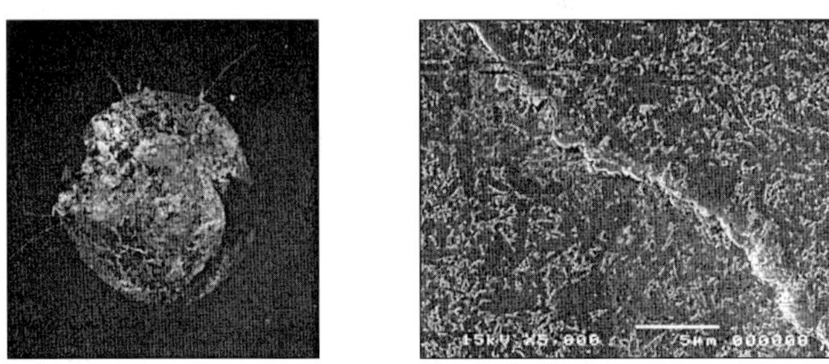

Fig. 4 Indentation and intergranular crack with grain pull-outs in material A after 200,000 cycles at 15 GPa in air (plasma etched).

When examining the tungsten carbide ball in a SEM after testing some shallow damage could be seen. However, this damage is too small to be considered to have had any influence on the indentations in the silicon nitride specimens.

4 CONCLUSIONS

Silicon nitride materials with the most uniform microstructure have the best cyclic contact fatigue performance in environments with low moisture content. The higher glass phase content material has the smallest damage due to a softening of the glass phase at increased contact temperatures, which smoothes asperities and prevents microfracture.

In distilled water the fine-grained silicon nitride with some elongated grains has the best damage resistance due to crack deflection. The material with higher content of glass phase has the largest damage, caused by stress corrosion.

The silicon nitride with a coarser and bimodal microstructure has deep and large wear scars.

5 ACKNOWLEDGEMENTS

The authors would like to thank Mr Erik Dahlberg, Chalmers University of Technology, for help with running the tests, Dr Håkan Björklund, Volvo Technological Development Corporation, and Dr Lars Kahlman, SKF Ceramic Bearings, for interesting discussions and finally Mr Lars Eklund, Swedish Ceramic Institute, for assistance with the scanning electron microscope. The project has been financed by the Industrial Ceramic Research Association and the Swedish National Board for Industrial and Technical Development (NUTEK).

REFERENCES

[1] Katz, R. N. and Hannoosh, J. G., "Ceramics for high performance rolling element bearings: A review and assessment", Int. J. High. Tech. Ceram., 1, 1995, 69-79.

[2] Burrier Jr, H. J., "Optimizing the structure and properties of silicon nitride for rolling contact bearing performance", Tribol. Trans., 39 (2), 1996, 276-285.

[3] Cundill, R. T., "Material selection and quality for ceramic rolling elements", in Proceedings of the Fourth International Symposium on Ceramic Materials and Components for Engines, Eds: Carlsson, R., Johansson, T., and Kahlman, L., Elsevier Science Publishers Ltd, London, 1992, 905-912.

[4] Cundill, R. T., "Impact resistance of silicon nitride balls", in Proceedings of the Sixth International Symposium on Ceramic Materials and Components for Engines, Eds: Niihara, K., Hirano, S., Kanzaki, S., Kommeya, K., and Morinaga, K., Japan Fine Ceramics Association, Tokyo, 1997, 556-561.

[5] Horibe, S., "Fatigue of silicon nitride ceramics under cyclic loading", J. Eur. Ceram. Soc., 6 (2), 1990, 89-95.

[6] Tikare, V. and Choi, S. R., "Influence of the Si_3N_4 microstructure on its R-curve and fatigue behavior", Ceram. Eng. Sci. Proc., 12 (7-8), 1991, 1437-1447.

[7] Xu, H. H. K., Wei, L., Padture, N. P., Lawn, B. R., and Yeckley, R. L., "Effect of microstructural coarsening on Hertzian contact damage in silicon nitride", J. Mater. Sci., 30 (4), 1995, 869-878.

[8] Lawn, B. R., "Indentation of ceramics with spheres: A century after Hertz", J. Am. Ceram. Soc., 81 (8), 1998, 1977-1994.

[9] Jacobs, D. S. and Chen, I-W., "Mechanical and environmental factors in the cyclic and static fatigue of silicon nitride", J. Am. Ceram. Soc., 77 (5), 1994, 1153-1161.

[10] Guiberteau, F., Padture, N. P., Cai, H. and Lawn, B. R., "Indentation fatigue - A simple cyclic Hertzian test for measuring damage accumulation in polycrystalline ceramics", Philos. Mag. A, 68 (5), 1993, 1003-1016.

[11] Sasaki, S., "Environmental effects on friction and wear of ceramics", KONA, No 10, 1992, 58-64.

[12] Kishimoto, H., Ueno, A., Fujiwara, Y., Kondo, T., and Kominato, I., "Influence of water vapour on crack propagation of sintered silicon nitride under cyclic load", JSME Int. J., Series A, 39 (3), 1996, 435-441.

[13] Gomes, J. R., Miranda, A. S., Silva, R. F., and Vieira, J. M., "Tribological properties of $AlN-CeO_2-Si_3N_4$ cutting materials in unlubricated sliding against tool steel and cast iron", Mater. Sci. Eng., A209 (1-2), 1996, 277-286.

[14] Chen, Z., Cuneo, J. C., Mecholsky Jr, J. J., and Hu, S., "Damage process in Si_3N_4 bearing material under contact loading", Wear, 198, 1996, 197-207.

[15] Williams, J. A., "Engineering Tribology", Oxford University Press Inc., New York, 1996, p. 468.

[16] Kahlman, L., Personal communication, 1997.

[17] Legin Eyzop, B. and Karlsson, S., "Cyclic contact fatigue of silicon nitride", Wear, 225-229, 1999, 1303-1308.

[18] Niihara, K., Morena, R., and Hasselman, D. P. H., "Evaluation of K_{IC} of brittle solids by the indentation method with low crack-to-indent ratios", J. Mater. Sci. Lett., 1 (1), 1982, 13-16.

[19] Laugier, M. T., "Elastic/plastic indentation of ceramics", J. Mater. Sci. Lett., 4 (12), 1985, 1539-1541.

[20] Wani, M. F., Prakash, B., Das, P. K., Raza, S. S., and Mukerji, J., "Friction and wear of HPSN bearing materials", Am. Ceram. Soc. Bull., 76 (8), 1997, 65-69.

Tribological behaviour of polyethersulphone-steel pairs

M D BERMÚDEZ, F-J CARRIÓN, J-J CERVANTES, I MARTÍNEZ-MATEO, and
G MARTÍNEZ-NICOLÁS
Grupo de Ciencia de Materiales e Ingeniería Metalúrgica, Universidad Politécnica de Cartagena, Spain

SYNOPSIS

Polyethersulphone (PES) is a thermoplastic with an unusual combination of mechanical, thermal and chemical resistance with ease of processing. Its range of uses covers from electrical/electronic to automotive and aerospace industries, including medical/surgical instrumentation or milking machine parts, due to its ability to withstand repeated sterilization. The purpose of this work was to establish the convenience of using polyethersulphone-steel contacts in the food processing industry. In this work we present the results of friction and wear tests carried out on a pin-on-disk tribometer using steel pins against PES disks. Influence of normal load, sliding speed and distance will be discused.

Keywords: Dry wear; Pin on disk; Polyethersulphone; Steel

1. INTRODUCTION

Polyethersulphone (PES) and other high temperature thermoplastics are increasingly gaining access to applications which have previously been reserved for graphite, ceramics or metals [1]. Special applications for those materials are maintenance-free sliding bearings with sliding surface temperatures (within the range) up to 300°C.

PES combines high thermal stability and mechanical strength with the ability to be melt-processed on conventional equipment at reasonable temperatures. Key properties [2] are thermal stability, creep resistance, electrical, chemical and flammability resistance. Thus applications cover electrical, electronic, automotive and food industries. PES is also used as lining of steel tubes and corrosion resistant coatings of automotive components [3].

The molecular structure of PES makes it suitable for many membrane applications in medical uses. It is used by drinks producers in equipment for removal of alcohol during production of low-alcohol beers and has been approved by the Food and Drug Agency [4] for food service applications.

The sulfone unit $[C_6H_4SO_2C_6H_4O]_n$ is responsible [5] for thermal stability and strength, while the ether linkage provides chain mobility, allowing easy processing. The material is amorphous and is transparent in its pure form. It has good metal-to-plastic bearing properties, and it has the ability to seal against all inner valve materials.

The use of thermoplastics, however, is hampered by a lack of established load-carriying and wear data, when compared with metals [6], because of their relatively recent arrival.

2. EXPERIMENTAL

Tribological experiments have been carried out using a real time pin-on-disk test machine ISC- 200 PC which continously records friction coefficients as a function of time. For the tribological tests, AISI 316 steel pins were used sliding against PES disks. PES is commercially available as Ultrason from BASF.

2.1. Preparation of PES disks.

PES disks were pressure molded from the melt according to the following procedure: PES pellets (8 g.) were introduced in a clindrical steel mold and compacted at room temperature under a pressure of 80 MPa for 30 minutes. This cold compact was then heated in an oven at 400°C for 40 minutes. The melt is then allowed to cool in air while applying a pressure of 50 MPa. In this way, 25 mm diameter and 10 mm height disks were obtained. Properties of PES are shown on table 1.

Table 1. Physical properties of polysulphone disks

Density (kg/m^3)	1370
Glass transition temperature (°C) (10°C/min.)	225
Humidity absorption (%)	0.7
Viscosity number (cm^3/kg)	56
Hardness (Shore D)	82
Roughness (R_a) (μm)	1.0

Volume loss was determined from wear track width, assuming no significant pin wear [7], using equation (1):

$$V = \frac{\pi(\text{wear track radius})(\text{track width})^3}{6(\text{pin sphere radius})} \quad (1)$$

Wear track radius = 9 mm.; pin sphere radius = 0.6 mm. Track width for each test is the mean value after at least twelve measures distributed along the wear scar and was measured by image analysis using Optimas 5.0 by the method already reported [8]. Volume loss is the mean value of at least three tests under the same conditions. Optical micrographs were taken with a Leica DMRM microscope and SEM micrographs with a JEOL 6100 microscope.

3. RESULTS AND DISCUSION.

Friction and wear of PES/steel pairs were measured under variable distance (load = 0.25N; speed = 0.10 m/s), speed (load = 0.98 N; distance = 500 m) and load (speed = 0.10 m/s; distance = 500m).

3.1. Friction tests.

Friction grows fast for short sliding distances. As distance increases further, friction increments slow down as shown on figure 1, to reach a maximum of 1.25 after 4000 m.

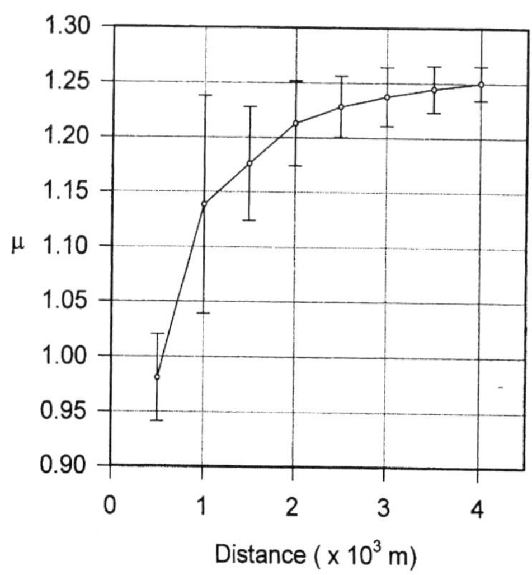

Figure 1. Variation of friction with sliding distance

Friction values remain low under increasing sliding velocity, until a strong increment is observed when passing from 0.10 to 0.12 m/s (figure 2).

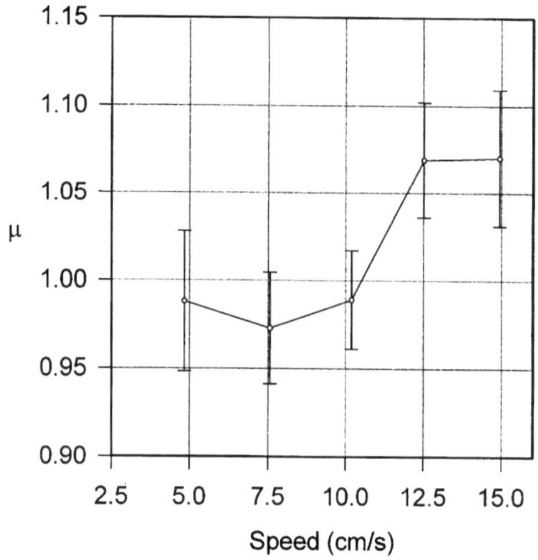

Figure 2. Variation of friction with sliding speed

After an initial peak under the critical load of 0.49 N, friction values decrease to remain practically unchanged around 0.98 from 1.47 N (figure 3). This behaviour has been previously observed for other polymers [9], the position of the maximum being a function of sliding speed.

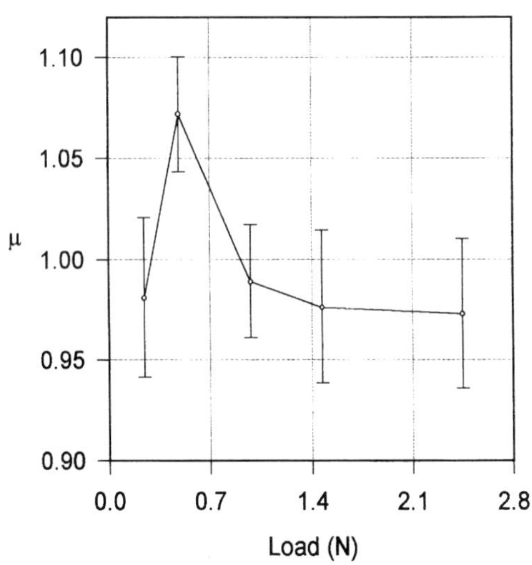

Figure 3. Variation of friction with applied normal load

3.2. Wear tests.

As expected, volume loss steadily increments with distance. More interesting results the analysis of wear coefficients as shown on figure 4.

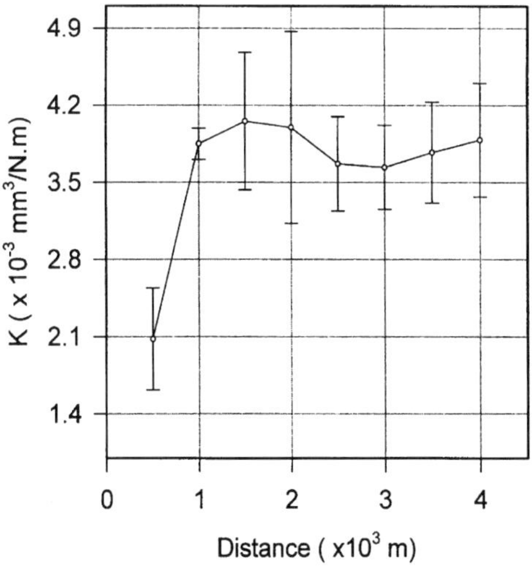

Figure 4. Wear coefficient vs sliding distance

After an initial increment in wear coefficient from 500 to 1000 m sliding distance, parallel to the increment in friction (see figure 1), wear coefficients are not significatively affected by increasing distance.

Wear coefficient (see figure 5) increments under the interval of sliding speed studied, reaching a maximum for 0.15 m/s, which also corresponds to maximum friction, and is associated to removal of wear debris from the contact path (figure 7b).

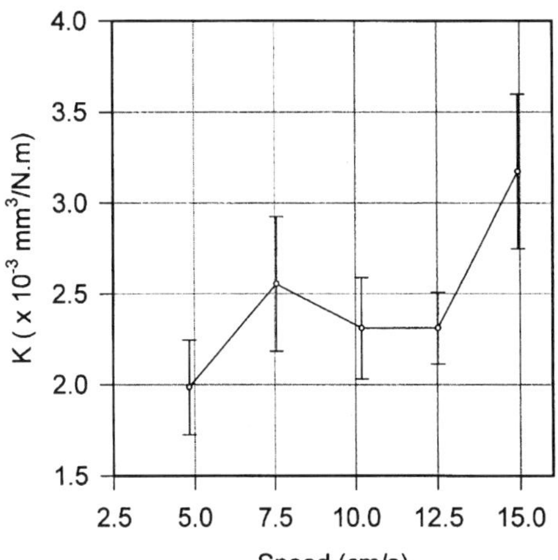

Figure 5. Wear coefficient vs sliding speed

At low loads, friction and wear coefficients show the same variation with load (see figures 3 and 6, respectively). From 0.98 N to 2.45 N, wear coefficient increases linearly with increasing load, while friction coefficent decreases for the same load interval. This could be due to thermal softening effects [10].

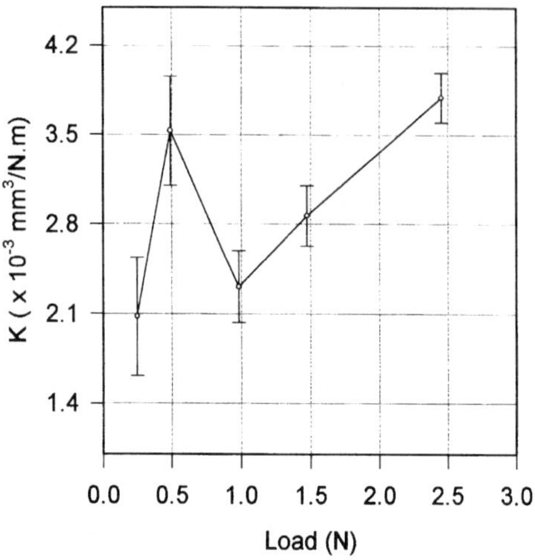

Figure 6. Wear coefficient vs applied normal load

As a consecuence, if low wear is required, as in the case of food processing equipment, normal loads higher than 1 N should be avoided in direct contact steel-PES. More important for this type of applications, is the fact that no wear of the steel pins were observed under the experimental conditions studied here.

3.3. Wear mechanism.

The polymer/steel contact is formed by a rigid asperity traversing a deformable polymer substrate. This configuraton involves relatively large volume deformations beneath the contact. Previous studies [11] have shown that for a metal slider on a polymer disk configuration, the wear rate is substantially lower than that observed for the polymer pin on metal disk configuration, under the same test conditions. These differences can be quite important from an engineering and enviromental point of view, and can be attributed to differences in contact conditions.

The large strains may cause rupture of the polymer close to the interface [12], provided that the transfer process is not hindered by the presence of a boundary film.

The metal pin on polymer disk configuration helps the wear debris to be trapped in the contact zone. As a result, the initial contact between the metal and the polymer changes to one between the metal and wear debris smeared all over the contact path.

The mechanism of creation of a single wear particle apparently involves creation of an adhesive junction that must be stronger in shear at the interface than the shear strength of the polymer. As a result, a thin, flakelike debris is created. For relatively low sliding speed (figure 7a) the majority of wear particles is trapped in the contact zone, particularly when the wear track reaches a certain depth. This gives rise to an adhesive and abrasive wear mode as shown by the wear tracks (figure 7). At higher speed (figure 7b) most wear particles are removed from the wear track, giving rise to a higher wear rate. A smoother worn surface is expected to correspond to a greater degree of thermal softening [13].

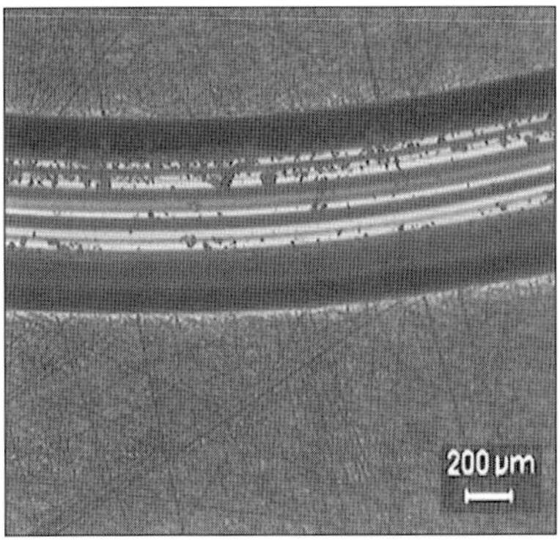

a

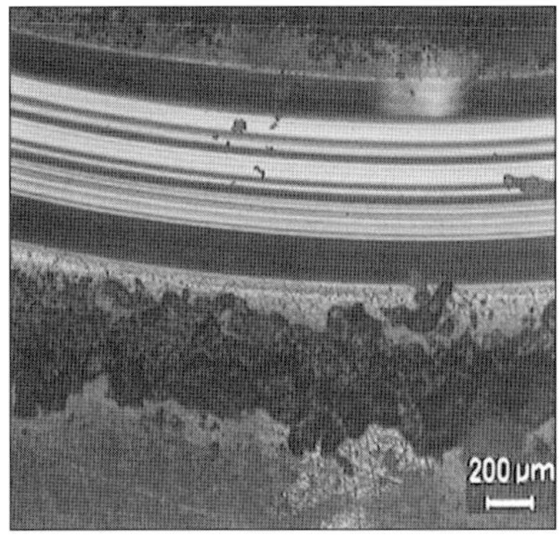

b

Figure 7. Wear tracks on PES disk. a) Sliding speed 0.12 m/s. b) Sliding speed 0.15 m/s.

Most wear particles have the appearance shown on figure 8, and are in accordance with the proposed mechanism. No fatigue cracks are apparent, although this mechanism prevails in wear of PES/steel contacts in the presence of liquid environments [14].

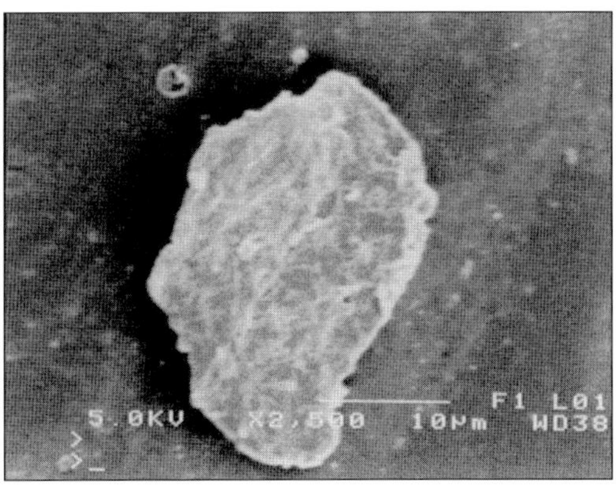

Figure 8. Wear particle trapped inside the wear track.

Aknowledgments

Authors wish to thank the Dirección General de Enseñanza Superior e Investigación Científica (PB97-1035) and the Fundación Séneca (Exp. 01738/CV/98) for financial support. We are also grateful to HRS Spiratube S.L. for their cooperation in this work.

REFERENCES

[1] Hufenback, W., Junghans, R., Kunze, K., Marx, S., "Friction and wear of maintenance-free sliding bearings of high-temperature thermoplastics". Tribol. Schmier.tech., 42(3), 1995, pp. 144-147.

[2] Maher, J., "Polyethersulfone: high-temperature performer". Adv. Mater. Process., 147(4), 1995, pp. 47-48.

[3] Harris, J.E., Johnson, R.N., "Encyclopedia of Polymer Science and Engineering", 13, 1988, pp. 196-211.

[4] FDA Regulations, Title 21, Vol. 3, 1999, pp. 318-319.

[5] ASM Engineered Materials Handbook. Vol. 2, "Engineering Plastics". 2nd reprint, 1995, pp. 159-162.

[6] Williams, E.H., Quinn, K.R., "Materials for plastic gears", Mach. Des., 6, 1995, pp. 50-52, 54.

[7] Standard test method for wear testing with a pin-on-disk apparatus. *ASTM G 99*, 1990.

[8] Yamaguchi, Y., "Tribology of plastics materials". Elsevier, Amsterdam, 1990.

[9] M.D. Bermudez, F.J. Carrion-Vilches, G. Martinez-Nicolas, "Wear of liquid crystal-additivated polymers against steel". J. Appl. Polym. Sci. 74, 1999, pp. 831-837.

[10] Lancaster, J.K., "Materials-specific wear mechanisms: relevance to wear modelling", Wear, 141, 1990, pp. 159-183.

[11] Hosseini, S.M., Stolarski, T.A., "Morphology of polymer wear debris resulting from different contact conditions", J. Appl. Polym. Sci., 45, 1992, pp. 2021-2030.

[12] Ovaert, T.C., Ramachandra, S., "Wear particle formation of polymers against smooth hardened steel counterfaces", Tribol. Trans., 40, 3, 1997, pp. 445-452.

[13] Haverstick, J.K., Ovaert, T.C., "Polymer-metal conformal sliding contact with forced convective cooling", J. Tribol., 117, 1995, pp. 1-8.

[14] Atkins, A.G., Omar, M.K., Lancaster, J.K., J. Mat. Sci. Lett., 3, 1984, pp. 779-782, cited by Hutchings, I.M. in "Tribology. Friction and wear of engineering materials" Edward Arnold, London, 1992, p. 127.

Authors' Index